A HORSE BY NATURE

MARY ANN SIMONDS

A HORSE BY NATURE

Managing Emotional and Mental Stress in Horses for Improved Welfare

TRAFALGAR SQUARE
North Pomfret, Vermont

FOREWORD BY **Julie Goodnight**
WILD HORSE PHOTOGRAPHY BY **Barbara** AND **Marty Wheeler**

First published in 2023 by
Trafalgar Square Books
North Pomfret, Vermont 05053

Copyright © 2023 *Mary Ann Simonds*

All rights reserved. No part of this book may be reproduced, by any means, without written permission of the publisher, except by a reviewer quoting brief excerpts for a review in a magazine, newspaper, or website.

Disclaimer of Liability
The author and publisher shall have neither liability nor responsibility to any person or entity with respect to any loss or damage caused or alleged to be caused directly or indirectly by the information contained in this book. While the book is as accurate as the author can make it, there may be errors, omissions, and inaccuracies.

Trafalgar Square Books encourages the use of approved safety helmets in all equestrian sports and activities.

Trafalgar Square Books certifies that the content in this book was generated by a human expert on the subject, and the content was edited, fact-checked, and proofread by human publishing specialists with a lifetime of equestrian knowledge. TSB does not publish books generated by artificial intelligence (AI).

Library of Congress Cataloging-in-Publication Data
Names: Simonds, Mary Ann, author.
Title: A horse by nature : managing emotional and mental stress in horses / Mary Ann Simonds ; foreword by Julie Goodnight.
Description: North Pomfret, Vermont : Trafalgar Square Books, 2021. | Includes bibliographical references and index. |
Identifiers: LCCN 2021012039 (print) | LCCN 2021012040 (ebook) | ISBN 9781646010448 (epub) | ISBN 9781646011827 (paperback) | ISBN 9781646011827(paperback) | ISBN 9781646010448(epub)
Subjects: LCSH: Horses--Psychology. | Horses--Health.
Classification: LCC SF285.3 (ebook) | LCC SF285.3 .S545 2021 (print) | DDC 636.1--dc23
LC record available at https://lccn.loc.gov/2021012039

All wild horse photographs by Barbara and Marty Wheeler (Barbara Wheeler Photography/barbarawheelerphotography.com). All other photographs courtesy of Mary Ann Simonds *except*: 7.1 (Raquel Mazur); 7.3 A, 8.1 (Lisa and Kevin Sink—Stormyranch Mustang Training and Learning Center); 9.13 A–C, 13.3 B (Nicole Harrington); 7.0, 10.11, 12.26 (Diane Petzel DeLano); 9.8, 9.42 (Gabriel de Matos Machado); 10.4, 11.8, 12.1, 12.5, 12.6, 12.7, 12.8 A & B, 12.28, 13.0 (Kim Suddaby); 12.9 (Orca Farm Foto); 12.18 B (Courtesy of Warwick Schiller); 12.18 C (Emma Whillans)

Book design by *Katarzyna Misiukanis–Celińska (https://misiukanis-artstudio.com)*
Cover design by *RM Didier*
Typefaces: *Forma DJR Text, PT Serif* and *Raleway*
Index by *Andrea Jones (JonesLiteraryServices.com)*

Printed in China
10 9 8 7 6 5 4 3 2 1

This book is dedicated to all the horses, wild and domestic, who have taught me how to listen to their subtle communication, to those true horsemen and women who take time to understand the real nature of horses, and to those who are committed to ethical and kind practices, putting the horse's welfare first.

CONTENTS

FOREWORD BY *Julie Goodnight* — XI

INTRODUCTION — 1

// PART ONE: HORSES WITHOUT HUMANS — 7

INTRODUCTION TO PART ONE BY *Dan Rubenstein, PhD* — 9
- How Wild Herds Function — 9
- The "Wild" in Your Domestic Horse — 10
- Horses Are Adaptable — 11
- Understanding the "Wild" in Your Horse — 11

Chapter [1] Why Understanding Wild Equines Is Important to Horse Welfare — 13
- Living the Dream — 17
- From Human-centric to Horse-centric—An Old Approach Refreshed — 22

Chapter [2] Horses as a Species in Nature — 25
- Fifty Million Years of Evolution, 6,000 Years with Humans — 27
- Free-Roaming Social Herbivores — 29
- Adapting Toward Group Cohesiveness — 31
- Adoptability vs. Adaptability — 42

Chapter [3] How Horses Sense Their World — 45
- Sight — 47
- Smell — 50
- Taste — 53
- Hearing — 54
- Touch — 55
- Subtle Energy Sensing — 57
- Integrating the Brain and Body — 59

Chapter [4] How Horses Communicate — 63
- What Do Horses Talk About, and How Do They Talk About It? — 65

Chapter [5] Understanding Horse Culture ... 85
- Adapting Toward Emotional Intelligence, Awareness, and Leadership   87
- Social Structures   89
- Space Games   91
- Family Life, Herd Rhythms, and Relationships   93
- Learning and Education   99

Chapter [6] Functional Horse Society ... 105
- Normal Horse Behaviors   109

// PART TWO: HORSES WITH HUMANS ... 113

Introduction to Part Two by *Dr. Med. Vet. Dorothe Meyer*   115
- Is the Horse a Prisoner?   116
- Behavior Caused by Stress   117
- Lowering Stress Equals Success   118

Chapter [7] Considering Our Impact on the Horse—How Domestication Has Changed the Horse (or Not) ... 121
- How We View Our Horses   126

Chapter [8] Giving Up Freedom, Finding Friendship ... 129
- From Wild to Domestic   131
- Adaptability Depends on Temperament   133
- Dominance, Leadership, and Friendship   136
- Interspecies, Trans-Species, and Equestrian Psychology   148
- What Kind of a Horse Person Are You?   154

Chapter [9] Horse-Human Communication and Relationships ... 161
- Feeling Safe Together—The Horse-Human Bond   165
- So...How Should We Communicate with Horses?   169
- Communication Channels   177
- Integrated Communication: Bringing the Channels Together   208

Chapter [10] Analyzing the Individual: Equine Personality, Temperament, and Training Assessments ... 217
- General Observation Assessment   221
- Specific Horse Personalities   222

- The SAICC Evaluation ... 224
- Benefits of Understanding Your Horse's Personality ... 244
- Training Readiness by Age, Gender, and Experience ... 244
- Assessing Trauma ... 247

Chapter [11] Understanding Equine Mental and Emotional Stress ... 257
- Consider the Research ... 260
- Types of Stress ... 261
- How Stress Affects Your Horse ... 264
- Primary Stressors ... 265
- How Horses Show Stress ... 268

Chapter [12] Managing Stress in Horses ... 283
- First...Manage *Your* Stress ... 285
- Understand How Horses Think and Learn ... 287
- Reducing Stress Related to Breeding ... 294
- Improving Early "Foalhood" Education and Initial Training ... 297
- Preparing Your Horse to Learn (Whether Young or Not) ... 307
- Less is Always Better: Tack and Training Devices ... 316
- Limiting Stress When Teaching Specific Skills ... 324
- Managing Riding Stress ... 327
- Habitat Management ... 335
- Shipping, Showing, and Selling Horses ... 349
- Solving Common Behavior Issues Related to Stress ... 358
- Our Humane Responsibility ... 363

// PART THREE: WHAT WE DO WITH HORSES: ETHICS, ECONOMICS & WELFARE ... 365

Introduction to Part Three by *Duncan Peters, DVM, DACVSMR* ... 367

Chapter [13] The Future of Horses ... 373
- How Do We Best Help the Horse? ... 377
- Industry Ideas for Improving Horse Welfare in Equestrian Sports ... 385
- Everyone Is Responsible ... 388
- Keys to Remember ... 391

Acknowledgments ... 397

Index ... 401

FOREWORD
BY
JULIE GOODNIGHT

> **Over thousands of years, horses have continually adapted to the needs of humans, although our views on horse welfare have not always kept pace.**

I was a horse-crazy teenager the first time someone paid me to ride a horse, and I couldn't believe my lucky stars! Nearly a half-century later, after an unimaginable, international career riding and training horses and their humans, I've seen an incredible revolution in horse training methods and how the science, economics, and even our societal view of horses have evolved.

I came of age at a time when horses were still transitioning from necessity to luxury, from utilitarian to sport; a time when few financial resources were dedicated to scientific research. The trainers of my youth were often retired U.S. Army officers who became horse trainers once the Cavalry was disbanded. They were tough and exacting trainers, and a perfect fit for a young equestrian like me. I was a determined hunter-jumper competitor, passionate about riding sports and completely consumed by horses.

Yet even as a teenager, I saw the harm that was routinely done to horses in the name of competition. Even as a willing participant, in my heart I refused to accept the misguided notions of the day that animals did not feel pain or form bonded relationships, and that heavy-handed training was the only way.

Julie Goodnight

Years later, as a budding young horse trainer, I was watching a local schooling show when I witnessed a lovely young mare being abused by a trainer who was jumping her in draw reins. In her distress, caused by the entrapment, she was biting her own shoulders until they dripped blood. Yet no one in authority called out the popular trainer for his abusiveness. It was the first, but sadly not the last time I saw a horse driven to self-mutilation to escape its own pain. This incident further cemented my determination to find a better way.

I've made it my life-long mission to understand more about horse behavior and to prove that not only can horses be trained in more humane ways, but that the end result is far superior when they are. While experimenting with techniques as a young trainer, I also turned to science and behavioral research for answers.

It was in this pursuit of a better way to train horses and my determination to share the results with others that my path first crossed with Mary Ann Simonds—a pioneer in equine behavior and practical applications. Strangers in all other ways, Mary Ann and I were united in our common mission to increase the awareness of the horse's natural behavior in all horse-lovers, and especially in the riders and trainers who require the most from their horses.

Throughout my career as a horse trainer and horsemanship clinician, I've come to see how well and how willingly horses can perform—*in any discipline*—when their hearts, minds, and bodies are healthy and committed to the cause. I've bonded with many horses to the point they eagerly perform complex reining maneuvers, flawlessly cut a cow, or attack trail-obstacle courses for me—all while riding without a bridle or any restraint. I've experienced firsthand, the thrill of having a deep, inter-species relationship with an animal that people once believed was stupid, obstinate, and incapable of forming bonds.

In nearly half a century of training horses to do our bidding, my understanding of their natural behavior and their capabilities has grown, as has my ability to draw out their eagerness to please me. I've formed deep athletic partnerships with many horses over the years and experienced firsthand the superior result of a performance horse that will give you everything he's got; horses that I can communicate with almost telepathically because we know and trust each other; horses that seek only my acceptance and will give me anything I ask in return. This is a thrill and an honor that many riders will never know.

A Horse by Nature has all the information you need to reach this level of connection with your horse *right now*—it won't take you 50 years to get there! Jump into it with your heart and mind to fully understand the nature of horses—their evolution, their adaptations to human society, how they think and learn, and how they perceive their reality. Armed with this knowledge and understanding, your horsemanship will soar to new heights.

Over thousands of years, horses have continually adapted to the needs of humans, although our views on horse welfare have not always kept pace. Economics and ethics do not always have to be at odds in horse sports, and modern scientific research into horses has proven it. *A Horse by Nature* allows readers to get deep inside the psyche of horses, to formulate a plan to better meet their needs, and to train them with logic and empathy, using methods that make sense and that get results.

Julie Goodnight
Trainer and Horsemanship Clinician
Poncha Springs, Colorado

INTRODUCTION

With thousands of books written on horses, why is this one needed? I asked myself this question and realized the answer was because I find myself repeating the same simple statements daily with trainers, veterinarians, and students.

Statements such as, "Learn to look with not at the horse," or "Mares will give a soft eye glance followed by a gentle nose bump when greeting, so pay attention," or "Horses are energetic, emotional, and social creatures who use feeling more than thinking, so help them feel good and be confident." Perhaps back when humans "lived with" their horses, they were more attuned to their feelings, but in the modern fast-paced world, the day-to-day and moment-to-moment behaviors and actions of horses go unnoticed by most people. And with so many horse-training programs claiming to be "natural," along with hundreds of devices marketed to give better "control" or "communication," and a vast array of supplements to help your horse perform better, it can be confusing for people to know what is best for their horses.

Relationships Rule

I'll tell you what is most important to horses: their social network, which provides them safety and comfort. *It is all about relationships.* Nature has wired horses to seek social bonds and maintain them. While horses may be prey and humans predators, it is our similar need for social bonds—for friendship—that brings us together.

Our hearts can connect us often more easily than our minds. While both are important, integrating knowledge with kindness and love will win over almost any creature, particularly horses, who seek to "belong" and value lifelong friendships. While science is slow to explore interspecies consciousness, I was fortunate enough to do my graduate research from 1987 to 1989 in this field. I worked primarily with horses and aimed to establish some parameters in equestrian psychology. The results of numerous clinical studies clearly showed that the most important source of "healing" and "happy, relaxed" states of being, in both horse and human, resided in the heart and not just the mind.

Being able to integrate knowledge with feelings is one of the main goals of this book: to educate and inspire "horse lovers" and equine professionals to use their knowledge, their hearts, and their voices to improve the mental and emotional welfare of horses.

Horses' lives interweave with humans as friends and willing companions. They exchanged freedom for friendship, trusting their human companions would care for them, value their social natures, and keep them safe and comfortable. While this may have been a good evolutionary decision for horses as a species, it came at a cost. Having a protected space, even if only 12 feet by 12 feet, with food and water that comes regularly and lots of friends nearby, may produce a little stress (along with safety and comfort) for some individuals. But when "wild" genes, which can be expressed in even the most domesticated horses, present themselves then those horses become *really* stressed by domestic life. Temperament and the ability to adapt are key factors in the horse's ability to handle mental stress, whether he is domesticated or wild.

While *physical* stress in horses has been fairly well researched, little attention has been paid to how stress affects their mental and emotional welfare. One of the purposes of this book is to help people understand the "drivers" of equine society—those traits that are inherent in horses as a species and will surface when "triggered." Although horses are "wired" to sense, communicate, and relate in species-driven ways, their interaction with humans can either help or hinder them.

One Shoe Does Not Fit All Horses

No one can say that *all* horses prefer to be wild, living in a group, or that *all* horses would prefer not to compete in sports. Certainly, many horses are very happy just hanging out being horses, but many horses enjoy the social aspect of horse-human interaction, including being our partners in sports.

Having studied behavioral adaptations in both domestic and wild horses for over 40 years, it became clear that there are:

- Wild horses who quickly realize the benefits of domestication and adapt well.

- Wild horses who never give up their wildness and merely "tolerate" human interaction.

- Domestic horses who enjoy working with humans.

- Domestic horses who are always "on guard" (vigilant), always worried when friends leave, and are stressed by confinement and changes. You could say they are not very "adaptable" as their wild genes are more expressed, causing a higher level of stress because of their higher sensitivity to both their social network and their environment.

By being able to accurately assess your horse's personality, your horse's ability to interact functionally as "a horse," and your horse's capacity to adapt to human wishes, you can design learning programs best suited to the individual horse. Thus you will greatly reduce emotional and mental stress, increase your horse's learning ability, and by virtue of having a well-adjusted horse, improve performance.

It should not take scientific studies to show the mental and emotional stress horses endure in their partnerships with humans. But horses have become an "industry," generating billions of dollars in the global economy. Because of this reality, horses' relationships with humans vary from being a commodity that is "bought and sold" to being a recreational partner to being a therapist. Believing horses are "property" that can be treated as commodities rather than valued companions puts horses at risk. Why? Because there is currently no guidance on the welfare standards of managing the psychological stress related to what humans do with horses.

It is my hope that by integrating wild horse knowledge from my own field studies and data from other equine ecologists/ethologists with current scientific studies being conducted (related to mental and emotional stress in domestic horses) together we can start to develop "Best Management Practices" in the care and welfare of horses and horse sport.

How to Use This Book

This book is designed to help any equestrian—from novice to expert—gain better insight into how horses think, feel, learn, and generally perceive our human world. I aim to empower readers to then feel confident in applying an understanding of the social ecology of wild horses in order to better manage and reduce stress in our domestic horses.

One of the most common statements I have heard from trainers or veterinarians after one of my clinics or lectures has been, "Gosh, this is so simple and makes sense. Why isn't everyone taught this?" I wondered the same thing, then realized the model used in teaching equine science is "human-centric" and based on horses studied in a clinic or confined environment.

The tips in this book, on the other hand, are based on a "horse-centric" model of equine behavior and communication. This model is built from my years in the field, first studying wild horses and then successfully applying the knowledge I'd gained to the domestic sport horse industry.

Before sitting down to write *A Horse by Nature*, I asked a number of top individuals in the equine industry what kind of book would be most helpful. Most answered they had little or no time to read, so said I should "give them tips." When I asked a number of my students the same question, however, they wanted more detailed information and examples of key knowledge points. So, I have tried to do both with this book: give simple tips, followed by a discussion of the science or justification of the tip with resources, exercises, and examples.

The tips you will find in the pages ahead are color coded:

★ **Essential (RED)**

★ **Important (BLUE)**

★ **Makes Life Better (GREEN)**

The colors should make the knowledge easy to embed in your brain. And, like when you learn a second language, hopefully, you will become progressively more "horse-like" in your thinking, feeling, and perception.

_ Part One

The first section of this book will ground you in the natural history and social ecology of free-roaming wild horses. The natural instincts, learned behaviors, gender difference, social roles, and importance of social bonds in horses will become clear as the driving factors in horse culture. You'll discover how adaptable horses really are, leading to variations in equine cultures and making them an ideal species to share their social lives with humans.

_ Part Two

The second section of *A Horse by Nature* focuses on horse-human relationships and how to apply the knowledge from Part One of the book to the management, care, and training of, and interactions with domesticated horses. You will be guided through discussions on:

- How to better assess your horse's personality.

- How to communicate verbally and nonverbally with horses.

- How horses learn compared to humans.

- How to establish trust and friendship through spatial awareness and equine social etiquette.

- How to reduce equine stress.

- How to apply all this knowledge in most horse-human interactions.

_ Part Three

The third section dives into why we have horses in our lives and all the things we do with them. It addresses the ethics of horse sports and asks the questions we need to be exploring to ensure horse welfare in all the disciplines.

While working on this book, I interviewed a number of respected sport-horse veterinarians and horse industry professionals. I asked them all, "What would be the best tip you could give to people to help sport horses?" The most common response was that there aren't many "horsemen" anymore. Many of those who compete—and this is across the disciplines, from Western to English—do not have time to just "be" with horses and find out what makes them happy. A number of the equine professionals I spoke with—involved in all different sports, from racing to jumpers—lamented the fact that horses have turned into disposable commodities that can be bought, traded, and thrown away when they do not "win."

Luckily, for most people who own and ride horses, this is not true. We love our horses as part of our families and would rather feed them than ourselves. But in looking forward as a society, we must deal with the ethical questions that dictate whether horses can continue to be in our lives. Can we maintain happy and healthy horses as open land and turnout space become more limited, and as competitions and disciplines expand to meet human demands for entertainment and excitement?

_ Come with an Open Mind

A Horse by Nature is designed to fill in the missing knowledge that is forgotten, ignored, or unknown and give readers further resources to explore. You may read a tip in the pages ahead and think, "Well, this makes sense," or you may read a tip and find it contradicts something you believe you have been doing correctly. Luckily, horses are very adaptable and will learn whatever they need to learn to stay "in good" with whatever social group they belong to, whether it is made up of horses or humans. However, this driving desire of horses to "belong" can also cause unnecessary stress.

Whether you are a professional trainer, a veterinarian, an aspiring student, or a horse-human interaction specialist, I hope this book will feed your appetite for horse knowledge and give you valuable tips you can use, regardless of your chosen equine discipline or activity. I encourage you to adapt a "horse-centric" model as you read. Be aware of your own "speciesism" (assumption of human superiority) and ingrained beliefs you may have about horses. Try and keep an open mind. Look for the bridges that bring horses and humans together, and become perceptive to the differences so you can adjust your own thinking and feeling to better understand and empathize with all horses.

PART ONE

HORSES WITHOUT HUMANS

← [1.0] A family of wild horses looks on with curiosity.

INTRODUCTION
TO PART ONE
BY
DAN RUBENSTEIN, PHD

Dr. Dan Rubenstein is Professor of Ecology and Evolutionary Biology, Emeritus, and former Director of Environmental Studies and African Studies at Princeton University. He served as Chair of the Department of Ecology and Evolutionary Biology for over two decades.

> **Understanding the 'wild' in the domestic horse is crucial, since people are now an essential part of the horse's world.**

How Wild Herds Function

Wild horses and domestic horses are very similar genetically—about 97 percent of their genes are the same. Over historical time, humans have selected from the natural traits exhibited by free-ranging horses, those traits which serve humans best.

Wild horses form bonds, motivated by their own self-interest, that facilitate living in groups to increase their survival and reproductive abilities. Staying together in large groups creates benefits: for example, more eyes for watching out for competitors or predators while also providing more time for feeding. Up to a point, living in larger groups can enhance the components of fitness that increase the horse's evolutionary success. Hence living in groups and forming social bonds is essential for horses to sustain themselves today and in the future.

Within horse family groups there are two types of leaders: Males provide females with material rewards—protection from harassment by other males seeking mating opportunities and protection from predators, both of which give females more time to spend on feeding and nursing of young.

Females organize movement and shape time budgets, which often differ depending on female herd members' reproductive states.

Whereas males are "showing" in their behavior, so dominance is easy for other horses and researchers to detect, females are more circumspect in what they do. Females typically reinforce dominance through subtle head nudges, ear movement, or simply stepping away. In general, interactions in females appear to take the path of least resistance since it conserves energy for the signaler and all others in the group. Really good friends in an equine herd almost "mind read."

The "Wild" in Your Domestic Horse

Although domestication has changed the relevant frequency of certain traits we see in horses, none of the traits have been abolished and canceled. The behaviors seen in barnyard or stabled horses will be part of the repertoire of wild horses even though some of the nuances that invoke them, or the frequencies at which they occur in their social repertoires are likely to have been changed by domestication. This primarily is because these horses now have to read and send signals to people as well as horse peers.

Horses are highly social in the wild, and what makes them different from other species is that they are *neolocal,* like humans. This means that their sons and daughters leave the herd when they are mature. So, social relationships that develop among adults do so with strangers—essentially non-genetic relatives—and what cooperative and altruistic tendencies they develop will be among strangers. Evolutionarily, this is hard to do, but it most likely pre-adapts horses to form bonds with people who are strangers as well.

Paradoxically, socialization traits may have gotten stronger with domestication even though horses have had to adapt to living in limited space, confinement, and unnatural settings. If people only understood this, they could utilize the enhanced set of traits associated with socialization and cohesiveness to increase the ease with which human-centric goals or behavioral expectations could be realized. By "getting inside the horse's world," a person can better understand what makes the horse "tick" and more easily evoke those behaviors in the horse that the person wants.

The behavior of wild horses is often context-specific—that is, wild horses make decisions and change behavior according to the situation. Moreover, one size does not fit all because horses have different temperaments.

Since cohesiveness is critical to survival, wild horses have found ways to meld individual difference into collective action that enables them to adjust to changing circumstances. Such nuanced flexibility is something that early humans capitalized on and is something that modern humans can draw on as well to meet specific needs. If people do not see the inklings of wildness in their domesticated horses, then their interactions with their horses are likely to be counterproductive.

Wild horses have functioning societies that minimize stress and maximize reproduction. Wild horses are really good at balancing tradeoffs associated with surviving, foraging, moving, mating, and parenting, and their solutions tend to better the fates of all members of the herd. Adding people's wishes and desires to the list of competing options is not likely to be a problem *as long as* people realize that horses are natural problem-solvers and reducing conflicts will help them get to a mutually advantageous outcome.

Horses Are Adaptable

The key is to understand the needs and abilities of a particular horse, especially the horse's state of being and personality. Then, creating a context that plays to the horse's natural intentions while having the horse freely accede to human desires reduces stress and enhances clear thinking. Horses change as they develop, and their relationships—governed by social networks and personalities influenced by past experiences—are generally malleable. When trainers, riders, and breeders understand these dynamics, they can modulate both physical and social environments that make it easy for horses to act in ways that fulfill their wishes.

Understanding the "Wild" in Your Horse

Knowing how horses navigate the world *without* the assistance of humans is key to understanding how to get the best out of horses. The deep relationships of fostering sociality, the movements to associate with strangers as youngsters, and the impulse to build new relationships are part of what makes horses special. These natural tendencies also open the possibility for humans to work *with* horses as opposed to against them.

Understanding the "wild" in the domestic horse is crucial, since people are now an essential part of the horse's world.

← [**1.1**] Horses have evolved to have strong social networks and adapt to a variety of habitats.

CHAPTER

[**1**]

WHY UNDERSTANDING WILD EQUINES IS IMPORTANT TO HORSE WELFARE

> **There is no fundamental difference between man and animals in their ability to feel pleasure and pain, happiness and misery.**
>
> Charles Darwin

[1]

During the early years of my wild horse research (1973 to 1993), I documented a number of functional social behaviors in wild horse populations across the Western United States. I was also able to collaborate with scientists and photographers around the world whose work has validated many of the behaviors identified in my early research. Three such photographers, Robert Vavra (of the book and film *Such Is the Real Nature of Horses*) and Barbara and Andy Wheeler (who have contributed many of the wild horse photos in this book), I am especially grateful to have met as they have captured a number of little recognized behaviors so intrinsic to the functional social lives of horses. Many of the behaviors observed in wild populations of horses can be observed in our domestic horses but often go unnoticed. Recognizing and understanding these behaviors can lead to better management and training of our domestic horses, as well as improved management of wild horse populations.

Because horses have adapted so well to domestication, they have been both penalized and rewarded for their cooperation. In exchange for food, water, and shelter they have given up their freedom and often their natural lifestyle and social networks. But they may have found mutual friendships with humans. This was not much of a problem in the past when humans

↙ **[1.2 A & B]** Stallions posturing in the wild **(A)** and stallion "C" rears as he plays in the paddock **(B)**. It is important for stallions to rear to stay fit and develop strength in their hind ends. Whether playing or fighting, stallions rear more than mares. You can often observe young horses rearing and "mock fighting" both in the wild and turned out in the pasture.

were living more closely with horses, day to day, and most horses had pastures to roam at will. But with land becoming scarce and more horses being confined to smaller spaces, mental welfare figures more prominently in the picture. While horses may have their physical needs met, their mental and emotional needs are often ignored. And mental and emotional stress can lead to physical ailments and behavioral issues.

Today with the work of such groups as the Equid Research and Conservation team from Princeton University, and other equine ecologists and ethologists around the world, scientists are starting to investigate animal thinking and decision-making and apply the data to improved horse welfare. Assessment tools such as the Horse Grimace Scale (HGS) and the Equine Facial Action

Equids or Equines?

In science, all species in the family *Equidae* are referred to as *equids*, which includes zebras, tarpans, Przewalski's horses, asses, and modern horses. For the purpose of this book, wild horses are referred to as "wild equines" rather than "wild equids."

Equine Facial Action Coding System and Horse Grimace Scale

- The *Equine Facial Action Coding System* (EquiFACS) is a systematic methodology identifying and coding facial expressions in horses based on facial muscle movement. Researchers have identified at least 17 various expressions in horses.

- The *Horse Grimace Scale (HGS)* incorporates Facial Action Units (FAUs) based on the Equine Facial Action Coding System related to pain. Both are useful tools to help better understand the emotional and mental states of horses.

Learn more

Learn more

Learn more

Coding System (EquiFACS) are providing tools to evaluate pain and feelings in horses that can be correlated to natural expressions in horses (see sidebar). This data can offer us insights into our domestic horses' minds and behavior.

★ **TIP:** / Domestic horses express similar behaviors to wild horses but are frequently misunderstood. /

Living the Dream

While not everyone can live out on the range and study wild horses, it was my dream to better understand horses by learning what they teach each other when humans are not around. As an equestrian, I felt that understanding more about horses *without* humans, in their natural habitat, would give insight into the complexity and intelligence of the creatures we partner with and ride (figs. 1.2 A & B). I was fortunate to be able to live both dreams—living out on the range, studying wild horses, and riding and showing domestic horses for fun and competition. If everyone had the same opportunities to experience really "living" with horses, they would also see the similarities between domestic horses in the barn and those living in nature with their families.

My early research was conducted while at the University of Wyoming as an undergraduate in Wildlife Biology and Range Management. My research targeted a whole-systems approach, investigating the influence of habitat on social behavior and ecology of wild horses in relationship to other species who inhabited the range. After graduation I was able to continue studying various equine populations while working as a consulting behavioral ecologist and range biologist. This gave me continued access to many of the herds I had studied in the early 1970s, and allowed me to investigate wild horse populations not just in Wyoming, but in Oregon,

California, Utah, Nevada, Arizona, Montana, Colorado, and New Mexico through 2002.

My research focus investigated the social ecology of wild horses, identifying functional behaviors that favored adaptation and sustainable equine populations. While a variety of factors were investigated according to behavioral ecology models, the information that I found most important to note is that *social bonds define horse culture* (fig. 1.3). My early field studies have since been documented in much more detail by other equid researchers, showing *the adaptive advantage of growing large social networks*. Large social networks give each individual more time to eat, raise young, and socialize, and reduce the time required for being on alert to threats and dangers. Vigilance goes down as herd size goes up, as the author of the Part One Introduction Dr. Dan Rubenstein's research has shown (see p. 8).

More Than What We Expect

Humans have a bias toward horses based on beliefs, attitudes, perceptions, and often false "knowledge." Be aware how ingrained old beliefs about horses can be unconscious and drive your behaviors. Similarly, there has been a bias in behavioral studies to interpret meaning of observations by researchers based on old models in science. Male-dominated models (where aggressive male behaviors have been noted) in social species, while common, have often missed the more subtle behaviors of females in the species who may also be communicating direction to the social group. An eye glance or a flick of an ear from a mare signaling danger, recognition, or action tend to go unnoticed compared to "snaking" of the neck and other more obvious herding behaviors of a stallion.

→ [**1.4**] Wild horses in the Red Desert, Wyoming, bunched together, curious but cautious. When you look "with" instead of "at" horses, you can see subtle communication you may have otherwise missed.

↙ [**1.3**] A group of young bachelors greet each other. Note the three-way conversation. Some stallions remain friends their entire life, even after getting mares to join them.

> **Human beliefs drive attitudes about horses.**

When researchers learn to look "with" instead of "at" horses, their observations may be enhanced. Looking "with" involves slowing down one's thinking to become a part of the natural system. By being more aware of what an individual horse is sensing—smelling, hearing, tasting, touching, seeing—the observer can become more "horse-centric" in recording behaviors. Looking "at" a system and observing behaviors from a human perspective often leads to false assumptions. A subtle shift in awareness of observation to become more a part of the whole system allows the researcher to see more clearly the dynamics of the situation and individual interactions being observed (fig. 1.4).

This was especially true when I began my research in Wyoming in 1973. Most of the ranchers and the Bureau of Land Management (BLM) personnel I interviewed to collect historic data on the horses and their diets had little real data but were convinced the horses were eating the same plants cattle were grazing on the range. Part of my research was analyzing manure in the lab to determine plants eaten by various animals. The data showed that horses were eating more Western wheatgrass, a native plant, while cattle preferred to eat crested wheatgrass, an introduced plant. The studies also showed that horse diets in many habitats overlapped more with elk than cattle. But the *paradigm,* or accepted belief, was that horses were grazing the same plants cattle were and therefore were in direct competition with them, even though the data said otherwise.

Diet studies on wild horses indicate horses are opportunistic herbivores and do develop food-taste preferences based on their habitats and learned

STORY FROM THE FIELD:

Is It a Stallion, or Is It a Mare?

When I was a young undergraduate conducting range- and wild-horse studies in Wyoming, I had a meeting with the BLM office in Rock Spring, Wyoming, where I was told that there was a wild bay stallion with a band of mares using a rancher's stock tank. Technically, this herd was off its allotted "Herd Management Area," so the BLM representatives asked me to document their movement and let them know if I saw the group while I was out conducting my studies.

All week long I observed a group of horses who matched the description and movement patterns the BLM personnel had described, but there was one problem.

So, I had the BLM manager come out to the field with me to indicate whether I had the right horses. When he spotted the herd in question through a spotting scope, he quickly said, "Yeah, that's them. See? There's the stallion I told you about, leading his mares right to the stock tank." "If that lead horse is a stallion," I replied, "then what was that foal sucking on all week?"

He looked puzzled, and I explained the big bay leading the group was a mare with a foal. The stallion was a smaller chestnut, moving in the back of the group. The bay mare would stand, wide-legged and head high, looking at what might be danger, and I understood how the BLM manager's bias misinterpreted the behavior and communication to be stallion-like (fig. 1.5). He never noticed the mare sending her foal back into the group of mares for protection, and then when at the water tank and "safe," bringing her foal out to nurse and be with her. All he saw was a "big bay stallion," leading a group of horses to water across the range. He saw what he was programmed to see. //

→ [1.5] Misinterpreted behavioral data concerning wild stallions and mares has led to the proliferation of incorrect information. Here, the lead mare of a herd races out to what she perceives as potential danger.

→ [1.6] A wild mare teaching her foal to be "aware" and on the lookout for unfamiliar stimuli.

knowledge about food. My field studies showed that while one herd of horses in the same habitat would dig up snow to find dried grasses in the winter, another more opportunistic group would head up the mountain and browse on juniper berries. Horses, whether wild or domestic, given free choice when it comes to food, are quite good at selecting what they need nutritionally, once they know how and where to get the nutrients. Given a chance, almost any stabled horse will try and find what his body needs, even if it means grabbing a weed or digging the ground and eating dirt.

_ Innate and Learned Behaviors

It is often stated that "domestic horses behave differently than wild horses." We must first understand that there are *innate behaviors*—those instinctual behaviors shared by all in the species that are direct reactions to stimuli and done without thought—and there are *learned behaviors*—those behaviors an animal learns through interactions with his environment and other individuals. *Innate behaviors between wild and domestic horses are virtually the same*, unless the domestic horse has overridden instinctual behaviors with learned behaviors. For example, horses, whether wild or domestic, will react to new or unfamiliar stimuli, unless they have learned the stimuli is safe (fig. 1.6).

Because innate behaviors of domestic horses are similar to those of wild horses, it is important for people to understand what is "normal" in both wild and domestic contexts. Many normal behaviors in domestic horses are

> **Innate behaviors between wild and domestic horses are virtually the same.**

[1.7] BLM facility director Fred Wyatt and I observe recently captured wild horses at the Palomino Valley holding facilities in 1992.

misinterpreted because most people do not know either what is triggering the behavior or what the behavior means. For example, a mare pinning her ears at a new horse stabled next to her often is interpreted as an aggressive act by a "bitchy" mare. But, in fact, ear-pinning in this scenario is a functional communication of a mare who is aware and simply communicating to a new horse to not enter her space without permission. If the new horse ignores this mare's message, then the new horse is not showing good horse etiquette and is the one not acting "normal." Although social greetings can be learned, most horses have innate behaviors like this to greet each other. And you will learn that mares are usually much better at equine social etiquette than male horses!

From Human-centric to Horse-centric—An Old Approach Refreshed

Science offers us great viewer/observer models for objective study. But the problem with viewer/observer models is that they place humans outside the system they are trying to study and create an energy emphasis of looking "at" the system. This energy can influence the systems or animals being studied, particularly in natural systems. Often wildlife who are "prey" species have developed heightened senses to "feel" when they are being "looked for" or "looked at." The wildlife species may then sense danger or feel threatened when being observed, which alters their behavior.

Hence humans will witness only a limited amount of natural wildlife behavior (non-human-influenced) when they are observing shy, sensitive creatures (fig. 1.7).

Humans and horses are both a part of nature. In order to better understand horses, try and imagine you are a horse living with horse friends, roaming the hills and grasslands. While humans may have evolved away from nature, horses have not, and although domestic horses may not be living out in a wild herd, they still are genetically wired for innate behaviors, including socializing.

Interestingly, socializing is also highly fundamental to humans. Both species are "social beings," and this fundamental core characteristic may be what has brought horses and humans together for thousands of years. Rather than focusing on the "prey-predator" relationship, it is our social friendships that may well be why horses enjoy being with humans and sharing in our activities.

★ **TIP:** / Learn to look with not at horses. Be "in the moment" with horses. /

STORY FROM THE FIELD: //

Looking "With" Not "At"

As a young field biologist studying wildlife while at the University of Wyoming, I designed an experiment to test whether "looking with" versus "looking at" would make any difference in observing the diversity and numbers of various species of wildlife in a particular habitat.

Wildlife observations were conducted by laying out transects—lines across a habitat or area of one—traveling them at different times of the day and evening, and recording what wildlife species were observed. My study was conducted while riding on horseback as I had found that there seemed to be more wildlife to observe when I was riding than when I was walking on foot. After the first week I noticed that every time I was just riding for pleasure and not collecting data for the study, I saw a high number of species, including badgers, foxes, ground squirrels, pronghorn antelope, and red-tailed hawks. But when I was conducting the formal surveys, and stopping and recording wildlife at the designated area, there were very few species to observe. I had designed the study to involve me riding morning, afternoon, and evening in hopes of learning more about wildlife movement patterns. But after two weeks, it became apparent that there was a notable difference in my observations, no matter the time of day, when I was riding for fun rather than riding while surveying.

I realized that when I was just having a pleasure ride and appreciating the sunshine, the smell of grass, and the open spaces, I was in a different brainwave state. My thoughts and feelings seemed to have a different effect on nature. I was looking "with" nature. And when I was on transects looking "for" and "at" wildlife, I did not see nearly the same number of animals or as much diversity. Over the course of the study I found there was over a 40 percent higher sighting of wildlife during the informal rides (for fun) than during formal surveys. This insight led to further research, which also showed significant differences between the two scenarios and was tested using other biologists. This led me to teach many natural resource field biologists, as well as horse professionals, how to be more "in the moment"—to look "with" not "at" nature and animals. //

[2.1] A bachelor stallion stands on the edge of a herd comprised of several family bands of mares with stallions. This stallion remained outside the herd as a loner but was not challenged or run off by the other stallions.

CHAPTER

[2]

HORSES AS A SPECIES IN NATURE

No one statement about horses can hold true to all horses.

[2]

Fifty Million Years of Evolution, 6,000 Years with Humans

The genus *Equus* includes modern horses, as well as tarpans (*Equus ferus ferus*), Przewalski's horses (*Equus ferus przewalskii*), zebras, and wild asses. While many wild subspecies have died out and most current equid species are threatened or endangered, "modern" domestic horses seem to be doing well with humans. (Note that while some argue there are no "wild" horses today, only "feral" horses, this labeling simply refers to nomenclature and not to the biology of the horses themselves. Whether horses are called "wild" or "feral" is irrelevant to those horses; they are all born into a natural world without humans and must rely on each other for survival, as they have done for thousands of years.)

The origins and distribution of modern horses (*Equus caballus*) continues to be debated as more and more data is produced through paleontological and genetic studies. Modern horses evolved in North America about two million years ago and migrated to Eurasia. Their amazing adaptability allowed them to acclimate to many habitats and climates, from the cold north to marshy subtropics to dry, arid rangeland.

★ **TIP:** / Understanding horses as a species gives insight into their physical, mental, and emotional priorities. /

Equus caballus seems to have increased its chances of survival by heading toward domestication. And since horses have been mingling with humans for thousands of years, most wild horses today have had some human influence. But that can be said of deer and elk as well, as most wildlife that shares habitats with humans alters behavior and genetically selects for individuals who adapt, taking the species toward cohabitation and domestication.

Instinctual or innate behaviors are still very much the same for domestic and wild horses. Both remain emotional and sensitive creatures, designed to be alert for danger, find safety in social groups, and learn behavior from their friends (fig. 2.2). This has not changed. Equine brains are also wired for having awareness of group status and leadership. They develop relationships with other group members through observations, vocalizations, and tactile interaction. Equine populations, both wild and domesticated, may have loose-to-strong hierarchies with status changing constantly or not much at all, depending on the stability of the social structures. Forming and maintaining social groups, whether composed of two horses or a herd of horses, seems universal. Horses may migrate toward larger group sizes when resources are plentiful and break off into small groups when resources are scarce, throughout appearing to be flexible with their social structure and friendships (fig. 2.3). These adaptations have helped domestic horses cope with losing friends, make new friends easily, and accept various types of habitats.

→ **[2.3]** A herd of Onaqui wild horses, spread out in smaller social groups and family bands. Observing interactions within the herd and smaller groups gives us valuable information about the behavior and social needs of our domestic horses.

↙ **[2.2]** Wild mares and foals watching something in the distance, alert and ready to flee danger if it is perceived.

Because domestic-bred horses have many of the same traits as their wild cousins, understanding horses in their native habitat, *without* humans directing their actions, can give us insight that helps us care for, train, interact with, and generally manage horses in ways that keep their physical, emotional, and mental welfare in mind (fig. 2.3).

★ **TIP:** / In the order *Perissodactyla* (odd-toed ungulates) and suborder *Hippomorpha* in the family *Equidae*, horses share their evolutionary history, as well as many traits, with rhinoceroses and tapirs. /

★ **TIP:** / Horse genetics have not changed much in their transition from wild creatures to domestic partners. /

Free-Roaming Social Herbivores

As we will discuss, social relationships are the core of functional sustainable equine society. Not unlike human cultures, there are a variety of "equine cultures" in the wild. Although there are common core behaviors or "drivers" among horses from every part of the planet and their biology and psychology backed by "instinct" and years of evolution remain very similar, each group can have a different culture based on their habitat, environmental factors, life experiences, learning, personalities, and genetics.

Learn more

↙ [**2.4 A & B**] Horses have adapted to a variety of habitats, from harsh, arid landscapes **(A)** to wet marshes **(B)**, evolving slightly different microbiomes of the digestive system. Horses living in harsh climates must spend more time and energy moving to find food and water, while those in more suitable habitats can spend more time socializing and reproducing.

★ **TIP:** / Just as there are diverse human cultures, so there are many diverse equine cultures. It is beneficial to have broad knowledge of various breeds, personalities, and genetics as they will give you background and insight into the behaviors and thus adaptability traits of different horses. /

Habitats direct herd rhythms and behaviors for eating, sleeping, moving, reproducing, watering, and socializing (figs. 2.4 A & B). Herds with large open grassy meadows and plenty of forage and water can spend more time eating, resting, socializing, reproducing, and rearing young. Herds in rugged habitats must dedicate more time to movement in order to find food and water. Some herds have to keep moving all day long in order to eat and drink sufficient amounts, and yet they still adapt. Horses are *nonruminant herbivores* that evolved to be continuous grazers, with specialized digestive tracts able to utilize high levels of plant fiber—forage only stays in the stomach for about 15 minutes, so they must keep eating small bits of food. They can then process a large quantity of forage with little nutrients and meet their needs. Typically, horses spend between 12 to 20 hours a day grazing, depending on habitat forage availability.

★ **TIP:** / Horses are free-roaming social herbivores, eating 12 to 20 hours a day while constantly moving. /

Primarily grass-eaters, horses will also eat broad-leaf plants, shrubs, trees, mosses, lichens, and dirt—basically almost any plant material that can be digested (fig. 2.5)! Depending upon their habitat and the available forage, they will eat various plants according to the season and preference. Having observed horses eating toxic plants on numerous occasions with no ill effects, it seems horses have adapted to handle certain toxicities. They will often dig up dirt to acquire needed minerals and soil microbes.

As already mentioned (see p. 19), during my undergraduate work at the University of Wyoming between 1973 and 1976, I conducted dietary overlap studies to compare wild horse diets to cattle, elk, and pronghorn antelope. Horses in most all habitats had more overlap with elk than with cattle. Various herds of horses, from the Red Desert to Green Mountain, roamed slightly different habitats, and while their diets overlapped, each group also seemed to have dietary preferences often unique to it. Having since conducted a number of other range studies of wild horse diets throughout the western United States, it is clear that horses are opportunistic grazers, seeking out a variety of plants to eat each season.

Adapting Toward Group Cohesiveness

★ **TIP:** / There is adaptive advantage in growing a large social network and being opportunistic foragers. /

Horses have evolved to handle physical stress. They can go all day without water; live on limited poor-quality forage; handle dry, hot conditions and

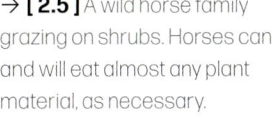
→ [**2.5**] A wild horse family grazing on shrubs. Horses can and will eat almost any plant material, as necessary.

← [2.6] A large group of Onaqui wild horses rests. Being together reinforces friendships and means more eyes are alert to ensure the herd's safety.

↘ [2.7] The mares and foal in this wild herd lean in the same direction as they gallop, sensing where each horse body is as the roan stallion pushes them.

cold, freezing conditions; recover from kicks and bites; run over rugged terrain all day long; and as long as they are all together, they handle life and death fairly well. But horses have not evolved to handle mental and emotional stress. The horse's brain is wired for group cohesiveness, and when under such stress, to come together to run. The *fight-or-flight* response is deeply ingrained. The endocrine response to stress is to secrete cortisol, increasing horses' ability to convert glucose for energy and run faster and longer. But horses will rarely run off alone, rather they will seek to cluster into groups when threatened. Having lots of eyes on the "look out" for predators gives horses in a herd more time as individuals to eat and explore nutrient needs. Hence, there is an adaptive advantage to growing large cohesive groups.

★ **TIP:** / Horses are emotional and energetic creatures. /

Although horses are cognitively aware and make decisions, their essence of response to stimuli is based on emotions and energy. As prey species, they have evolved to be highly sensitive and alert to danger, and safety is associated with being in their group, close to their friends (fig. 2.6). Being able to sense danger when a herd member's energy changes is again an adaptive advantage. Those who are not as aware and sensitive to energy changes may end up getting eaten.

★ **TIP:** / Spatial awareness and respect is learned for safety and structure. /

Thus horses in cohesive herds are very sensitive to the "energy dynamics" of the group. They have great *proprioception*—sensing where their bodies are relative to their environment and other horses. If one horse moves slightly and starts to run, the whole herd can move in rhythm (fig. 2.7). This is similar to fish and other social animals used to moving together. There is an energetic connection, which helps them move in unison. To a casual observer, a group of horses may look unorganized and disconnected, but although they may not express *murmuration* (a term specifically used to describe how starling flocks move as a group to confuse predators and keep warm) as well as birds and fish do, a well-connected herd will stay together without horses running into each other. This may be because horses learn *spatial awareness* at a young age.

★ **TIP:** / Herds with the strongest social bonds tend to have the best ability to adapt. /

More Important Than Food or Water

In fact, the greatest motivator for horses is not food or water, but their friends. As you have just learned, there is a strong survival reason for this behavior. For the last three to five million years, horses have developed a strong ability to form functional social societies because their ability to form social bonds and friendships within their family bands, social groups, and herds has allowed them to adapt to a variety of habitats and environmental stressors.

PART ONE / CHAPTER 2 / Horses as a Species in Nature

[2.8 A–C] The two stallions in Photo (A), perhaps brothers, may acquire mares together rather than separately, while staying together in a single larger group. Mares, like those in (B), often make friends for life from outside their band. The gray stallion in Photo (C) has a band of mares and foals, and the black stallions are from a bachelor band, but they all seem like friends or family.

Interestingly, social bonds are not limited to family members or even to other horses within a familiar group. Horses have the ability to seek other friendships outside their own herd. You see this with large groups of 100 to 150 wild horses spread out in intermingling social groups and not all from the same band. The stallions may socialize with other stallions, foals may play with foals from other bands, and mares from various families may interact (figs. 2.8 A–C).

Generally, the herds with the strongest social bonds tend to have the best ability to adapt because they spend little time establishing social structures and more time reinforcing actions like eating together or buddy-scratching each other (fig. 2.9). Even small family groups with a strongly bonded male-female pair and perhaps two offspring can do well in the wild. Thus, it does not appear that the size of the group matters as

→ [**2.9**] Horses spend a good deal of time socializing and touching each other when they are not eating. This interaction allows them to stay constantly emotionally connected to their friends.

> **In however complex a manner this feeling may have originated as it is one of high importance to all those animals which aid in and defend one another, it will have been increased through natural selection; for those communities, which included the greatest number of the most sympathetic members, would flourish best.**
>
> Charles Darwin

much as the ability to form social bonds and make good decisions related to the group.

_ Social Disruption

Herds with the most functional social bonds seem to have good parenting skills and supportive social networks across ages and genders. Prior to repetitive human interference caused by gathering herd members and removing functional leaders, wild horses I observed in Wyoming had maintained bonds between stallions and mares, stallions and stallions, and mares and mares for many years. While horses are often good at restructuring and reorganizing themselves after gathers, a number of "pair-bonded" individuals were removed, leaving behind dysfunctional social structures. Bachelor stallions preyed upon young

[2.10 A–C] A caring father "talks" to his crippled foal and encourages him to walk **(A)**, then supports the colt, keeping his nose on the foal's flank **(B)**, staying behind the band to encourage the youngster up the hill **(C)**. The stallion helped the foal along for hours. Note the father's concerned expression.

females as there were no stallions to defend the youngsters, which caused reproductive rates to increase quickly.

Since horses by nature seek cohesive groups, when there are individuals willing to restructure the group, then herds will reform. But in cases where there aren't individuals willing or capable of functional restructuring, social dynamics can shift in ways like I've described—to "predator bachelor" bands in some cases or smaller stallion-mare groups. Other herds may increase the integration and migration of stallions and mares as they struggle to redefine stable social networks.

_ Altruism

What stood out in my time observing wild horses, and perhaps is the reason horses touch such a deep chord in

> **Besides love and sympathy, animals exhibit other qualities connected with social instincts, which in us would be called "moral."**
>
> Charles Darwin

humans, were the individuals who demonstrated altruistic behaviors (behavior that benefited others at their own expense)—the bachelor bands that adopted an orphan foal and raised it, or the stallions that stayed behind with injured youngsters to help them walk, sometimes going without food and water for hours (figs. 2.10 A–C); the group of mares who were strongly bonded to a stallion and protected him from relentless "predator bachelors" when he was injured (figs. 2.11 A & B). Observing these

special nurturing qualities made me wonder why we often don't see such caring behaviors in our own species.

★ **TIP:** / Horses have evolved well to handle physical stress but not mental and emotional stress. /

_ Nurture vs. Aggression

Functional herds usually have one or more *social facilitators*—the horse or horses that the others seem to interact with most. The social facilitator is often not the "leader" but the horse that is most aware and social. These behaviors can be interpreted as *emotional intelligence* (the capacity to control and express emotions, and to handle relationships with others empathetically) because these individuals seem connected to the most herdmates in

↑ **[2.11 A & B]** The stallion known as "Golden Boy" maintained strong bonds with his mares for years. Fertility treatment and roundups disrupted the social structures of area bands, and Golden Boy was injured fighting off a number of stallions, seeking to breed his mares. He is seen here in the middle of a circle of his mares, who surrounded him the day before he died **(A)**. The mares would charge and try to run off other stallions who approached him. He and his favorite mare kept soft eye contact on their last day together **(B)**. She would not leave his side.

STORY FROM THE FIELD:

Stallion Friends

I have seen deep friendships override traditional social structure and behavior. One example was the case of two stallions born about the same time in the same herd (fig. 2.12). As two-year-olds they were run off and joined a bachelor band of other young stallions. Several years later, I observed the same two stallions with a group of young mares—still friends and still enjoying each other's company, sharing herd responsibility and breeding rights. While this behavior is not unusual in functional herds, it is less common today with the impacts of population stress and social-structure disruption.

→ [2.12] Two stallion friends enjoy grazing together.

Another case of stallion friends involved a group of horses gathered out of Arizona off a ranch donated for wildlife conservation. The BLM had handled the gather and I was consulted on unusual behavior in one of the stallions—a pinto. The horse had been rounded up and placed in a large arena but now would not move. He stood like a statue, and if a horse and rider came close, without looking at them, he charged. I was told that all the foals and mares in the herd, except for this stallion's band, had been killed by mountain lions. It became clear the pinto stallion had likely adopted unique behavior—standing still and silent and then suddenly charging—to protect his family group.

This same pinto stallion also charged at people to get them out of his way so he could reach a corral where another stallion was lying on the ground, sick. The pinto stuck his head through the fence and licked the injured horse, and then later protected his "friend" from the other stallions when they were hauled to a new facility. The pinto stallion had developed special bonds with other stallions and eventually with people too. He was a truly intelligent horse with excellent social skills. //

→ **[2.13]** An injured and weak foal is surrounded and protected by a bay mare, a buckskin two-year-old friend, and his mother, in the back. The group of horses would not leave the foal and stood watch over him until he eventually died.

a friendly, nurturing manner. (Note that emotional intelligence often goes unacknowledged because humans usually feel communication among horses is all about body language.)

While you can indeed observe horses using physical calming signals such as yawning, head lowering, rubbing forelegs, sighing, licking, eye blinking, and buddy scratches in an effort to relax and bring those in their group back into "harmony," these behaviors need to be put into context of what is happening in the herd for accurate interpretation.

★ **TIP:** / Within most functional herds, there is at least one *social facilitator*. /

Horses can appear aggressive, but most of their body language reflects their gentle nature of nurturing. As with all creatures, individual differences can demonstrate both aggressiveness and gentleness, but those herds with the most aggressive behaviors also have weak social bonds and unstable groups.

By studying a number of wild horse populations in various habitats and during various seasons, I was able to make broad observations of equine interaction during breeding, foaling, movement, feeding, watering, and other daily activities. Although each herd varied, depending upon the social structure of the group, the herds that were considered "socially functional" at the time (meaning they included stable, bonded pairs of stallions and mares with foals and other young horses) had ten times the nurturing behaviors to every

> **Behaviors need to be put into context for accurate interpretation.**

[2.14] The orphaned yearling (on the left) was adopted by a bachelor group when his mother died. The pinto protectively resting his head on him is the lead stallion, and the two remained friends for years.

[2.15] A palomino band stallion greets two bachelor stallions. Note the three-way conversation. Stallion friends will often either stay close with their individual bands or become "lieutenants" for each other, sharing the responsibility of a single band.

Interpreting Ears Back

I find that on many occasions people interpret a gentle "ears-back" expression as aggressive, but in horse language there are many positions of the ears, and they may vary according to age, gender, and herd. A gentle tilting back of the ears while eating is considered a greeting, like, "Hi, I see you, but I don't want to stop eating." Or it might be, "Hi...don't get too close to my space, but don't go away." Often horses in a herd that were friends would greet each other with a gentle look of the eye and a tilt backward of the ears.

aggressive behavior observed (fig. 2.13). In stable herds, there was actually very little aggressive behavior noted, aside from stallions posturing or defending their mares, or an alpha mare running off a lesser mare before she was bred and settled. Even in bachelor groups I observed, friendship played a central role and "mock" fights were not always aggressive but instead "play" behaviors. The bachelor groups had at least four times the number of nurturing behaviors to every aggressive behavior noted. On several occasions when foals had been orphaned or lost, bachelor males protected and adopted the young horses (fig. 2.14).

★ **TIP:** / In functional herds, nurturing behaviors (across genders and ages) are far more common than aggressive behaviors. /

STORY FROM THE FIELD:

Curiosity Overcomes Fear

Curiosity surfaces when fear is suppressed. This is especially true for horses. Curiosity can be a motivator when working with horses because they use smell and touch as nonverbal communicators to determine if something in their environment is safe.

One example was in a field on top of Green Mountain, Wyoming, with the herd I had been studying all summer. I knew the horses' routine, but they had been difficult to get close to in the daytime, so I was sleeping in their "nighttime meadow." I was only half asleep, snuggled up in my sleeping bag, when two foals approached me. Their mothers were dozing and somewhat unaware of what the youngsters were doing as one of the foals started to tug at my sleeping bag. He tried to lift it up with his mouth, but of course I was a little too heavy for this maneuver. I peeked out of my bag and wanted to say, "Boo!" but I also wanted to observe how far this play could go (fig. 2.16).

I stayed quiet, just enjoying and watching as one foal brought his nose to my face, sniffing me. He seemed unaware that I was a living, breathing human being. The second foal lay down next to me and then, within a short time, a whole nursery of foals showed up, lying in the safe, quiet energy around me. From then on, I became a "normal" part of their habitat and was able to get closer, even in the daylight (as long as I did not stand up!). It was clear my "sleepy time" energy was relaxing. Similar situations happened with elk walking around and foraging over me at night when I was sleeping out on the range. //

→ [2.16] A curious foal pesters his sleeping mother.

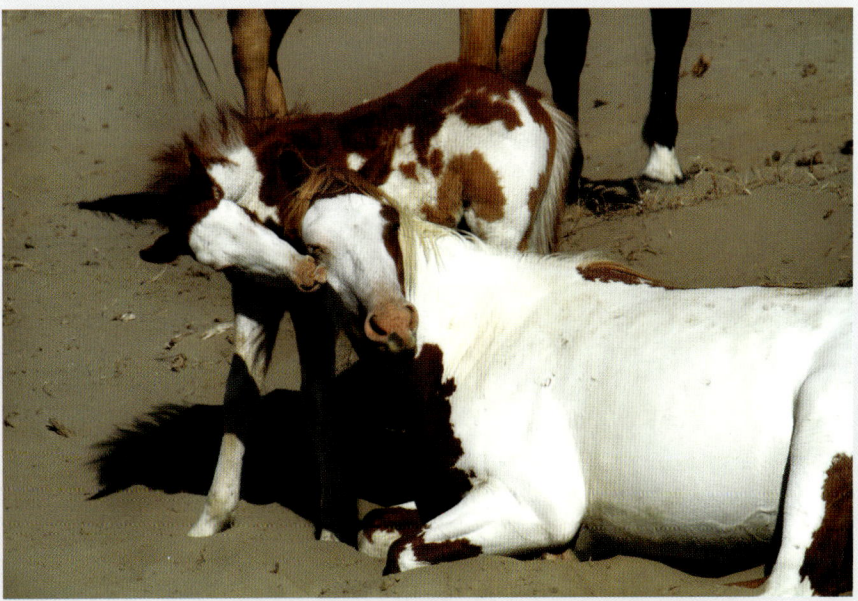

Nurturing behaviors are easy to see if you know what to look for—such as nuzzling, nudging, head bumping, shoulder touching, scratching each other, wrapping heads and necks around or over each other, gentle nips, nickering, and soft tail swishes—but at the time of my work with wild horse herds they were not well documented. Most observational studies investigated and focused on stallion behavior and the aggressive fighting that took place during breeding season. However, the fact was that many stallions formed lifelong friendships with each other, and those who were friends showed little if any aggressive behaviors toward each other (fig. 2.15).

_ Curiosity

★ **TIP:** / Horses are curious when they are not fearful. /

Most people know about the *fight-or-flight response* as humans have a similar reaction to stressful stimuli. But people often overlook the trait of curiosity, which has made the horse a great domestic friend (fig. 2.17). When a stressor is *not* perceived to be frightening, then curiosity may be stimulated. It can be learned from others, but some consider that it may be more related to genetics and hence a cultural marker in particular herds of horses.

★ **TIP:** / Relaxing and sleeping energies act as "calming signals" to the herd. /

Adoptability vs. Adaptability

Recent studies have shown that equine genes play a role in whether a horse will be more likely to be "vigilant" (highly sensitive and reactive to stimuli) or more "curious" (more interested in discovering what something is than running from it). A Japanese study found temperament traits related to polymorphism in

Learn more

the dopamine D4 receptor gene strongly influenced this behavior (scan code for more information).

Although no genetics tests were available at the time of my work in the field, equine cultural differences in wild herds no doubt would have reflected these genetic differences if they were tested. For example, the Beatys Butte herd in Oregon was made up of very sensitive and alert horses. When they were gathered and put into pens with horses from a different area, such as Stinkingwater or Warm Springs, which were more curious and less sensitive, the Beatys Butte herd often got pushed around. The horses who were less "vigilant" could just look at the Beatys Butte horses and get them to move out of their space.

★ **TIP:** / Genetics can play a role in equine temperament and culture. /

This was, in fact, something I was looking for. An observer could see that the Beatys Butte horses were more "sensitive" and "reactive"—traits that helped them in the wild but not in a domestic situation. My study identified cultural and behavioral traits that would allow horses to be more "*adaptable* in nature" versus those that made them "*adoptable* into a domestic lifestyle." One of the primary traits for *adaptability* was the horse's high level of "alertness" to his environment and sensory stimuli such as novel objects, new smells, sounds, and being touched. Horses who had more adoptable traits were more curious and less reactive toward these things.

Because of horses' ability to adapt quickly, they have survived all over the world. They have been domesticated and were able to re-adapt to their habitats. Their innate ability to form strong social bonds and work collaboratively provide a wonderful model for survival.

← **[2.17]** The white stallion is talking to another band's stallion (the buckskin) with youngsters next to him who seem curious and like they want to "help." The yearling and two-year-old palominos, as well as the roan, followed the white stallion around, learning from his exchanges.

[**3.1**] This pinto stallion looks at something he heard or saw as he decides whether he needs to respond.

CHAPTER

[**3**]

HOW HORSES SENSE THEIR WORLD

> Whether it is an ability to sense electro-magnetic fields, highly sophisticated proprioception, or something else, science is only just starting to recognize a 'sixth sense.'

[3]

Evolving as a reactive species that "runs when in doubt" has proved to be an asset for equines. Their senses have played an integral part in this success. Horses, like other social mammals, use all forms of senses, including sight, smell, taste, hearing, and touch. But as we've already touched upon in the previous chapter, horses also use subtle energy perception—a "sixth sense" of sorts. Whether it is an ability to sense electro-magnetic fields, highly sophisticated proprioception, or something else, science is only just starting to recognize this ability in animals and study it (fig. 3.2). Let's look at each of the senses in more detail.

★ **TIP:** / Horses tend to use vision primarily for determining danger, and smell and touch for social recognition, foraging, and breeding. /

Sight

With eyes on both sides of their heads instead of in front, horses have mostly *monocular vision* (both eyes are used separately), which means they see differently through each eye. This is important when you spend much of your time grazing with your head down and need to observe your herd's movement out

Learn more

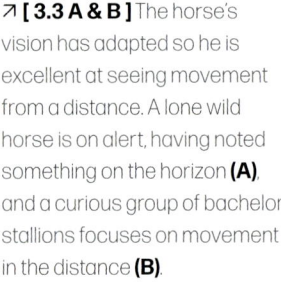

[3.2] As we touched upon in chapter 2, horses use subtle energy perception to enable them to move safely in large groups "as one."

of one eye while keeping the other on the horizon, looking for danger. Although they do not have great depth perception compared to humans, horses can sense even slight movement at a distance very well (figs. 3.3 A & B). They are wired this way because this is where predators may lurk.

★ **TIP: / Horses have a keen ability to see slight movement at a distance. /**

When horses raise their heads and look forward, they have *binocular vision* (both eyes used together with overlapping fields of view), and better depth perception, but since horses use their other senses to determine danger, their vision is focused on sensing movement. With their eyes positioned on the sides of the head as they are, horses also have about a 340-degree view around them.

↗ **[3.3 A & B]** The horse's vision has adapted so he is excellent at seeing movement from a distance. A lone wild horse is on alert, having noted something on the horizon **(A)**, and a curious group of bachelor stallions focuses on movement in the distance **(B)**.

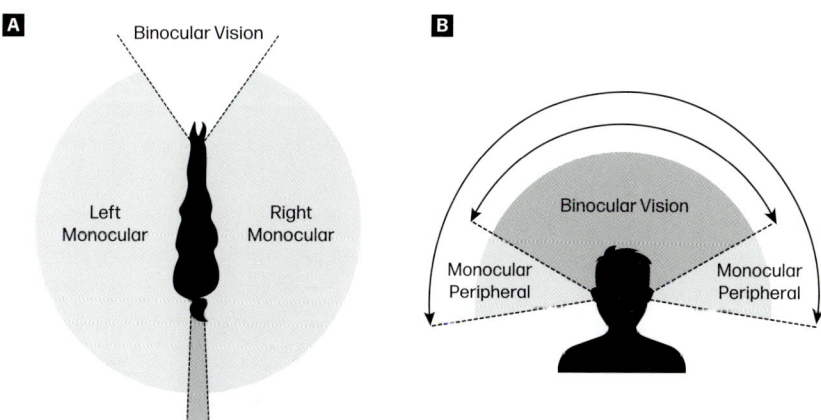

[**3.4 A & B**] Differences in visual perception between horse **(A)** and human **(B)**.

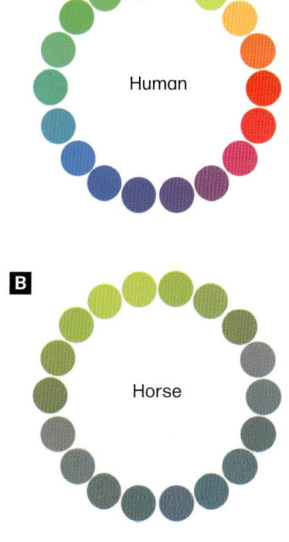

[**3.5 A & B**] Horses see color but not the same as humans. Horses may not have the ability to see red and orange, based on current research.

(In comparison, humans have about a 90-degree view.) Horses have blind spots directly in front and directly behind them, which is why they need to move their heads from side to side to see clearly and feel safe, and why you will often see them grazing together head to head, or head to tail (figs. 3.4 A & B).

Research has shown horses do not see color the way humans do—they lack the red cones in their retinas that humans have—but they do *sense* color (figs. 3.5 A & B). Their retinas are also packed full of cells compared to other animals, and research suggests most horses can see more like 20/30 to 20/60, compared to humans seeing 20/20. (That said, not all horses see well, and behavioral oddities may be partly due to poor vision.)

While horses can adapt to sleep cycles of the social network they are in, they are more nocturnal than diurnal and have very good night vision—better than humans and many other animals. This adaptation is an advantage because predators may hunt at night. However, equine eyes do not adjust quickly to changes in light, as compared to humans. Horses can navigate down a rugged mountain at night with no trouble seeing, *but* they would have trouble seeing if a bright light were suddenly shined around them. Research shows it takes horses anywhere from 15 to 30 minutes to accurately adjust to light changes. It is important to keep this in mind—when going from a dark barn to outside sunshine or from bright daylight into a dark trailer, horses only see blurry objects.

★ **TIP:** / It takes horses 15 to 30 minutes to adjust to changes in light, but they see better than humans at night./

The horse has adapted well for natural changes in his environment, and his vision acclimates as the sun goes down or comes up. But his eyes have not adapted well to human spaces where light changes rapidly. This is seen

PART ONE / CHAPTER 3 / How Horses Sense Their World

[**3.6 A & B**] Two stallions greet each other nostril to nostril, exchanging friendly information **(A)**, and a gentle greeting between a stallion and a mare **(B)**.

when captured wild horses are hauled to adoption sites and chased off dark trailers—they run into the pole fences. Their eyes have not had the opportunity to adjust to the light, and they are frightened. If instead they are allowed to walk off the trailers at their own pace, they will put their noses down and use their sense of smell to direct them and guide them to safety.

Smell

Little research has been done involving equine smell, and it may well be the horse's most important sense. It is used for finding food, noting familiar habitats, participating in reproduction, and gaining social recognition (figs. 3.6 A & B).

> **Smell appears to be the dominant sense for social interaction and finding food and water.**

Horses have two olfactory systems: One that has two olfactory branches extending from the nose over the surface of the *nasal mucosa* (the *vomeronasal organ*, once known as the *Jacobson's organ*), and another that senses odor in the nasal passages and sends messages directly to the brain. There are two distinct olfactory areas in the brain, located at the front on each side of the cerebrum and connected to the main olfactory nerves and receptors in the nasal passages. The left nostril leads to the left side of the brain and the right to the right side.

★ **TIP:** / Horses have a sophisticated sense of smell, which appears to be the dominate sensing organ for social interaction and finding food and water. It is the only sense that has direct access to the cerebrum (the uppermost part of the brain—the right and left hemispheres). /

While the horse's sense of smell is not as good as that of a dog, it is over 50 times better than most humans, with over a hundred million olfactory receptors (humans have only five or six million receptors). Equines also have over 1,000 olfactory receptor (OR) genes (the receptor genes responsible for smell), while humans only have about 350. Long-nosed mammals, in general, have long nasal passages allowing more olfactory receptors to pick up smells and linger.

Another great adaptation of horses is the *nasal diverticulum*. Often called a "false nostril," the pouch extends about 5 to 6 centimeters in each nostril. Under strenuous exercise,

STORY FROM THE FIELD: //

Blind Horses Are Calm Horses

On several occasions I have observed unusual behavior in wild horses, both in the field and when captured. One such horse was an albino mare with a foal. She was unusually quiet during the gathering and loaded well with her foal by her side, following the other horses. Her lack of anxiety during the process was unusual, and she actually appeared to have a calming effect on the horses around her.

The mare and her foal were adopted in Palomino Valley, Nevada, shortly after they had been gathered. But this time a team of men had trouble loading her and her foal. I have witnessed many wild horses choosing to run into trailers when being chased or pushed by handlers as the trailer offers safety and cover. But this mare just stood quietly, simply refusing to go in. Her foal eventually lay down, as he was not upset either.

After 30 minutes watching, I could tell the mare was blind as I had observed blind horses in the field—often noting they were quieter and when in doubt would stand and call to a friend to come and support them. Although I was consulting with the Bureau of Land Management at the time, I was not very welcome when it came to giving advice to others. But after watching eight men with two long ropes try to pull and push the mare onto the trailer, I rode over and suggested they pick up the foal or lead him into the trailer first, allowing the mare to smell and follow her foal forward, as she was blind. When encouraged, the foal moved into the trailer, and the mare touched her foal's rump with her nose and walked right in behind him.

While not common, blindness did seem to arise in some wild herds more than others. Generally, if the blind horse had a friend, then he seemed to do well, exhibiting calmer behavior than the rest of the herd, and when brought into captivity, often adjusting well when his vision limitations were recognized and he was placed with a friend or scent markers (see p. 186) for doors and objects. //

← [3.7] New smells and reproductive or hormone smells often elicit *flehmen response* in horses, as seen in this stallion, as they curl their upper lip back to hold an unusual or interesting odor in the nostrils. With over 1,000 receptor genes for smell, horses have an amazing ability to differentiate odors.

the pouch opens up to allow more air intake, then closes down again when horses are at rest. This adaptation allows wide-open nostrils (when running, for example) to take in more air, but reduces to minimal openings when the horse is feeding or relaxed.

★ **TIP:** / **Horses use their sense of smell to recognize friends (via their manure), find food and water, assess reproductive readiness, and know the health status of other members of the herd, as well as whether they are relaxed or stressed in general.** /

In studying wild horses, it became clear that smell (along with taste and touch, which we'll discuss next) was used far more than other senses in the daily activities of the herd. Smell was often the primary form of communication and necessary for social functioning, plant selection, finding water, and of particular importance when it came to determining the health and welfare of the other horses in the group through the smelling of manure and urine (fig. 3.7).

Taste

Taste and smell go together in horses. You can observe horses smelling dirt, manure, or a novel object, and then they will often taste it with their lips or their tongue. This ability to smell, taste, and use the flehmen response and vomeronasal gland gives horses multiple ways to process olfactory information.

★ **TIP:** / The ability to taste, smell, and use the flehmen response and vomeronasal gland gives horses multiple ways to process olfactory information. /

Horses seem to have or to cultivate taste preferences. This is true both in the wild and the barn. As I've mentioned, horses are opportunistic foragers and will seek plants to meet their nutritional needs—but sometimes they choose food just because it tastes good. Those living in the same habitat but from different social groups may eat different plants. One example I've already shared is the herd I studied on Green Mountain in Wyoming enjoyed browsing more with the elk, while another herd preferred grasses and staying at lower elevations. Horses' ability to find and survive on a variety of plant material, from lichens to grasses, is one reason they are so adaptable.

When domesticating wild horses, you can observe the various tastes and how long it takes

STORY FROM THE FIELD: //

Seasons of Larkspur

One of the most interesting behaviors I observed in the field was watching a mare teach her foal about what to eat when. In early summer, I observed them, along with several other horses, eating a number of plants that later in the season become toxic. One of these plants was larkspur, which was blooming at the time. Later in the summer, observing the same herd, I watch the mare push her foal away from eating the larkspur. She did this more than once, laying her ears back and physically pushing her foal when he tried to take a bite of the plant. At first, I thought the foal had done something—misbehaved in some way—to elicit such a dramatic response from his mother, but after watching the same event play out several times when the foal attempted to eat the larkspur, I realized his mother must have known the plant was becoming too toxic to eat. (Larkspur can increase in alkaloid toxicity as its pods form. //

[3.8] This pinto is using both sight and hearing to assess whether danger lurks in the distance.

certain horses to learn to consume new foods. For example, horses coming into confinement and seeing alfalfa for the first time often will not eat it until they see another horse doing so. They might be familiar with dried grasses and will eat grass hay right away. Once they acquire a taste for alfalfa, they will munch it down just fine.

★ **TIP:** / **Horses have a keen sense of taste and use it to back up their sense of smell. Using the two senses, they can identify the nutrients they need and select the food they need or want to eat.** /

Hearing

Horses have well-developed ears and hear separately in different directions. Hearing studies in horses have not provided clear results, but researchers think horses' hearing is similar to humans in the lower ranges but that they hear higher pitches than we do. Studies have shown humans hear in the frequency range of about 20 to 25,000 Hertz, while it has been reported that horses hear in the range of 20 to 33,500 Hertz. Separate findings indicated horses responded to even lower frequencies—in the 14 Hertz range. Regardless of what they hear, response to noise can vary from horse to horse.

★ **TIP:** / Horses can hear noises much farther away than humans at both higher and lower frequencies. /

Horses can rotate their ears about 180 degrees without turning their heads. They have 10 ear muscles (humans have three and most people can't move them). This adaptation allows them to listen while they eat with their heads down, constantly moving their ears. Combined with their vision, they can sense what is going on all around without lifting their heads, and they have the ability to track the direction of the noise. During my wild horse studies, I noted that some horses always seemed to sense or hear my truck coming from miles away and would be gone before I got close, leaving just hoof prints and manure behind. Other groups did not seem to hear (or were simply not alarmed by) the sound of a vehicle.

When horses hear something, they look in the direction of the sound. Likewise, if they see something on the horizon, they track it with their ears, trying to pick up sounds that may give them more information. Horses with limited hearing living in groups seem to do well by relying on friends, just as we see with horses with vision impairments. I have found those who appear to be overly sensitive to sound often have limited sight. In the field, I found on several occasions that one horse might pick up his head, startled by a sound or movement, and then nudge his friend to look up and check. This kind of "cooperative sensing" provides safety to all horses in the herd.

★ **TIP:** / Horses like to understand and associate sounds with actions. /

Touch

Horses have developed an acute tactile sense. The extensive neural network in the skin allows them to feel exactly where a fly lands and, using the *cutaneous trunci* muscles that cover most of the dorsal and lateral walls

> **Horses' hearing and vision 'cooperate' to help them sense what is going on all around them.**

[3.9] A wild foal leans against his mother. This is a comfort to him and allows her to continue grazing while maintaining connection.

of the abdomen and thorax, twitch the fly away. This means the horse can continue grazing while using his tail to switch and his *trunci* muscles to twitch the areas he cannot reach with his tail.

Touch is also extremely important for social bonding, nurturing, and connecting in horses (fig. 3.9). From full-body touching and mutual grooming to gentle nose bumps and head rubs, horses use touch constantly when interacting in social groups. Touch is often overlooked in field observations because it can be subtle. A gentle nose bump to greet a foal to acknowledge it is okay to nurse or a head rub on a buddy's shoulder can go unnoticed but are forms of communication (fig. 3.10). When threatened, such as when they are being herded or chased, horses who usually show respect for each other's spatial distancing, will bump shoulders into shoulders as they run or cluster together for protection.

↑ [3.10] Touch is critical for comfort and communication.

Horses use their facial whiskers to sense their environment, but they can actually feel through all the hairs on their body. In addition to whiskers, you can observe small hairs on the front of the horse's nose. These help horses eat in places they cannot see and navigate objects in the dark.

Horses have a prehensile nose and lips that allow "grasping," giving them the benefit of being able to manipulate things in their environment. It also helps in finding food, grooming, digging, and defense. For example, a horse might use his nose (and his great sense of smell—see p. 50) to move a prickly plant out of the way so his long tongue and prehensile lips can grab a bit of hidden green grass.

★ **TIP:** / The horse's highly sensitive prehensile lips are used to grab, touch, scratch, and interact with the environment. /

Subtle Energy Sensing

Because humans do not yet know everything about how the horse's senses work, we don't have a dependable way to measure his ability to make a decision or to react to something that we humans cannot explain. We may generally chalk it up to their five main senses, which of course are keenly important, but they do not explain unusual scenarios, such as the blind horse that lived capably for years, as his herd traipsed up and down mountains and across rugged terrain, simply by listening, smelling, and occasionally touching a friend. The horse's "sixth sense" may well be used as often as the other five senses.

Henry Blake was a horseman and an "equine naturalist" who wrote several books outlining his studies on equine awareness, including the well-known *Talking with Horses*. In one study he isolated two horse friends over 20 miles from each other and had observers record their behaviors. When one horse was threatened and alarmed, the second horse reacted similarly, even when separated by distance—for example, running around the field. Blake documented several experiments in which horse friends who were separated from each other in this way showed an "empathetic pairing." Blake used domestic horse subjects, but this ability to sense danger or stress between strongly bonded individuals has also been observed in the wild. As Dr. Rubenstein stated in his introduction to Part One, "Really good friends in an equine herd almost 'mind read,'" referring to closely bonded individuals within an equid population (see p. 10).

Awareness is rewarded in wild horses as it is an adaptive advantage to being able to assess sensory input and make cognitive decisions about safety and food. Horses have evolved to integrate this sensory input from their environment, and from other horses, in order to make rather rapid decisions about safety. The fight-or-flight instinct (which most often takes the direction of "flight")

↓ **[3.11]** The bay stallion in the front is highly alert. He may have sensed something before the other horses in the group, but his reaction is then felt by them. It pays off to be sensitive and aware in nature.

is wired deep in horses, and they have not yet evolved away from quickly reacting to perceived movement or sensed danger. Like other prey species, they seem to have the ability to sense danger even when no apparent stimuli has acted as a trigger, and often make decisions based on feelings not thinking (fig. 3.11). Still, they are able to maintain their awareness in the moment and stay focused on subtle stimuli.

Horses are good at "reading energy" for safety. My undergraduate research demonstrated that they, like some other animals, have the ability to sense mental and emotional energy around them. They also have the ability to synchronize their breathing, heart rate, and heart rate variability (the measure of the variation in time between each heartbeat, which has been the subject of recent studies). After the stress of being gathered and separated from their bands, horses experienced elevated heart and respiration rates, which quickly dropped when they were rejoined by friends and herd mates. This observation was made many times with various groups of horses, leading to the conclusion that being around friends created a sense of safety even in stressful situations. It is likely horses—and other social mammals—have the ability to calm and relax each other by slowing their own heart rates and synchronizing their breathing, developing a sort of "heart coherence" (an empathetic connection that is also called "heart coupling"). This may also be why horses are such valuable teachers in horse-human interaction programs.

★ **TIP:** / Horses are capable of sensing subtle energies. /

Integrating the Brain and Body

Living in groups has allowed horses to rely on each other—the more eyes and ears and noses tuned in to the environment, the better the survival rate for the group. Not every horse has to have perfect vision, hearing, smell, or proprioception when they can "buddy up" with friends and find safety in their social group.

While all mammals share similar shapes and functions of their brains and nervous systems, horses have some unique differences from humans.

Because horses need to be able to get up and move quickly to avoid predators, they have evolved fast-developing *myelin,* which acts like an insulator to the nerve fibers that transmit information. This allows a foal to be up and running around in minutes after birth. By stretching many times in the first

"
Horses often make decisions based on feelings, not thinking.

STORY FROM THE FIELD: //

Ten Horses, One Breath

In the 1980s a friend and I were taking a load of ten wild horses to a sanctuary in Northern California. We had three stallions in the front part of the trailer and mares and foals in the back of it. The trailer was fairly closed up, except for windows at the top. When we stopped for gas, the station attendant asked us what we had inside. We said, "Wild horses," and he just laughed, as there was no sound or movement of any kind coming from within. Normally, when you stop with a trailer full of domestic horses, you hear them moving around in anticipation of getting off or food or water. But we heard nothing. My friend and I put our ears on the side of the trailer and looked at each other in awe. All we could hear was one deep breath in, a moment's hold, then a breath out. The horses were breathing in synchronicity—even the foals—so it sounded as if only one horse was in the trailer, not ten. Was this a way to connect and relax in a stressful situation? Or was it an adaptive behavior based on calming each other in their previous habitat, which was populated with mountain lions? We did not know. //

few days of life, the foal can quickly strengthen the *motor pathways* (neural pathways that originate in the brain or brainstem and descend down the spinal cord to control the motor neurons, and thus posture, reflexes, muscle tone, and conscious voluntary movements), causing rapid myelination to take place. Foals may eat their mothers' manure to gain not just important intestinal bacteria, but also *deoxycholic acid,* which assists in proper myelination.

The *cerebellum* (the part of the brain that coordinates and regulates muscular activity) becomes more developed as it acts to integrate sensorimotor information as a foal learns important physical information about his body, balance, and proprioception. This is essential so the foal does not run into other horses when the group takes off galloping.

Repeated series of motor movements in the young horse—such as bucking, running, stretching, and getting up and down—also increase myelination, which starts in the brain, developing what is most important for survival, and then moves toward the primitive frontal lobe associated with attention and focus. Hence, it is critical for young horses to get up and moving, as it helps ensure development of their cognitive ability later (fig. 3.12).

Equine brains are wired for sensory input and motor learning from an early age. As they develop, they are able to make cognitive choices and decisions, but they are still primarily associated with group cohesiveness. Most everything in the horse's brain and body has evolved to learn how to move in balance and coordination *relative to their environment and other horses* while integrating sensory input.

↑ [**3.12**] Foals can get up and start running within minutes after birth.

← [**4.1**] This mare strike is a clear communication to the stallion to "Back off!" Note the eyes, ears, and tail providing emphasis.

CHAPTER

[**4**]

HOW HORSES COMMUNICATE

The careful observer will note that sometimes equine communication is in the silence—the subtle gestures.

[4]

What Do Horses Talk About, and How Do They Talk About It?

Communication and the transfer of information is key to the survival of any species. As social mammals, horses are no exception. What humans deem "communication" is usually limited to a signal we observe and an action it may have related to, but while some forms of communication are well defined, such as a horse "screaming" for his friend or a mare kicking out to define her space, more subtle communication is not always noticed (figs. 4.2 A & B). Often missed are the meaningful tail swish, eye glance, or head bob. The careful observer will note that sometimes the communication is in the silence—the subtle gestures.

So, what do horses have to talk about and what channels of communication do they use to exchange information? As a prey species, horses tend to be quieter than social predators, but come upon a herd during breeding and foaling season or a group of bachelor stallions engaged in play and mock fighting, and you'll find horses can be quite vocal.

As I've already indicated, horses use various channels of communication, depending upon the situation, and they often integrate more than one.

STORY FROM THE FIELD:

Learning to Listen

Spending days alone in the field with no cell phones or other people, just animals, one learns to listen to the silence and become more horse-like and thus able to observe and be a part of the subtle communication they use. The language of silence slows the mind and opens the intuitive channels, allowing clearer interpretation of what kind of and how communication is being exchanged within a group of horses.

Many times when the horses I was studying would react to something I did not see or hear, I would assume it was a smell or something beyond my normal range of sensing. However, the more time I spent alone in the field and as I became a part of the habitat, the more I realized I had initially missed many subtle gestures.

→↘ **[4.3 A & B]** A group of worried mares takes off after a bachelor band chases a foal from the side of the mare in the middle **(A)**. The mares pursue as a group but do not seem to know how to get the foal back. When the mare in the middle finally does get her foal back, you can see more signs of the worry she and her bandmates experienced: note her eyes and ears, and those of the mares in the back, still watching the bachelors **(B)**. This cooperative behavior is known as prosocial.

At first, when watching from a distance through binoculars, I could not observe the animals' eyes well. But when I was close enough to clearly see the facial expressions of the individual horses, particularly those making decisions and those in the roles of social facilitators, I realized the horses were in constant eye contact with each other. Their ears tracked and acknowledged this, and the expression in their eyes acknowledged this conversation, even when their heads did not move. //

↑ **[4.4]** A nurturing mare nuzzles her foal. Note the soft eyes in both mare and foal.

The following are the dominate forms of communication associated with specific activities. Some I have already touched upon in this book, but I will discuss each of them in more detail as pertains to equine communication on the pages that follow:

- Vocalization
- Body Language
- Smell
- Touch
- Subtle Energy
- Spatial Awareness
- Play

★ **TIP:** / Horses use a variety of methods of communication, depending on the situation and including body language, smell, touch, taste, vocalizations, and subtle energy. /

↑ [**4.2 A & B**] What an equine conversation looks like: The roan mare in this family group bites the stallion to reprimand him for responding to a flirting youngster **(A)**. The mare then reprimands the filly for flirting with the stallion **(B)**.

_ Vocalization

Horses are constantly communicating. While vocalization often takes the form of the familiar gentle nickers of greeting or longer, higher-pitched whinnies when calling to a friend, there are different calls, primarily used for alarm, greetings, and expressing emotions. Grunts, snorts, squeals, and blows are also forms of vocalization. Grunts and snorts can be heard when horses are unsure of new objects or curious about something. Some horses use snorts or blows through the nostrils as greetings. Squeals are often sounded by youngsters expressing joy and by mares engaged in courtship with stallions. During breeding season, squeals can be heard as a stallion gently nips a mare, either on the flanks or the

↑ [**4.5 A & B**] Young foals will nicker and "clack" their teeth together as a sign of submissiveness to show older horses they know they are young and pose no threat. Here a filly nibbles at her older sister, trying to engage her to wake up and play **(A)**, and a colt clacks at his father, showing submissive respect **(B)**.

top of the tail, as a way to test for acceptance. Squeals may also be used after a social greeting, sometimes combined with a leg strike. Ever notice how when your horse greets another horse and one of them squeals, only the humans are startled by it? That is because your horse understands what is being said: Sometimes a squeal can represent a challenge to another horse, or it can simply mean they are just mutually coming to an agreement. Individual vocalizations can mean different things, depending on the individuals and circumstances (figs. 4.5 A & B).

When horses are separated from a group, they will use vocalizations—usually a long whinny—and then wait for a response to guide them. Horses who are gone from a group and then return use gentle nickers as soft "hello" greetings. This communication also occurs

[**4.6 A–C**] Stallion posturing: In **(A)**, note the eye roll and hyperflexed neck in the pinto. This is usually only seen in stallions. The buckskin in **(B)** shows confidence while the roan observes and backs off. The tails, ears, and eyes in both indicate that neither wants a fight. In **(C)** we see two stallions, most likely brothers, mimicking each other's body language.

when a foal who has been sleeping wakes up and wants to nurse–both mom and baby exchange a gentle nickering dialogue.

Again, various herd cultures use communication differently. Not all horses communicate exactly the same way because they have learned different signals.

★ **TIP:** / Vocalizations are primarily used to sound alarm, call herd mates, participate in courtship, express feelings, or greet another. /

Body Language

Body language includes eyes, ears, facial expressions, tail, body postures, and movement, which are all constantly used to communicate among herd mates and offspring. From calming signals to alarm, body language can be subtle or loud, depending upon the herd's culture. It can also mean slightly different things, depending on the herd. Hence, horses tend to combine eye contact, smell, and their "subtle energy awareness," as well as body language, to assess meaning of each other's communication. My field observations often found two horses, usually stallions, coming upon each other, then standing apart and staring at each other, both waiting for the other to move or take action. Sometimes this initial meeting would lead immediately to "posturing," greeting, or chasing (figs. 4.6 A–C). But in other instances, the two horses would stare at each other, then grab a scent in the breeze before moving in for a friendly greeting or simply trotting off with no other observable body language.

Common Body Language and Subtle Gestures

- **Ears forward:** Alert, attentive, focused on something.

- **Ears neutral:** Relaxed and at rest.

- **One ear back and one ear forward:** Scanning, resting, but paying attention to what is going on.

- **Ears back slightly:** Says, "Don't push me out of my space" or "Please move, as I want your space," or acknowledges the presence of another horse.

- **Ears flat back:** Aggressive gesture—that is, "Move out of my way, now!"

- **Eyes soft:** Relaxed.

- **Eyes wide:** Alert, worried, curious.

- **Eyes rolled back:** Stressed, fearful, angry.

- **Tail soft and relaxed:** At rest and comfortable.

- **Tail swishing tightly up and down or back and forth:** "Something is bothering me."

- **Tail swishing gently back and forth:** Keeping flies off oneself or another, relaxed but attentive.

- **Tail up and out:** Happy, confident, playful.

- **Tail tight to body:** Timid, worried, stressed, fearful.

- **Tail slightly out from body, even under stress:** Relaxed, confident, demonstrating leadership.

- **Tight lips:** Stressed, worried.

- **Relaxed lips:** Comfortable.

- **Head shaking or "snaking":** "Move out of my way," "Get going."

- **Gently moving head up and down with ears forward:** Asking to enter another horse's space.

Horse body language that communicates reproductive readiness, courtship, or respect for space is easy to observe and very familiar to most horse people. However, humans tend to observe the obvious but not the subtle communications. Smell or eye contact is often the initiating trigger for the body language we do notice, such as a stallion posturing or a mare raising her tail and urinating. (Both stallions and dominant mares may posture and either defecate or urinate as a way to show their strength and health.)

Facial expressions, especially related to the eyes, are critical to communication in horses, but often missed by humans. The horse's ears may be obvious, but they are often misinterpreted (figs. 4.7 A & B). Mares typically have more ear positions than male horses, and since horses can move their ears independently, they may be "talking out of both sides of their mouths"—in other words, one ear may be paying attention to the other horses or "looking" where the horse is going, while the other ear might be communicating to a foal to keep up.

As I've mentioned, horses can misunderstand each other if they have learned different or the "wrong" communication signals. For example, a foal might learn to put his ears back flat as a social greeting. This is not uncommon when mares pin their ears back if nursing hurts, although they still want the foal to nurse. The foal sees his mother's expression and ears, but then feels her push him gently back to nurse, so the foal thinks ears flat back is an appropriate way to greet other horses. He has now learned the wrong communication signal, and may be misunderstood by other horses, as well as humans.

Smell

Smell may well be the dominant communication channel used in horses, and the one most underestimated by scientists. As I discussed in the section about the senses on p. 50, smell employs the largest and most direct neuropathway in the brain, and horses use the sense to communicate just about everything, from determining the health and well-being of other horses to displaying

and noting reproductive readiness, to finding friends, to marking familiar trails with urine and manure.

★ **TIP:** / **Taking up the largest and most direct neuropathway in the brain, smell may be the dominant sensory organ in equine social communication.** /

↑ **[4.7 A & B]** Two examples of ears flat back: In **(A)**, a stallion snakes his head and neck with his ears flat back and neck stretched flat. This is common body language for herding, telling others, "Get going." In **(B)**, the weanling on the left is telling a visiting colt to back out of his space with his flat ears. Note the facial expression and body language of the bay colt as he stops himself.

In my studies of a number of different wild herds in the Western United States, I found that much of the day was spent grazing and smelling. This behavior was not limited to stallions, who do spend a lot of time creating "stud piles" (see p. 74) and smelling each other's and their own manure and urine, as well as that of any mare who left a mark along the way; it also included social greetings and nostril blows while horses grazed next to friends. Mares smelled foals, foals smelled mares,

STORY FROM THE FIELD:

A Scent in the Wind

I remember watching a palomino-pinto mare in 1977 on Green Mountain in Wyoming who quickly brought her head up from grazing and looked into the wind, smelling the scent she caught. After a few seconds of processing, she trotted over to another mare with her foal at her side, and with a glance of her eye and a nose bump for validation, the two soon began alerting the herd to the possible threat. Foals were guided to their mothers' sides, and the family stallion began "snaking" his head and neck at the mares and foals that did not want to stop eating, saying, "Take note, we are leaving." The palomino-pinto mare and her friend trotted off first, with the other horses following single file behind and the stallion in the rear, and the group disappeared into the woods surrounding the meadow where I still stood. This was a scene I would see many times with many different herds in many different habitats, demonstrating alert sensing, clear communication, and cooperative organization. //

→ [4.8] A palomino-pinto mare with her weanling and yearling offspring on alert.

↑ [**4.9 A & B**] A foal makes a flehmen response (see p. 52), most likely after smelling a strange mare's urine **(A)**, and another youngster investigates something interesting, employing his sense of smell **(B)**.

mares greeted and smelled each other, and on and on—sense of smell was constantly used in social interaction in the daily life of wild horses (figs. 4.9 A & B). It appeared to be like a form of "text message" or social media post—horses would check back in the same spots over and over again to see if any smells changed.

Smell interest and sensitivity varied in the wild horses I observed, particularly around the amount of time spent smelling manure. Some herds spent much of their time walking around, smelling each other's and their own manure, while horses in a different herd walked away from the area where they were grazing to pass manure and showed little interest in it after.

_ "Smell Dance"

Stallions leave "stud piles" of manure to loosely mark territories so strange stallions know they are there as well as to show other stallions their degree of health. The stallion who has the most manure is usually the healthiest and most "fit" stallion to breed. Mounds of manure grow during breeding season as stallions "mark" on top of other stallion manure piles.

Communication between stallions seems to have a similar routine all over the world and goes something like this: Approach, paw, eye glance, strike, paw, back up, pass manure, turn around and smell, then walk away while the other stallion smells the first stallion's manure and proceeds to poop on top. This "stallion manure smell dance" can keep the boys occupied for

[**4.10 A–C**] Smelling manure for horses is like going on social media for humans—lots of information can be transferred quickly. The two stallions in **(A)** are marking "stud piles" by defecating on top of each other's manure. In **(B)**, two other stallions are smelling each other's manure to see who is more healthy. A father teaches his son how to use his smell to read a pile in **(C)**.

hours, especially if they are "posturing" and not breeding (figs. 4.10 A–C). Stallions will also urinate and pass manure on the area where their mares have urinated to communicate that the mares are "taken."

★ **TIP:** / **Stallions use manure to create "stud piles" to mark dominance, while mares use urine to communicate both dominance and reproductive readiness.** /

While mares use smell more for social recognition and checking on their foals, they too can initiate a "smell dance." This is sometimes seen with dominant mares who may be ready to breed and will "wink" and "squirt" urine near another mare to signal that the dominant mare gets covered first. Mares are usually not as demonstrative as stallions, but I have observed this behavior when a new mare entered a herd. It can be used to establish a clear social structure and clarify which mare is in a "leadership" or dominant ("alpha") position. Since not all herds have obvious hierarchies, it is less common to see mares demonstrate smell dominance in this way.

★ **TIP:** / Smell and touch are primarily used in communicating information about the health and well-being of other horses, as well as for exploring an environment to determine if it is safe. /

Remember, smell is like an open book in "horsedom." Manure and urine "have no secrets." This is why horses may refuse to enter a familiar trailer if it smells like a stressed horse was in it before them, and yet they may walk right into what we might consider a "scary" place if it smells like friends have been there first and were relaxed.

— Touch

The horse's nose and whiskers are primarily communication tools using touch to gather information. As I discussed on p. 57, the whiskers allow the horse to assess spatial awareness while grazing and identify objects even in the dark in his environment. Along with their prehensile noses filled with neural proprioceptors, horses have an amazing way to communicate and gain information from touching and sensing as they walk along the ground foraging and exploring.

STORY FROM THE FIELD: //

A Pile of Information

Once as I was watching horses on the Red Desert head to water—a familiar activity done almost single-file every day as the horses came off the mountain—the leader, a well-built bay mare, stopped at a pile of manure left on the trail. The mare smelled the manure. Then her friend joined her, and the two spent some time over the pile, nostril to nostril. The rest of the horses stood patiently or wandered around a bit, but no one continued down the mountain.

The stallion in the back of the herd was trying to keep the group organized and seemed slightly impatient. I watched as the two mares nosed each other, smelled the manure again, then appeared to make a decision. They turned up the hill and followed a different route instead of heading down the mountain as usual to water. What did they smell? What information did they receive from the manure? Was it another friend and herdmate? Was it a strange horse? Another stallion? I could only guess.

As the herd headed up the other side of the mountain, following the bay and her friend, some horses stopped to casually smell the pile of manure while others seemed to not care and were happy just going with the consensus. The stallion stopped and smelled several times, pawed the manure to get a fresher scent, made a flehmen response, and followed the herd up the mountain as well. Did they discover another water source? Another group of horses? I still wonder, but having seen this decision-making behavior based on smell alone, I realized smell may well be the strongest sense a horse uses in decision-making. //

Horses also use touch to relax each other, whether just brushing shoulder to shoulder, wrapping head and neck over another's withers, or engaging in mutual massage (fig. 4.11). In fact, touch is used constantly in a well-functioning herd to reinforce social well-being, as well as in the form of a bite or a kick to communicate, "Move!" or "Get out of my way!" or even to initiate courtship or play.

Subtle Energy

Subtle energy perception could be said to stem from an "emotional field" that can indicate fear, worry, pain, or a nurturing impulse. As I mentioned in my earlier discussion, we know little about it in horses because it is difficult to study. But it has been studied in some species, such as elephants. Caitlin O'Connell-Rodwell of the Stanford Center for Conservation Biology found that elephants can sense very low seismic frequencies with their feet and communicate through them. Horses may also have this ability.

[4.11] Grooming reinforces social bonds. Here a youngster helps his mother scratch her friend.

→ [**4.12**] Horses learn spatial awareness as a form of communication. It enables them to gallop as a group without running into each other.

What is easy to observe are the emotional connections between horses, with those closely bonded seeming to think and feel in "one mind" (see Henry Blake's study—p. 57). They sense when their friends are stressed and want to be with them to provide comfort. Emotions, particularly strong emotions such as fear, take on an electromagnetic energy that most animals can sense. Not only can horses "smell" fear, but they can "feel" fear.

_ Spatial Awareness

Body language can vary among horses, but the concept of spatial awareness is woven throughout equine culture (fig. 4.12). Horses who do not understand spatial awareness and respect are at a disadvantage and will often get kicked and hurt until they learn to pay attention. As we discussed in chapter 2, spatial awareness is the energetic proprioceptive consciousness that is used to reinforce respect of social status and a group's structure and hierarchy as well as maintaining alertness and safety (I will discuss much of this in more depth in the next chapter—see p. 89). This energetic language may involve charging, biting, and kicking, but is also often much more subtle—just walking into another horse's space (space-taking) or incorporation of other body language, such as an eye glance or move of an ear, can establish whether a horse can approach or should move away (fig. 4.14).

Spatial communication is taught and reinforced by the herd. It is more critical with mares, no doubt because they must be aware in order to protect

> **“**
> **Horses sense when their friends are stressed and want to be with them to provide comfort.**

STORY FROM THE FIELD:

Greeting Communication

Horse etiquette and social greetings were well-defined in most of the herds I studied, and usually occurred as a series of steps, which went as follows:

1. **Eye contact**, a request or casual acknowledgment, leading to a decision whether to proceed with the greeting or to be cautious.

2. **Approach**, done cautiously and with spatial respect, especially with strange horses.

3. **Smell and touch** in the form of an exchange of breath nostril to nostril. This "chat" often ended with a gentle nose bump and maybe a squeal and a strike. Depending on what was said, at this point, the horses might go back to grazing, vocalize, blow, walk away, or if friends, go in for a mutual buddy scratch. If the exchange was between rival stallions, this greeting might lead to posturing and fighting.

4. **Buddy scratch**, usually starting at the base of the neck and withers but often leading up the neck or down the spine, depending on how each horse communicated what he wanted. I found that, typically, the more dominant horse would guide the action by scratching his friend where he wanted the friend to scratch him.

↑ [4.13 A–C] A chestnut stallion greets a small bay mare in (A). Note the soft eyes in both. A more cautious dialogue takes place in (B): Nose to nose, two stallions greet, with the pinto careful in his backed-up body language. The eyes and ears of both are relaxed, however, indicating a friendly greeting. In (C), two stallion friends greet and move in for a mutual buddy scratch.

their foals as well as themselves. In bachelor groups, spatial respect is often not practiced as much—I liken them to bumper cars as they engage in physical activity with each other, practicing their "moves" for gaining mares. But a good male horse will watch breeding herds and learn how close he can get to a mare before she kicks or runs him off (fig. 4.15).

★ **TIP:** / Spatial awareness and respect is a form of communication, providing status, safety, and structure, particularly in breeding herds. /

When a foal is first born, he learns to understand, "My space, your space" (see p. 101 for more on this subject). The foal learns to stay close to his mother's flank when running and remain out of the space of other horses. This is critical to the safety of the entire herd so others are not tripping over a slow foal or another horse who is not spatially aware.

↑ [**4.14**] The roan stallion kicks at a bay stallion who is trying to protect his mares. Note the horses' ears, eyes, and tight lips. The peaceful band was surprised by the unprovoked attack by the roan.

Spatial awareness communicates one's status in a herd. The horse who can demand the "biggest space" often is the horse that other horses look to for guidance and leadership. A more dominant or respected horse will often walk into the space of a friend or less dominant horse. This social hierarchy allows the group to always have spatially aware horses on guard, looking out for the others' welfare and safety. Spatial awareness and boundaries do vary according to the activity in the herd. Friends can be "buddy scratching" or peacefully grazing next to each other, touching shoulders, but when social organization is suddenly necessary, each horse may demand the other horses respect their space.

_ Play

Primarily used by foals, young horses, and bachelor males, play is used for learning about other horses and the environment. Play can help teach respect in spatial awareness games as well as polite

↑ [4.15] The bay mare kicks out at the stallions who are fighting too close to her foal.

STORY FROM THE FIELD:

By the Waterhole

Several times I watched a group of horses come down off Green Mountain, Wyoming, to water. There were several groups of horses who used the same water source, and the horses in the second group would stand waiting for the first group to finish before they would come in to drink. There was spatial respect for the first band of horses. As they stood, licking their lips, you could observe the second herd of horses was thirsty. Clearly there was communication understood by both groups.

But in one instance something happened that took me a while to understand. The herd that was waiting to drink had one mare wander off from her standing position within her group. The "lead mare" saw her and quickly chased her. I wondered if she was running her out of the group, as I had seen both stallions and mares do so in the past, but the lead mare kept after the wanderer until the young mare finally went back to where she had originally been standing in the group. At that point, the lead mare stopped and went back to her own former position.

I realized the communication was all about maintaining spatial awareness; the wandering mare was supposed to stay in her original spot until the herd went to drink. Now, of course wild herds are not that regimented, but the lead mare—through eye contact, facial expression, and spatial awareness (chasing the young mare)—made the rule crystal clear: "Stay in your place until it is time to move." This key observation validated the importance of respect for and awareness of space in horses as a language all its own. Science may call it "dominance hierarchy," but it appears to me more about respect and awareness as the space "bubble" around a horse changes, depending upon what he is doing. //

→ [4.16] Each herd takes a turn at the waterhole, showing respect and awareness for other groups' space.

→ **[4.17]** A young foal nibbles on his older brother in an invitation to play.

← **[4.18 A–E]** A buckskin youngster from another band asks the roan stallion if he can play with his son, the palomino colt on the right **(A)**. The buckskin allows the stallion to smell him as he is assessed to see if he is acceptable **(B)**. The roan stallion walks away and starts to graze, giving the "okay" for the "boys" to play **(C & D)**. The two colts are joined by a third youngster **(E)**.

social greetings (see p. 78). Sometimes in young stallions "horse play" can get rough and turn into fighting, but play communication is usually friendly.

Among social species in general, play amongst youngsters is an important way to communicate and have fun learning. In functional herds with time to socialize, foals will split their time between sleeping, nursing, and playing (fig. 4.17).

While play is a juvenile language in most social mammals, horses, particularly males, continue to play when they have time and plenty of resources. Mares conserve energy compared to males, but certain mares enjoyed playing with siblings and others when "life is good." A statement of good health and relaxing times, I observed wild horses playing in the spring after eating and naps were done, or when the bachelor males had nothing to do (figs. 4.18 A–E).

★ **TIP: / Play is an important developmental language helping horses learn essential life skills. /**

PART ONE / CHAPTER 4 / How Horses Communicate

← **[5.1]** Horses have developed emotional intelligence, allowing them to form cooperative social networks.

CHAPTER

[5]

UNDERSTANDING HORSE CULTURE

For a species to evolve over millions of years before humans entered the picture, it must have been doing something right.

[5]

Adapting Toward Emotional Intelligence, Awareness, and Leadership

Horses are one of the most adaptable social species. Their ability to grow large social networks and maintain strong bonds has not only helped horses survive but also given them opportunity to form friendships with humans and other species. Horses are rather unique because, like humans, they usually move their sons and daughters out of their family band when they are old enough to form family units of their own. These family units may still live within the larger herd population.

While it is clear horses have high emotional intelligence, little research has been conducted to investigate their awareness, cognition, and learning tendencies. Horses evolved over 50 million years ago from a small dog-sized forest-dwelling *Eohippus* into the modern horse, *Equus*, and domestication seems to have taken place in multiple areas with humans only about 6,000 years ago. For a species to evolve over millions of years before humans entered the picture, it must have been doing something right. Surely, horses have had to be aware, cognitive, and be able to teach their young how to stay alive.

★ **TIP:** / By making good decisions for the safety and welfare of the whole group, horses successfully evolved for over 50 million years before humans entered the picture. /

When wild horses roamed the western grasslands of North America in the hundreds of thousands, many social groups lived together. But because horses are so adaptable, they have adjusted to their shrinking habitat and reduced populations. Their social structures have been altered, and we only have sparse herds to study today. In the early 1970s, before the United States government began removing horses from public lands, there were still many herds with minimal human influence to study. Thus, my research was able to identify behaviors related to awareness, cognition, good decision-making, nurturing, and ability to teach others in the group.

★ **TIP:** / Functional horses are aware and have good emotional intelligence, allowing them to exist in peaceful, cohesive groups. /

Those horses who were able to maintain long-term social bonds between reproductive pairs as well as other friendships in the group—whether in bachelor bands or "harem bands"— demonstrated the fewest injuries and were able to balance their level of reproduction with their habitat. Given plenty of habitat and forage, wild horses will grow their social network. Average herd sizes today, on adequate habitat, are 100 to 150 horses made of up 10 to 12 harem bands (mares and offspring with stallions) and bachelor bands (all stallions) that live on the periphery of the harem bands.

The stable social structures I noted were based on good leadership, which could be demonstrated in both fluid and rigid hierarchies, and the emotional intelligence of the individuals—that is, whether they all got

Key Points of Equine Culture

- Free-roaming social herbivores.
- Adaptive to a variety of habitats.
- Social structure based on spatial awareness and respect.
- Sustainable herds have functional social bonds.
- Mares are primary educators and "social facilitators."
- Many males live in bachelor bands their whole lives, never reproducing.
- Culture varies based on habitat and the behaviors of the individuals.

↗ **[5.2]** Some mothers form strong bonds with their offspring and allow them to stay in the herd as long as they like. Here, the mare in the middle grooms her two-year-old daughter while her yearling rubs on her.

along, established "equine etiquette" when it came to respecting others' space, and formed cohesive bonds. This conserved energy. The rhythms of the social groups were clear: there were times for eating, times for sleeping, times for moving to water, times for rolling, times for playing and learning, and times for re-establishing and maintaining social structures. We will discuss all these things in more detail in the pages ahead.

Social Structures

Social organizations in horses vary. Overall horses appear to have loose hierarchical structures that can range from strongly dominant to shared leadership. While some groups appear to be *stallion-dominated,* other groups tend to have more *matriarchal* influence. While not territorial, some groups of horses adopt protective behaviors concerning resources, particularly when they are limited. But generally, horses' free-roaming, social nature drives them to form cohesive social bonds with a seeming awareness that social order and organization is important for the welfare of the group.

Social structures in groups of 12 or fewer horses most often consisted of a bonded male and female pair, and other mares and offspring. The group would include foals, yearlings, and two-year-old—and often even three- and four-year-old—family members. Although horses are *neolocal* (younger family members live or relocate away from both their father and their mother) like humans and typically pressure reproductive offspring to leave their family bands, some family groups maintain social relationships with their offspring. I have observed both mature males and females living with their own families in their original family bands. It is not uncommon to see a mother, her daughter, and her granddaughter remaining together if their social bonds are strong.

★ **TIP:** / **Leadership in equine society can be stallion-dominant for some specific tasks and matriarchal for other group decisions.** /

The *alpha* or *lead* mare was usually the most alert horse in the group and would notify the other group members when there was danger. The stallion would then gather the band from behind, and as the lead mare took off, the stallion would protect from the rear. In larger herds with several bands of horses, the social structures adapted slightly differently: frequently, there would be a second stallion, a friend of the dominant stallion who would assist in protection and guarding. (Sometimes, there could even be two assistant stallions.)

As you already know, much of a group's energy and risk is reduced by keeping cohesive units at peace. Although ingress and egress is not uncommon with horses coming and going, group dynamics that include being able to make friends quickly and "agree" to leadership decisions are essential for the welfare of the whole group. In my research, I noted individuals who often seemed to regulate conflict and make decisions regarding actions and movement, and I called them *social facilitators* (I mentioned them briefly in chapter 2—p. 37). These were also the horses that other horses would most often check in with through social contact. They were the "rhythm keepers" of the herd. They were the "aware" horses who kept the peace. Sometimes they were the "leaders," but sometimes they were somewhere in the middle of the herd hierarchy and just more social than others. It could be that this trait is one that has allowed horses to become so social with humans.

The terms *social dominance* and *hierarchy* are often used regarding horses, and while some herds I studied demonstrated clear social hierarchy, others did not. Equine culture can vary, as I've mentioned, depending upon individuals and habitat, but "horse etiquette"—those skills needed to stay safe in a group and get along—seem universal and simple. The driver in horse society as a prey species is "to stay alive," and horses are still wired with the motto, "One for all, and all for one," thinking and learning as a "one-mind species." By this I mean they often appear to learn as a social group, adapting behaviors through a variety of learning styles (see p. 290) and driven by changing habitat and circumstances.

→ [**5.3**] Onaqui wild horses graze in large groups of 75 to 150 horses. Their healthy herd culture has allowed them to maintain large groups, which increases their social networks and provides more time for grazing, raising foals, and forming and strengthening bonds within those networks.

Space Games

We've discussed how horses use space as a form of communication (see p. 77). It is key in influencing social structure. In bachelor bands, spatial awareness does not seem to be as key as as in mare or family groups. But in well-organized herds, lead mares and more dominant individuals are able to walk into the space of lower-ranking individuals. While there is a strong tendency in functional herds to "get along" with each other, spatial awareness and respect for the social structure is obvious (fig. 5.3). From the moment they are born, all horses learn to understand the social dynamics and importance of reading the energy and intentions of herdmate leaders who dictate movement of the group. Horses within a herd learn to read the energy of other horses. When a dominant animal walks toward a less dominant horse, the less dominant individual must recognize that he either needs to *stand still* or *move*. Stallions may do this to younger stallions or offspring who are allowed to live in the herd, or a mare may do this to a younger mare who in turn may repeat the same behavior to a younger animal, or one of lesser status. Once "awareness" and status has been acknowledged, then the individuals often return to grazing, relaxing, and nurturing social bonds. If the dominant horse wants the space, his energy "pushes" the other horse into moving, often with eye contact, ears slightly back, and a head bob. Stallions will often use "snaking behavior" to move a sleeping foal or a mare who wants to continue grazing. If the dominant horse wants mutual grooming, then he can walk into the less-dominant horse's space, greet him, tell him to stand still with his energy, and ask for mutual grooming. This behavior can be seen in stallions and mares.

[**5.4**] A pleasant family photo: A pinto stallion with one of his mares and her foal.

Even young foals will try taking space from each other. These "space games" are all a part of leadership development and training. Because the most aware horses usually end up being the decision-makers and leaders, horses learn to respect each other's space. This awareness requires diligence as well as good proprioception. Some horses just want to stay out of the way of others and not cause trouble, while others may constantly play space-taking games to challenge the awareness of a leader. (It is interesting to note here that most often lead mares have foals who also become leaders.)

More dominate horses can demand the biggest space, take space from other individuals, or tell them to stand still. When you spend time watching a group of horses, you will observe that it is pretty common for one horse to walk over to another horse and into his space, and then that horse will walk away. This may appear to be subtle show of dominance, but it is actually a useful awareness game that ensures all horses are paying attention to each other. Mares commonly wander over near various friends and graze, but when they are posturing, the more dominant mare may ask the lesser mare to move away.

Stallions also play space games; theirs often involve chasing each other out of an area. Even friends do this with each other, again as a way to ensure both stallions are aware and paying attention to each other. In bachelor bands, there is more "bumping into each other" until young males mature, and you can observe space games like one stallion running by and biting another stallion as he grazes or two males suddenly having a rearing match. If a horse isn't aware of space and where other horses are, he may get kicked, struck, or bitten.

★ **TIP:** / Understanding spatial awareness—"My space, your space," and "Stand still" or "Move"—is critical to horse society. /

Family Life, Herd Rhythms, and Relationships

It may be difficult for people to imagine the complexity of equine family life and social culture if they only have observed their domestic horses in the barn. But given a chance, horses living in pastures together often do form close friendships and surrogate family units, as they might have in the wild (fig. 5.4). It is not uncommon to see geldings hanging out together as if in a bachelor band, or a mare and gelding pairing up and demonstrating mating behaviors. Horses seek social relationships, and even with our human influence, they still try to achieve functional social groups.

Most wild horse groups I studied maintained a peaceful existence, with some horses developing deep friendships for life. Conflict was often avoided under natural circumstances when food, water, and reproductive opportunities were abundant. Confrontations usually ended with posturing and keeping respected spatial distances.

STORY FROM THE FIELD: //

Human Influence on Wild Horse Society

Many wild herd social structures were destabilized or destroyed when roundups and "gate cuts"—whereby a specific number of horses are pushed into corrals and the gates are closed (dividing the herd) when the quota is reached—were implemented in the 1970s in the Western United States as a method to remove horses off public lands. A more recent development has been the use of fertility control, which allows mares to cycle but not become pregnant. The result of all these factors is many long-term bonded pairs have been dissolved.

The leadership of many herds was disrupted due to removal or injury of mature stallions or dominant mares during roundups, and so young stallions were able to breed younger mares. This destroyed the stable social structure, leading to unnaturally high reproductive rates. The youngest stallion with a harem in Wyoming at the time of my research in 1973 was 13 years old, and many were in their late teens. Today's wild horse herds appear to have much younger stallions breeding younger mares. Reproductive rates in the herds I studied were 2 to 4 percent, consistent with other equid rates globally in stable populations. There were also few mares in foal younger than four years old and most mares only had a foal every other year. This, like maintaining a cohesive herd that "gets along," conserves energy for the whole group and is consistent with other equid populations. But today, this is not the case in most wild horse herds in the United States. //

← **[5.5]** Breeding is a social activity in the wild, and in functional herds, it is a time for young horses to learn all the nuances of courtship. Here two foals watch, and one colt makes a flehmen response.

↓ **[5.6]** The black-and-white pinto stallion has formed a new band with the little mare near him. He postures with a mature stallion, as if to say, "Look, I have a mare now, too!"

— Courtship, Breeding, and Raising Offspring

Courtship between stallions and mares is critical to reproduction: Behaviors may include hyperflexion of the stallion's neck, strutting, posturing, nipping and licking the mare's flanks, and running together. In functional social groups, mares have a strong say in who they allow to breed them and may become very aggressive toward unwanted stallions. The lead mare has an adaptive edge as she is usually bred first, thus her offspring will be delivered first, which gives both behavioral and resource advantages. Herdmates, particularly bachelor stallions and youngsters, watch and learn from courtship behaviors (figs. 5.5 and 5.6).

Most wild horse groups I studied maintained a peaceful existence, with some horses developing deep friendships for life.

★ **TIP:** / Courtship between stallions and mares is critical to successful reproduction. /

Reproduction in herds typically occurs in the spring after and during foaling season, although stallions who are fond of their mares may "cover them" even when they are already in foal to reinforce social bonds.

★ **TIP:** / Reproductive success in wild horses is closely tied to strong social bonds and nurturing parents. /

Family life starts when foals are born (fig. 5.7). Often mares leave the herd and have their foals on a gentle slope, dropping the baby while standing up. The foal is usually on his feet and nursing within minutes. The mare may remain away from the

↑ [**5.7**] This brand new foal has returned to the herd with his mother. Foals stay very close to their moms the first few days of life.

herd for a day or two, but returns quickly to be greeted by the stallion first, followed by the mare's other offspring, social friends, and then rest of the herd. The other horses share in the care of foals, even the stallion (figs. 5.8 A–D).

★ **TIP:** / Horses have family and friends who all may share in rearing the young. Stallions commonly are involved in raising their offspring. /

↑ [**5.8 A-D**] Stallions often "babysit" their offspring when the foals' mothers have gone off to graze.. They are also important "teachers," educating youngsters on being alert and learning about their habitat.

⎯ Food and Water

The rhythms of the equine family are dictated by the herd's habitat. Horses living in dry grassland habitats often have to travel long distances to water. They may

→ **[5.9]** A mare and her two daughters groom each other after a visit to the waterhole. Youngsters learn by watching and helping.

↓ **[5.10]** A group of horses sips water in the late afternoon sunlight. I found that mares typically decided when a band would travel to water as they were aware of their foals' needs.

spend the night eating and traveling, and in the day head to the waterhole where they often rest, roll, and spend time grooming and renewing social bonds (fig. 5.9). On the other hand, in habitats where food and water is plentiful, adult horses may spend more time socializing and youngsters playing.

The mares with foals are often involved in encouraging the herd to go to water (fig. 5.10). And many times the "social facilitator" (see p. 90) pesters the others until they cease eating and move toward a water source. Mares with foals need water sooner than "dry" mares or stallions. As a rule, mares conserve energy while stallions expend more energy, so mares are usually the "rhythm keepers" in the herd—they know when it is time to eat, to sleep, to find water, and to safely travel.

★ **TIP:** / Horses, both wild and domestic, have natural rhythms for eating, sleeping, moving, rolling, and playing. /

⎯ Rolling

Horses will roll 10 to 12 times a day during the dry months when they have the opportunity (fig. 5.11). Rolling in a particular and preferred spot frequently is done in hierarchical order, with more dominant horses getting to roll first. Horses roll to help their coats dry (domestic horses like to roll after their baths, too), for insect protection, to rebalance their muscles and skeletal structures, and just because it feels good. Even in wetland habitats, horses will roll in shallow water, but not as often as when a nice dry area offers itself.

[**5.11**] Rolling is a daily activity for horses. Some will roll and roll, while others just do a couple and are quickly back up, eating or resting.

[**5.12**] A pinto mare babysits napping foals.

★ **TIP:** / Rolling and mutual grooming are important for mental and physical welfare in horses. /

Sleeping

In most wild habitats, horses sleep more in the day than the night, but they can be both *nocturnal* and *diurnal* sleepers. Horses see well at night, and as a prey species, it is better to be awake at night when predators are more of a threat.

In order for horses to get the deep sleep they need, they need friends watching over them, "on guard." They will alternate eating and resting in the herd so everyone is never asleep at the same time. Foals will sleep together with other foals, and mares may "assign" younger horses or other mares in the herd to "babysit" the nappers. Sometimes, there is a particular mare who just loves to babysit, and you will regularly see sleeping foals, all around her (fig. 5.12).

Several times during my field work I came across horses lying flat out on the range (fig. 5.13). Thinking someone had shot them, I would get almost close enough to touch them only to find the horses, always stallions, still very much alive. Relaxed in the warm sun and feeling safe, these stallions had no doubt gone into a deep sleep and been left behind when their herds moved on. Horses usually can only go into a deep sleep (*REM* sleep, *paradoxical* sleep, or *desynchronized* sleep) when they feel safe and are lying down, and most horses need at least 30 minutes of this deep sleep a day.

★ **TIP:** / Horses can only go into a deep sleep (REM) when they feel safe and are lying down. /

Learning and Education

As social mammals, most behavior in horses is learned. Mares teach their foals most everything they need to know to fit into the herd, but both parents may take an interest in educating offspring. Stallions often spend time with their colts, teaching them how to "patrol" and keep an eye out for danger, or regulating young male "horse play" that may get too rough. While "mother" mares are usually the primary educators in the herd, the functional groups I studied often had strong participation from other siblings, relatives, and herdmates.

Education of young horses may vary depending upon the herd's culture, but how horses learn and process information is fairly consistent. As *associative learners,* horses may associate objects, individuals, sensory input, and experiences with "positive" or "negative" reactions. Equine genetics also play a role in how horses learn, as well as their personalities. The same stimulus that causes one group of horses to flee, feeling threatened (*vigilant gene allele*) may trigger another group of horses to investigate (*curiosity gene allele*). While curiosity is usually associated with learning in mammals, vigilance may keep the horse alive longer in the wild. (I talk more about how horses learn on p. 290 and about vigilance and curiosity on p. 131.)

Top priority for horses is how to stay alive. So, what is most important for horses to learn is how to get along in a group, make friends, respect social structure, be alert to danger, and be able to move cohesively in a group without running into others—in other words, know how to use their bodies correctly.

[5.13] Often horses that look dead, lying flat on their sides, are actually just in a deep sleep.

[5.14] Foals learn the difference between "Be still!" and "Move!" at a young age. This stallion tells his foal to move with strong, snaking energy. The rest of the herd has already left.

★ **TIP:** / **Horses are *associative learners*—they learn through associations with sensory input attached to situations.** /

Fear is no doubt the strongest emotional driver that has kept horses alive—far stronger than physical pain. Fear of separation from friends, which represent comfort and safety, is a much greater stressor than threat of fire, for example. Horses are wired to stay together when danger presents itself, and most do not have the deductive reasoning that can analyze a situation and determine another course of action. Most will gallop on with the herd, even if in a lot of physical pain.

Because much of what horses need to learn has to do with how to fit into their group, make friends, and keep cohesive bonds, all in an effort to grow a large social network and stay free of fear and safe from being eaten, horses have developed their emotional awareness in order to "keep peace" and have developed strong associations to "safe places" and "safe individuals." Regardless of a horse's culture, there appear to be "Golden Rules" of functional horse society. Every young horse learns these almost as soon as he is born. And those horses who do not learn these skills have trouble fitting into a well-functioning herd. They are often run off until they *do* learn the appropriate skills. It should be noted that these same "Golden Rules" can be taught to domestic horses and work well in establishing horse-human communication.

★ **TIP:** / **Social greetings in horses appear universal across various populations.** /

The Golden Rules of Horse Etiquette

❶
My space, your space.

❷
Stand still or move.

❸
Always pay attention to the leader (unless you are the leader).

❹
Never move into the space of a more dominant horse without permission.

＿Teaching Youngsters

Mares typically teach their foals everything they need to know to fit into the herd and stay safe. Thus, a mare's temperament and behaviors can have a strong influence on the foal. In the wild, horses can also be influenced by their fathers, uncles, aunts, siblings, and other herdmates (fig. 5.14). Sometimes horses even form friendships with other species who choose to stay in or near the herd. As social mammals, horses can learn from other social mammals. I observed this in a male bighorn sheep who bonded with two harem stallions and chose to stay with the herd for a number of years. The two stallions and the bighorn sheep patrolled together and kept watch together while exhibiting similar posture. Who was learning from whom was hard to say.

It appears that education beyond mother/foal is determined by the interest and social bonding of the other horses to the youngster. All kinds of relationships are formed, based more on personality than age, rank, or gender. Equine culture is *context specific* and can evolve as horses adapt to their environment and situations. When a foal wanders too far from his mother, another mare may guide the foal back where he belongs. Two-year-olds often are required to "babysit" while their mothers go off to eat grass. (In this case, both ages learn how to have patience and stay where they are supposed to stay—fig. 5.15.) When foals grow old enough to be apart from their mothers, other horses in the herd, particularly fathers and older male siblings, may teach young males how to "patrol" and be aware of herd dynamics and exterior threats (fig. 5.16).

[**5.15**] Siblings often spend a lot of time together, with older colts and fillies teaching and entertaining younger siblings while mothers graze. This two-year-old is babysitting his little brother, who keeps pestering him to get back up and play.

[**5.16**] A father (bay stallion in the lead) and uncle train a young colt on "patrol." Note their eyes, ears, and body language, all of which is being mimicked by the foal. The older stallions are telling the colt to stay back and keep his spatial awareness with them.

★ **TIP:** / Most horse behavior is learned and most of what horses need to learn to stay safely in a group they learn between birth and two years old. /

Horses can demonstrate competitiveness when learning in a group; you can observe young (mostly male) horses racing around, trying to outrun each other, bucking and rearing to see who is strongest (fig. 5.17). Play is the dominant language used among young horses to learn tasks they will need to survive. This innate motivation to demonstrate athletic readiness may be why horses seem to enjoy learning sports alongside humans. They are designed to use their athletic bodies to run, jump, rear, and race.

Horses can learn through mimicking behavior and through exploring and discovering "cause and effect" on their own. Foals that learn the Golden Rules of Horse

> **Play is the dominant language used among young horses to learn tasks they will need to survive.**

→ [**5.17**] Two stallions (the lead stallion and his friend and assistant) mock fight while mares and foals casually ignore them. When real fights take place, mares will move their foals away from the action. Stallion friends, particularly young stallions, often mock fight to stay in shape and hone their defensive skills.

↑ [**5.18 A & B**] Foals can learn skills, like how to jump and rub itchy spots, through communication and watching others

Etiquette (p. 101) tend to learn other tasks faster as they appear more aware and responsive to being in a group and motivated to learn from others, and others in the group seem more likely to be willing to teach them (figs. 5.18 A & B). (How to identify key traits critical to learning and fitting into a group, such as sensitivity, awareness, intelligence, confidence, and cooperation, will be discussed in more detail in Part Two—see p. 113.)

Bottom line? Horses use their emotional intelligence to learn as their survival depends upon staying connected to group members. A horse alone stands little chance of surviving.

★ **TIP:** / Education of the young horses in a herd is a responsibility shared among individuals in the social group, although mares usually are the primary teachers. /

→ [6.1] Two young friends from different bands, bonding. While it is unusual to see foals this young from different bands come together to play, their fathers or mothers may be related and be friends, as well.

CHAPTER

[6]

FUNCTIONAL HORSE SOCIETY

> There are no two horses born the same in nature, as there are no two humans or any other creature born the same.

[6]

Horses, whether wild or domesticated, are functioning from very similar genetics, behaviors, and intelligence. Therefore, to understand domestic horses, you must understand the horse in nature. To review some key points:

- Horses are very adaptable creatures, developing an integrated form of communication.

- Horses form strong social bonds with other horses. This evolutionary strategy gave the horse protection with numbers. As an herbivore, the horse had a better chance of surviving a predator attack in a large group or herd than he would by himself. Having strong social bonds ensures cooperation and minimizes fighting and injury in social groups. Respecting the horse's instinct to form friendships with other horses remains key to understanding the horse's nature.

- Like humans and other social creatures, horses have various temperaments and personalities. More timid horses often end up at the "bottom" of the herd, unless they have a friend of higher status. Brave, intelligent horses often end up as leaders. Horses spend much of their day in a functional herd maintaining leadership and social bonds, as they are the foundations for a sustainable group.

- Leadership is critical to the survival of a herd and a group of horses without leadership or good decision-makers usually does not survive long in nature. A leader can be a mare, a dominant stallion, two stallions, a mare and a stallion, and sometimes even a young decision-maker, especially if the herd has been disrupted in some way. Bachelor bands of males often share decision-making.

- Many functional herds have "social facilitators" that maintain strong social connections. The social facilitator is most often a mare who checks in with other mares and has a relationship and awareness of all the individuals in a group.

There are no two horses born the same in nature, as there are no two humans or any other creature born the same. Each animal develops a unique personality based on genetics, environment, experiences, and learning. But certain behaviors have been observed that support the adaptation and evolution of the equine species. These behaviors are considered *functional adaptive* behaviors and should be used as "baseline" behaviors when considering the domestic horse.

While many behavioral scientists identify behaviors related to biological activities—such as foraging behaviors, reproductive behaviors, and elimination behaviors—and these are important to note, the purpose of this book is to expand the awareness of the reader to better understand the emotional and cognitive behaviors of horses. Science has done a good job of dissecting the anatomy and physiology of horses, but it has a long

[**6.2**] Horses have not only physically adapted to a variety of habitats, they have learned to change their behavior to live with other horses and with humans.

way to go in understanding the complexities of cognition, awareness, learning, and emotions in horses.

What we've already seen is that not only are horses good at *physical adaptation* to be able to live in varied habitats on different diets, but they can learn to behave differently to fit into various social structures, too (fig. 6.2). They seem to possess the ability to transfer their emotional connection from horses to humans, allowing them to adapt to living with people in the same way they would adapt to live with a group of horses. This ability, which is sometimes linked to domestication, is seen in many social creatures.

What needs to be understood is that the horse has adapted to living with us, but his "normal" and natural equine behaviors are often considered annoyances in domesticity. By better understanding these normal behaviors, humans can make life more pleasant and less stressful for their horses. A little acceptance, understanding, and explaining to horses goes a long way as they will transfer their loyalty and override their innate behavior when their desire to fit in and bond with a human group is strong. The following outlines some of the most normal daily behaviors for horses. As domestication has reduced the need to be as alert as is necessary in the wild, you may not note these behaviors as often in confined horses. But understand they still exist, and when their expression is limited, they may be a cause for mental and emotional stress.

Normal Horse Behaviors

- **Forming social friendships/greeting:** No matter how well they know their friends, horses seem to like to go through greetings when another horse has been away or out of view for any amount of time. Greetings allow smell, touch, and assessment. Lots of information can be transferred with a simple greeting and a few nostril blows.

- **Making eye contact**: Horses are constantly keeping an eye on each other and looking for movement and danger.

- **Staying connected and aware of other horses:** If one horse spots danger or is agitated, he will alert other horses, and they will respond by forming a group.

> **The horse's 'normal' and natural equine behaviors are often considered annoyances in domesticity.**

↑ [**6.3**] Two friends in the same band, reinforcing social bonds. Horses choose their friends and may stay friends throughout their lives..

- **Smelling:** Used for social recognition, reproduction readiness, finding food, and exploring the environment, smell appears to be the dominant sense used in social recognition and decision-making.

- **Touching:** Functional social groups spend much of their time reinforcing social bonds through touch with greetings, nose bumps, and mutual scratching, for example (fig. 6.3). Horses use their noses like we use hands, constantly touching and smelling with them. This tactile activity is a dominant behavior in free-roaming horses.

- **Sleeping next to friends:** Because horses need to feel safe, deep sleep usually only takes place when horses know another horse or horses are watching for danger. Horses can spend relaxing time head to tail, resting with each other.

- **Posturing:** Position in a social group can be important and both mares and stallions will "posture" by defecating, urinating, and using body language to get noticed and respected, and then to maintain their social structure. Most posturing does not lead to fighting but rather to spatial respect. However, when another horse does not give the needed respect (such as spatial distance), then the posturing may lead to physical kicking. You may notice this in particular with mares in domestic situations: Kicking, which might be posturing in large open fields, can be misinterpreted as aggressiveness in confined spaces.

- **Following:** Horses from a young age seem to have a "follow reflex," as in, "Follow the tail in front of you." This social behavior creates safety and organization in the herd, and each horse seems to know where he fits in.

- **Defining spatial respect:** Space awareness is key in a functional herd for safety of the individual members, but with horses, spatial respect is also a language. Horses with the largest "space bubble" are usually leaders.

- **Mutual grooming:** This nurturing behavior, conducted daily by horses living in groups, is important for their physical, mental, and emotional well-being.

- **Reproducing:** In spring mares are going to come into season, whether there are stallion prospects or not. They are wired to breed and have babies.

- **Playing:** All young horses play, but as they age, male horses continue to play in bachelor bands, while mares tend to take life more seriously.

- **Rolling:** Rolling is often done several times a day in natural living situations. Rolling is also a learned behavior and a social activity—when one horse starts rolling, they all want to roll, and usually in the same dirt pile.

- **Moving:** In the wild, horses have free range of motion and are moving most of the day. This is critical for optimum physiological functions.

- **Exploring:** Horses either explore all day in search of food, or when full, they explore their environment out of curiosity.

- **Decision-making:** Although many horses prefer to follow the behaviors dictated by "whatever the group is doing," daily activities related to who to graze with, when to nurse (if you are a foal), who to have a mutual buddy scratch with, when to go to water, and what plants to eat all involve cognitive decision-making in horses.

- **Foraging/eating:** Horses are opportunistic grazers, eating a variety of plant materials, from grasses to shrubs. They have different food preferences that change with habitat and season, and even from herd to herd.

- **Eliminating/defecating/urinating:** Like leaving a "calling card" behind or an update on social media, manure and urine share important information with other horses.

PART TWO
———

HORSES WITH HUMANS

———

← **[7.0]** Because horses seek social friendships, most are willing to override past experiences and trauma with a person who is patient and emotionally understands their feelings. Here Diane DeLano, Director and Founder of the Wild Horse Rescue Center, approaches a new horse to the center, assessing his "safe space." Note he maintains awareness but is not moving as Diane is calm and talking quietly to him, asking him to accept her touch.

INTRODUCTION
TO PART TWO
BY
DR. MED. VET. DOROTHE MEYER

Dr. med. vet. Dorothe Meyer is a lifelong equestrian and equine behavioral nutritional consultant, and Managing Director of IWEST Animal Nutrition.

"

From an evolutionary-biological point of view, the horse's genes are still about 97 percent the same as those of his ancestors from 6,000 years ago.

The development of human civilization globally, especially in the last two millennia, was significantly enabled by the horse. Nevertheless, the horse is—in terms of evolution and biology—still a "young companion" to mankind. We note signs of domestication related to the horse of about 6,000 years, which is a really a very short period of time, say, compared to the dog, who has been at our side for at least 30,000 years, and possibly even much longer. The dog has been able to successfully adapt to life in human communities during this significantly longer period. Our dog today, for example, unlike his wild ancestors, can digest starch, and even though he is still a highly social creature, he has become an easy pet to keep.

Our domestic horse, on the other hand, is no pet at all—he is really still a "wild" horse. From an evolutionary-biological point of view, the horse's genes are still about 97 percent the same as those of his ancestors from 6,000 years ago, regardless of his breed. Our domestic horses, therefore, still have the same needs and innate responses to their environment as their forebears. Their requirements for fiber-rich nutrition, freedom to exercise and enjoy social contact with other equines, and their ways of learning and interacting are still the same as those of their free-roaming, social-herbivore ancestors.

Dr. med. vet. Dorothe Meyer

It is just the horse's genetically distinctive social nature and adaptability that allows him to form close bonds with humans, and live and work together with us.

This is exactly why it is so important for us to learn more about how horses "are"—how they experience their environment, how they communicate, how they learn, and what they require in order to have functioning social behavior and a sense of security. This is the only way we can minimize negative stress related to domestication that can make them ill, and the only way we can achieve a mutually beneficial cooperation.

Horses have an extremely high need for social contact because being part of a herd offers protection and security. The horse by nature, therefore, is an extremely social creature, and at the same time very adaptable and cooperative. This not only makes it easy for us to work with the horse now, but is ultimately the explanation for the great influence the horse has had on human culture over time. Even though today we have tractors to work the fields, cars and airplanes to take us places, and tanks, missiles, and drones to fight our wars, without the horse's social nature and adaptability, the rapid development of human civilization around the globe in the last 2,000 years would not have been possible.

Is the Horse a Prisoner?

What does the horse still mean to us today? Most people have horses as recreational companions or as athletic partners to compete with in equestrian sports. We enjoy contact with them and time spent with them, to the extent that horses are now a part of various therapies and personal development training.

And what do we mean to our horses? They live (by human standards) in safety, have regular access to food and clean water, are exercised to stay fit, and are provided treatment when they get sick.

But is that enough?

Our horses seldom have a choice. We determine the food they receive and when, how they can move and where, who they may have social contact with and for how long. Quite a few horses live a life most similar to that of a prisoner: They must eat what the prison offers, socialize only during a yard walk with other prisoners, finish uninteresting and dull work without complaint, and spend most of the day locked in a small cell. For basic survival, this is enough, but such an alienated life is truly not a *good* life.

It is a life full of stress and without any freedom and self-determination. Yet, still a massive number of horses throughout the world live such a life, working for us as ordered with unsuitable equipment and hard training, but still trying to do their very best.

We determine everything in the lives of our horses. Consider the breeding industry: we choose the stallion to cover the mare, and perhaps we do an embryo transfer and do not see how often the mare suffers in this process. Often at only four months, foals are separated from their mothers and turned out in a herd with peers, who all have one thing in common: too little experience with other horses and life in general, and therefore, usually, too much stress.

We regularly sell horses because they do not meet our expectations or because we simply do not want them anymore, and often we do not care who the new owner is or what kind of life the horse will have with that person.

None of this happens because people don't love their horses. Most people, in my experience, *do* love their horses—they just don't know enough about horses to give them a better life and minimize the stress they experience in a domesticated existence.

Behavior Caused by Stress

Horses that are difficult to train, or even "dangerous," are often just the result of their experience and the stressful circumstances of their lives. For example: The smart, clever, and strong-minded mare who, because the human in her life can't give her security, makes her own decisions (or even goes against the human's wishes) is rarely popular. But with horse people who can provide security for this mare, who can listen to this mare and make friends with her, and then reaffirm this friendship every day—well, for those people, this mare may be an outstanding horse.

I once had the good fortune to have such a mare in my life, and right at the beginning of our relationship, she explained to me in no uncertain terms that humans—with our lousy sense of smell and being almost completely deaf and blind, as well as incapable of quick flight—were, at best, a safety risk. And so, as a clever mare, it made sense for her to pay little attention to me and instead concentrate on her surroundings. It was my good fortune to be able to listen to her, to have enough empathy to understand her message, to not expect gratitude for the great stabling, but instead to set out to show

Dr. med. vet. Dorothe Meyer

her that I desired her friendship, that I could certainly make wise decisions, and that I could be the source of surprising and exciting interactions. We became very good friends, but for as long as she lived, that mare challenged me in wonderful ways. She taught me horse etiquette, how to look out into the world together *with* her and not just look *at* her. She taught me to think slowly and with more "feel." I had always talked to my horses before, but with her I talked *all the time*, and I slowly began to discipline my thoughts.

What is the personality of your horse? Do you have a stallion, a gelding, or a mare? And what is *your* personality like? What is your motivation for having one or more horses? Success in a sport perhaps? If you are a competitor, you have probably heard the opinion, still common among riders, that mares are difficult as athletic partners. No, they are not. It is just perhaps much more difficult to convince a mare that what you are asking her to do makes sense and is worthwhile.

Lowering Stress Equals Success

After reading the following chapters, you will gain a deep understanding of why horses are the way they are, how you can lower your horse's stress level with simple tools, and how you and your horse can grow together happily in whatever you choose to do.

I had the pleasure of first meeting Mary Ann Simonds many years ago. I asked her to work with German dressage rider Heike Kemmer and her horse Bonaparte before the 2008 Olympics in Beijing. Bonaparte was an extremely shy horse who at the same time always wanted to please and did not want to do anything wrong. This desire was so strong that he often breathed in an irregular and a shallow manner, "crawling into himself," and thus remaining far below his capabilities. Mary Ann showed Heike how to practice breathing together with Bonaparte and how to make him feel joy and pride in his movement and their partnership. The result for the pair was a team gold and an individual bronze medal, which hardly anyone would have believed possible before.

And consider Dalera, who won two gold medals at the Tokyo Olympics with Jessica von Bredow-Werndl—a mare who definitely has her opinions and requires certain things of her environment. To be successful with her, her rider must know, understand, and satisfy these requirements. For example, in Tokyo, Dalera preferred to jump and play on a long rein to schooling her upcoming tests. It was not easy for von Bredow-Werndl to defend that as the right warm-up for Dalera the day before the Grand Prix Special. After all, von Bredow-Werndl was riding for Germany in Tokyo, and all the other riders on her team prepared themselves and their horses as directed by their country's trainers. However, the results at the end of the Games proved her right: Dalera just needed to feel her body and have fun during schooling. It was *her* time; the competition was another time, and one where she would focus.

When we learn more about the individual personalities of our horses; when we learn to communicate with our horses nonverbally; when we learn to give them security; when we understand how to build trust, friendship, and social togetherness through spatial awareness; when we grasp the importance of smell and touch for horses; when we learn to minimize negative stress in their lives; and when we integrate *all this* into our everyday "togetherness," then we not only succeed in giving our horses really good lives, but we give ourselves the greatest gift: Friendship with a creature that is social as only a horse can be.

Your dog will always love and adore you, and be uncritical in his loyalty, but—even though you may not like to hear this now—your dog is definitely selfish, and also (partly) purposefully manipulative, almost always looking for his personal advantage. This is because your dog finds dominance normal.

In the nature of your horse, on the other hand, you will find nurturing behavior, communication, and cooperation. Your horse has a much more multifaceted, and yes, *caring* interaction with you. He takes you into a world of deep empathy with a sincere and respectful friendship across species boundaries. Horses are often our mirrors, telling us the truth about ourselves (which might be difficult to accept—like, "You are being a lousy leader right now"), but it is these qualities that can bring out the best in us, too, if we listen.

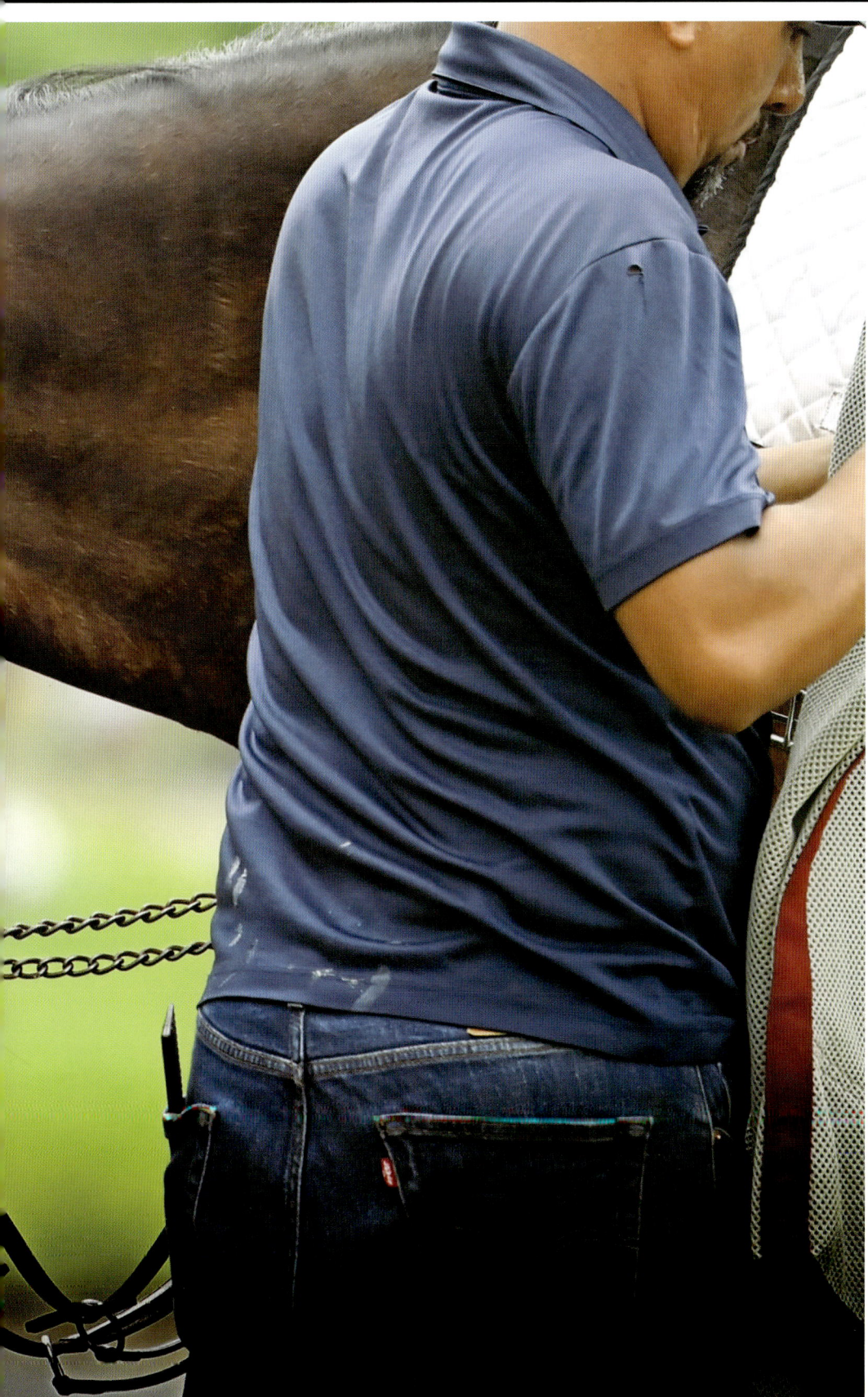

← [**7.1**] Horses willingly submit and often enjoy the activities we do together but many common devices cause pain and limit the horse's ability to communicate with us. Because of established equestrian cultures, few people know how simple, clear, and profound communication can be, or how sensitive horses can be in listening when they enjoy being with us. Here, Radar and Raquel have a bonding moment.

CHAPTER

[**7**]

CONSIDERING OUR IMPACT ON THE HORSE—HOW DOMESTICATION HAS CHANGED THE HORSE (OR NOT)

❝

As intelligent creatures, horses have made decisions about their lives for millions of years, passing on knowledge to each new generation.

[7]

Horses have been intertwined with human history for a long time, partially because horses are very adaptable and can form friendships with other species, including humans. And though domestication has limited the horse's natural ability to freely roam while foraging with friends, it has provided the horse with consistent care, food, water, and safety. Horse-human relationships could be categorized as *mutualism*, where each species can benefit the other in some way without harm. Part Two of this book explores ways we can evolve our mutual relationship with horses as well as gives guidance in areas were humans often unknowingly cause harm.

As intelligent creatures, horses have made decisions about their lives for millions of years, passing on knowledge to each new generation. I wanted to learn this knowledge, which is why I chose to study the behavioral and social ecology of wild horses. By doing so, I learned what we discussed in Part One of this book. To recap:

- Wild horses have a variety of cultures within equine society based on habitat and individual temperaments, but there are common practices across all: greetings, spatial awareness and respect, stallion posturing, play, and mutual grooming, for example.

- Most behavior is learned and starts at a very young age when the foal is born. Young horses are taught by both parents as well as siblings and friends about social etiquette, scratching, finding food and water, and the rules of general interaction (fig. 7.2).

- Functional social bonding between mares and stallions, mares and mares, and stallions and stallions supports the growth of strong social networks, creating sustainable herds with reproductive rates that match their resources.

- Horses are generally gentle creatures and invest significant time in nurturing behaviors toward friends and offspring.

- Smell and touch are dominant senses used for social recognition, support, and comfort.

- Horses are capable of making cognitive and altruistic decisions regarding the welfare of horses (and humans) they care about.

[7.2] Horses invest significant time in nurturing offspring. In the wild, mothers, fathers, and siblings can all influence the young horse.

- Many stallions live their entire lives in bachelor herds if they do not have enough confidence to attract mares. Stallion friendships can last a lifetime. Stallions may adopt orphaned foals, protecting them and raising them.

- Mares are the social facilitators—the "rhythm keepers"—in the groups, deciding when to go to water, when to sleep, and when to move as they are always aware of the needs of the young herd members. Mares will help raise and care for other mares' offspring if they are their friends. In functional herds, mares may adopt an orphan foal if the mother dies. Mares seem to work toward cohesiveness within groups.

- Horses have adapted to undergo tremendous physical stress, but *not* mental and emotional stress. I have seen horses willing to follow friends and fall in treacherous environments have their bodies heal rapidly, but when separated from friends and captured, colic and die (even when provided plenty of food and water).

- Horses are wired to make new friends and adapt to new situations *as long as they have a social network* (or even one friend) they can feel safe with and trust.

- Horses are good communicators, integrating facial expressions, body language, eye contact, vocals, touch, and energy. Herds may develop cultural differences in their learned communications.

As you read Part Two, keep in mind all the tips from Part One. They will help you connect the dots yourself. Horse people who spend the time to get to know their horses will recognize many of the traits, behaviors, and tips in the pages ahead and why they are important. But what was once assumed knowledge about horses is now becoming rare in the sport horse world. Winning, performance, and money often dominate over good horsemanship.

Since horses seek friendship, safety, and comfort, and because horses are willing to participate with us in a variety of ways, why not build strong friendships *first* before asking them to do more? Because they are willing, horses have allowed themselves to be overused and pushed beyond their mental, emotional, and physical capabilities, all in an effort to please humans and fit into our society. People, particularly those using horses for financial

> **Horses have adapted to undergo tremendous physical stress, but not mental and emotional stress.**

[7.3 A & B] Horses can be your "best friend": Here Daisy gets a hug from Elie at Stormyranch Mustang Training and Learning Center **(A)**. Or horses can be a business: While many sport horses can stay with their people for years, a large number are trained for sale **(B)**.

gain, owe it to horses to meet their mental and emotional needs. It not only will improve the horse's psychological welfare, it will also improve performance.

How We View Our Horses

The pages ahead will address the impacts humans have on horses, from breeding and handling, to training and selling, and how to minimize equine emotional and mental stress in all of these scenarios. As explained by Dr. Meyer in the Introduction to Part Two (p. 114), while horses and humans have been interacting for about 6,000 years, it is key to realize that no matter what breed of horse is in your barn, that horse retains approximately 97 percent of the same genes as his wild ancestors. Hence, he has the innate drivers to respond to his environment, experiences, and interactions as a "horse"—not as a "domestic pet."

One of the socioeconomic issues of keeping horses is their *classification*. Their classification is associated with human beliefs, which in turn relates to how people treat them. Horses do not fit the standard model for our relationships with animals as either "livestock" or "pets." They may be "livestock" for a tax write-off and "pets" when you are riding. Then there is an "equestrian lifestyle" that figures horses as the centerpoint of how some people manage their lives, and there are sports where horses are "commodities" or "investments" to be bought, sold, and traded. Now there is also a growing field of horse-human interaction that defines horses as "therapists," "facilitators," and "educators."

So how do you view horses (figs. 7.3 A & B)? As livestock? Pets? Friends? Exercise equipment? An investment? A statement of who you are? A therapist? For most people it is a combination of these classifications, as horses fulfill many roles for humans. It is important to understand your beliefs while reading this book. If you view horses as investments, commodities, or livestock to be sold or traded, the tips ahead will help ensure your investment has less risk by showing how to look out for the horse's welfare. If you consider your horse your best friend, athletic partner, or therapist, then hopefully you will find tips that take your relationship to the next level.

★ **TIP: / Become aware of your beliefs about horses, as this will influence how you communicate and relate to them. /**

← [**8.1**] Santana and Kevin sharing a bonding moment, reinforcing their friendship at Stormyranch Mustang Training and Learning Center. Santana, an ex-wild horse with a number of injuries from his life on the range, found safety and comfort with human friends.

CHAPTER

[**8**]

GIVING UP FREEDOM, FINDING FRIENDSHIP

> **Highly sensitive horses (and people) have trouble adapting to certain social situations. This is particularly true for horses that have been bred to be sensitive for performance.**

[8]

From Wild to Domestic

Living naturally in the wild is not stress-free. In fact, nature constantly dishes out stress to encourage all of us to adapt. But while a little stress is good, too much stress causes breakdown. Stress is also relative to the temperament of the individual. Highly sensitive horses (and people) often have trouble adapting to certain social situations. This is particularly true for horses that have been bred to be sensitive for performance in racing, dressage, or jumping. Along with the sensitivity comes the worry. Being highly sensitive and aware while living out on the range in a group of horses is a distinct advantage, but being highly sensitive and aware while living domestically in confined situations can be stress-inducing (figs. 8.2 A & B).

While selective breeding for domestication has usually favored traits such as *friendly, curious, playful,* and *wanting to please* (juvenile behaviors), breeding for high-performance equestrian sport has often gone the other way, with *talent, determination, courage, alertness,* and *sensitivity* being desired qualities. Recent research has helped us gain insight into how genetics affect equine temperament. Studies have shown that horses who are homozygous for the G allele (G/G) display both *higher curiosity* and *lower vigilance,*

[8.2 A & B] In **(A)**, recently captured wild horses stay connected through touch for safety and comfort. In **(B)**, two wild yearlings smell the ground for safety as they have their first experience with a human. A horse will always feel safer with another horse until he gets to know you.

while horses with one or two A alleles (A/A and G/A) show *lower curiosity* and *higher vigilance*. Horses with high curiosity seem to accept change and domestication more easily than horses with high vigilance, who have a tendency to be on "high alert." While more studies are needed, this gene seems to be displayed in both wild and domestic horses. This means we can intentionally breed for high curiosity and low vigilance (better temperament for domestic living), but it is often the case that horses bred for specific disciplines are highly vigilant horses. Understanding that these traits may, in fact, be genetic, can enable you to better manage your horse's stress.

★ **TIP:** / Genetic testing can give insight into equine temperament types, showing genetic alleles for *curiosity* or *vigilance.* /

Learn more

Adaptability Depends on Temperament

There are advantages and disadvantages to the domestication of horses. From a species standpoint, they ensured their survival by becoming partners to humans, who seem to be dominating the planet at a rapid pace by evolutionary standards. But individually, not all temperaments, and perhaps genetic types, are adaptable to domestication.

Just like people, there seem to be "city horses" (those who adapt well to stabled life as long as they have lots of friends) and "country horses" (those who do better living outside in large open spaces with friends). Both genetics and early foal-hood education play a strong role in how an individual horse will perceive the stressors related to domestication. Even horses with a highly vigilant genetic type can adapt *if* they learn from other horses (and humans) to feel safe when confined or faced with certain environmental stimuli.

★ **TIP:** / **Individual temperament plays an important role in how horses perceive stress.** /

Whether "city" or "country," horses need to be horses. They need free time without humans supervising their every move and making all their decisions for them. They need to play, move, and explore. Even if they seem content, when horses are not allowed to think, make decisions, and have choices, they may not continue to develop and learn, which is essential for performance horses. With that said, many horses seem to adapt well to living with humans in limited environments and enjoy playing our games as athletic companions.

People need to be aware that, as Dr. Meyer explains in the Introduction to Part Two (p. 115), horses are not dogs. Nor are they humans—although they can mimic and take on the characteristics of close friends, and they adapt well to various environments and situations. Horses are unique in that they bridge freedom and wildness with their ability to form deep connections and partnerships across species. Because horses seem to adapt well and seek our friendship, we must not overlook their needs—and that is, foremost, to be horses, *not* humans (fig. 8.4). For example (and I talk about all of these things in detail later in this chapter), horses want to feel safe and use their eyes, ears, and noses to investigate their environment, so limiting a horse's ability to engage his senses to evaluate his surroundings can cause undue stress. Good horsemen allow their horses free time to explore, roll, and

STORY FROM THE FIELD:

A Study of Two Temperaments

I studied two wild yearling foals, both taken off "lead mares"—one who was curious and gregarious and one who was shy and spooky (vigilant). The two fillies were adopted together and kept in the same environment until they died in their twenties.

The curious, friendly horse, Cedar, loved everything about people (once she realized people had water and food). She was actually motivated by water as she turned out to have a thyroid condition, no doubt caused by toxicity from the habitat where she was gathered. Her desire for water attracted her to sprinklers and people with hoses (who quickly became her best friend). There was little she did not enjoy with people—from being ridden to pulling inner tubes in the snow, she sought out interaction and novel activity. She thrived in the domestic environment.

Shamber, on the other hand, was shy and extremely sensitive. She would break out in a sweat just seeing humans and spent her first year always leaving her stall and going out into her paddock when people came into the barn. Then, one day, Shamber decided she wanted to be ridden. She had not even had a saddle or halter on yet, so this was a "big ask." I groomed her on her terms, letting her walk up to me when she was ready. She stood still in her stall, looking at the other two horses in the barn, staring at her, while I showed her what a saddle pad

→ [8.3] Cedar (left), Neptune (middle), and Shamber (right) living as a little herd. The Mustang mares turned Neptune, a Thoroughbred gelding, into as much of a wild stallion as they could make him! Although a gelding, he covered both mares whenever they asked, having very little idea what he was doing. This behavior continued throughout their 20 years together. Neptune provided companionship and taught Cedar and Shamber about domestic horse life.

would feel like on her back. She shivered and tensed when the pad touched her but turned and looked at me as if saying, "Go ahead, put the saddle on." She smelled the saddle and I placed it on her back. She acted interested and turned around and smelled the saddle again, licking it and biting it. I visualized what the girth would feel like around her middle before tightening it, and she stood absolutely still, even though she wasn't restrained and could leave me at any time.

I thought that was a good place to quit for the first time having a saddle on, but I also felt she was telling me she wanted me to get on her. It is moments like this where you want to be able to trust your communication skills with your horse. I used a mounting block, and before easing onto her back, I visualized what my weight would feel like to her. She remained still but tense. Then she turned her head, smelled the toe of one of my boots, licked and nibbled it, and walked out of the stall into the arena, where she continued to give me a lesson on how horses like to have humans on their backs on *their terms.*

Shamber "allowed" me to ride her after that, and she seemed to enjoy teaching people about horses (especially veterinary students), but she was always looking off into the distance, toward the mountains, as if wondering where her herd had gone. She tolerated domestication. I found that providing contact with close friends, in particular a Thoroughbred gelding who she taught to "breed" her all year long, helped reduce her stress (fig. 8.3). She clearly wanted to have foals and be a part of a large social network but had to settle for a small social group of horses, humans, dogs, and cats. //

↑ [**8.4**] Coco playing in the paddock, expressing herself freely.

smell another horse's manure, whether a horse is being turned out in a paddock or pasture or just walking around at a horse show, experiencing all the sights, sounds, and smells. Your horse does not change how he assesses his environment—his "system" is the same wherever he goes—and allowing him to establish that he can "feel" safe is core to building a positive relationship with him, and to his overall mental welfare and emotional welfare (fig. 8.5).

← [8.5] A young horse who showed nervousness being ridden in the ring was allowed to explore on the ground with his person standing next to him and telling him what everything was in the arena. Here he smells and picks up the ladder as he sees it as a curious "toy" (novel object) when he is not under pressure so he can learn not to worry.

Dominance, Leadership, and Friendship

Relationship models for horse-human interactions have been primarily based on training methods and so have mostly relied on theories of *dominance* and *leadership*. That is changing in some circles, such as in horse-human interaction programs that have a particular focus on friendship, but as money drives the sport horse industry, "dominance models" for training horses to perform are unfortunately still growing. Horses are not really interested in teaching and learning "models," but humans like to learn logically and follow a system. That is why there are many "horsemanship systems" and training programs teaching various steps and stages for developing relationships with horses. Keep in mind your horse may not have read the same books you have, and therefore you must be ready to adjust what you do to fit your horse. For example, while some horses enjoy learning games and repeating where to put their feet, others may find it annoying and react negatively. (This may be the case in particular with mares, as instinct tells them to not waste energy doing things that are not meaningful to their lives.)

Relationships with horses ideally start with friendship, but you should not confuse friendship with being weak. Horses may challenge you because they want you to "defend your space" and prove you are a worthy friend and partner. Remember, with horses, it all goes back to feeling "safe," and if you cannot defend your space with a horse, well, that horse recognizes that you probably won't do well with an attacking predator.

This dynamic is why even somewhat abusive training techniques have been able to get results with horses. Horses just want to "fit in" and there is "safety in numbers." This is true in nature and in domestic situations. If the horse figures out how to move away from pressure or to do what he is told with restrictive devices, he at least can be with another social creature for protection. However, this model fails when the stressors are stronger than the safety the horse feels.

For many horses, including wild horses who have been adopted, domestic life limits or reduces sources of stress, and so they happily adapt to whatever training and relationship model a human wishes to use. Keep in mind that it is not the training method, the tack you use, or the system you follow that creates strong horse-human bonds. It is the time you take to listen and understand your horse's needs, the attitude you have toward your horse, and your willingness to adapt your communication and training style to fit him.

> **Relationships with horses ideally start with friendship, but you should not confuse friendship with being weak.**

★ **TIP:** / **The more people can learn about how functional horses interact, communicate, and form strong trusting relationships with other horses, the more likely people are to have strong trusting relationships with their horses.** /

＿ Establishing Friendship

Since people have horses in their lives for different reasons, from a horse's perspective, there is no consistency in communicating and developing relationships with humans. Horses have to figure us out. But assuming we are not causing them harm or scaring them, horses will take the time to

[8.6 A & B] I make friends with a wild horse during a BLM demonstration **(A)**, and Gabriel, a trainer, introduces himself to a new sport horse (Radar) in a paddock **(B)**. Keep it fun. Laughter and happy energy are attractive to other social species.

try to understand our requests and make friends with us (figs. 8.6 A & B).

Horses value the time you spend with them just "being" and doing "horse things," like eating, resting, or exploring together. Do not limit horse interaction time to just grooming and riding; extend your relationship-building time. Sometimes this is forgotten as people focus more on performance or training goals than their horses' simple needs. Even the sales horse you are considering or the riding lesson horse you see once a week wants to know more about you. You may have had the benefit of hearing all about the horse from another person, but the horse has little or no knowledge of you. So think "horse-centric" and consider what you learned in Part One: Horses will always recognize a strange

> **Horses value the time you spend with them, just "being" and doing "horse things," like eating, resting, or exploring together.**

[8.7] Many mares do not like their faces touched, but Shamber would ask to have her forehead rubbed. I allowed her to guide my hand to where she wanted it, which gained a relaxation response from her and reinforced our friendship.

horse and run up to find out whether that horse is a threat or a potential new friend. And horses invest daily time *reinforcing* social bonds that already exist. We can use that as a bridge in developing positive relationships with horses. Instead of starting off in a round pen, chasing a horse in circles, or putting a bitting rig on a horse and longeing him, start with just getting to know each other (fig. 8.7).

Depending on your horse's experience, he may or may not know how to make friends with people or other horses. While you can use food as a motivator, often it is better to just start off with a simple "horse-type greeting" or a nice body scratch. Hopefully, you will recognize how natural and important it is for horses to engage in this regular greeting ritual. (You will also read in more detail later in this book about horse greetings, and the importance of eye contact, smell, and touch—see p. 209.)

[**8.8**] I talk to a new horse, asking him what he is thinking and feeling. A horse may not understand your words, but he will understand your intentions.

Here's how to make friends in a "horse-centric" way:

- Talk to your horse in a soft gentle voice and ask how he feels (fig. 8.8). He does not have to understand your words to understand your intentions.

- Create a "safe space" for him by slowing your energy down and relaxing into "horse time." Learn to just "be" with your horse; spend time hanging out. You will learn much more about him just observing and allowing him to safely interact with you. Remember horses have millions of years of adaptation allowing them to sense your feelings and energy. Use this to your advantage by showing your horse he can trust you.

- Once your horse makes soft eye contact with you, offer your hand for a sniff. Don't force it, but invite the horse to greet you through smell. Smell is used for social recognition and communication.

- Watch your horse's eyes, and if they remain relaxed, gently put a hand on him where you feel your horse would like to be touched.

- If your horse accepts this touch, ask him where he would like to be scratched or gently patted. People are often surprised when a horse turns around and guides them to the place he would like touched. Horses enjoy a "buddy scratch"; this practice develops a channel of communication and builds trust.

_ Listen to the Horse

No one model for horse-human relationships is the best, and you will have to discover the most suitable for you and your horse—that is, the one that causes minimal stress for you both while also keeping you and your horse safe. However, keep in mind that human-centric models tend to focus on achieving the desired goals of the human, not the horse. Horses have simple needs–to feel safe, to be comfortable and without pain, to have adequate food, to have the freedom to move and make choices, and to have the freedom to choose their friends, which may, in fact, be you (fig. 8.9).

★ **TIP:** / **Horses have simple needs: to feel safe, to be comfortable and free of pain, to have adequate food and water, and to have the freedom to move and make choices.** /

Horses spend much of their time in nature being "aware" of danger and the movements of their fellow herd members. This is obviously for their "safety." The most aware horse is often the one other horses turn to for guidance. (This is the "social facilitator" I discussed in Part One—p. 90.) This horse will often have the largest space, if it is requested, and this awareness is reinforced at various times. This can translate to horse-human relationships as well by teaching awareness and spatial respect with your horse who may or may not have learned it when he was young. Mares

→ [**8.9**] This person is keeping eye contact, talking to her horse, and using her body energy, clearly communicating both connection with and an awareness of her horse. Note the horse's ears are relaxed to the sides, his eyes are soft but attentive, and he is engaged with his person.

STORY FROM THE FIELD:

Establishing Spatial Awareness and Respect—"My Space, Your Space"

Stallion "C" is a sweet boy and lives with mares and geldings in a jumper barn. But like all smart horses and reproductively active males, he can become "pushy" and opinionated because there are decisions he wants to make that may not be aligned with what his humans want him to do in a particular moment. Ava and I worked with C in the paddock, using the following exercises to reinforce her position as leader based on the spatial awareness and respect that horses are already pre-programmed for learning (figs. 8.10 A–F). Horses in nature learn to either stand still or move as part of their social etiquette, so doing these exercises with your horse reinforces his nature to be aware, sense energy, read body language, keep eye contact, and cooperate. //

A

C

E

→ **[8.10 A–F]** Ava creates more energy around her space while maintaining eye contact and verbally tells C to "not to challenge her space." She opens up her body and sends energy toward C so his sensitive proprioception can pick it up. Ava is using her eye contact, her voice, her body language, and her energy to clearly communicate. C understands and backs off **(A)**.

Ava directs C to keep moving by maintaining eye contact, using her voice, opening her shoulder, and softening her energy in the direction she wants him to go, while strengthening and pushing her energy from behind to keep him moving and protect her personal space **(B)**.

By centering and calming her energy, softening her voice, and maintaining eye contact and body language, Ava asks C to halt and stand still. Ava visualizes C standing still while gently using her voice, telling him, "Good boy." C softens his eyes, and relaxes his ears and body language as he understands Ava's signals **(C)**.

Ava walks up to C to reward him with a gentle touch to the nose so he can smell her hand. She uses her voice to let him know he did exactly as she asked. Note his ear positions processing her touch and smell for recognition **(D)**.

Ava places her hand on C's withers, reinforcing that she is rewarding him for standing still and that they are friends **(E)**.

She then walks away, leaving C standing still and relaxed, waiting for her next assignment **(F)**.

are usually the teachers of this skill, but when foals do not have good moms or are weaned too soon, they often miss this—one of the most important lessons of being a horse.

Human awareness of horse awareness often goes unnoticed because horses have rhythms and their communication around spatial awareness can be subtle. You may notice it most at feeding time or if you ever walk into a polo barn full of mares. They stand resting, but then as you walk by, maybe point their ears back as acknowledgment that you moved through their space. Not wanting to wake up fully, they communicate their awareness nonetheless. Spatial respect and awareness is a basic horse skill and a simple lesson you can teach any horse, and he will feel safer and more well-adjusted once he understands it.

Keep in mind no matter what training method you use with your horse, it should follow a logical and safe path for both of you. The method must consider the horse's emotional reaction and response, and always represent the horse's best interests.

Many of the most dangerous horses I have worked with were simply smart horses who were annoyed at the repetitive stupidity of the person working with them, continuing to use a particular training method to get a desired response that a horse was either not able to give or felt was completely wrong from a horse's perspective. The

STORY FROM THE FIELD:

Good Intentions, Bad Information

The lovely palomino Berber stallion stood in a small stall at the Equitana event in Essen, Germany. He was still except for twitching frequently at the loud sounds of the speakers and hundreds of people walking by. I was very impressed with the stallion's apparent intelligence, alertness, awareness, and overall presence, and requested he and his owner participate in one of my demonstrations to show a highly functional horse. Unfortunately, they were already scheduled to appear in another demonstration. I went to watch.

Sadly, the trainer working with the Berber stallion had good intentions but appeared to have learned some techniques that were counterproductive. He stood in the middle of the demo ring, and as the horse trotted around, trying to make eye contact to help him figure out what he was supposed to do, the trainer dropped his head so the horse couldn't see his eyes. He did this every time the horse tried to look at him. Then he shook a bucket of grain to gain the horse's attention and have him come to him, but he again dropped his head so the horse couldn't make eye contact. And when the horse came toward him in response to the grain, the trainer made a "shsssss" noise into the microphone, which the horse learned to mean he should back off.

As the trainer did the demonstration several times that day, I watched the horse become more and more annoyed—shaking his head, rolling his eyes, and even pinning his ears—but the trainer continued. I warned the owner that the stallion was irritated by so much stimulus in the huge venue, that he had probably had no sleep, and now was dealing with very confusing and annoying demonstrations.

The next day, the Equitana management asked me to come meet with them as the Berber stallion had kicked the trainer, and the man was now hospitalized. They wanted me to verify the horse was dangerous. But quite the contrary, I argued that I felt the stallion was a very functional horse who had tried numerous times and ways to communicate to the trainer that his methods were not only not working but were also setting up a negative relationship between the horse and human.

When the trainer got out of the hospital a few days later, he came to me and asked why I thought the horse was not dangerous. So I informed him of all the misinformation he had adopted into his training method:

1. ***He never looked the horse in the eye or allowed the horse to see his eyes***. It is key for social mammals to see your eyes in order to read what you are thinking and feeling, even if they do not completely understand your body language.

2. *He used food as a manipulation trigger to lure the horse to come to him, but then when the horse tried to approach, he used loud sound to push the horse away.* This approach-and-retreat type of technique was not just confusing, it was annoying to a smart horse as it had little to do with the rules of spatial awareness and respect from a horse's perspective.

2. *He chased the horse in a circle for no particular reason other than to show he could make the horse go around him while he talked.*

The trainer told me he had learned from another international trainer that male horses never kicked, but only used their front ends, and that you should never look a horse in the eye. Obviously, he'd been given bad information. Although serious fighting may involve rearing, stomping, and biting, male horses do plenty of kicking, and the fact that the horse just turned out of annoyance and kicked him was lucky for the trainer—it could have been worse.

This story is a reminder to empathize with and listen to your horse—do not blindly follow a method when evidence suggests it isn't working. //

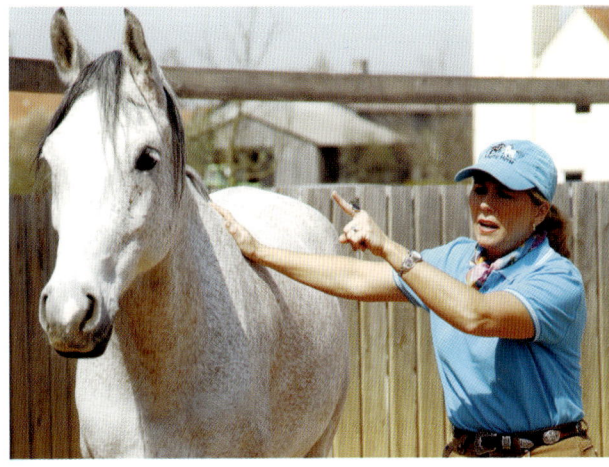

↑ [**8.11**] This mare directs her attention to something she sees in the distance. By putting my hand on her withers, I address her question, and she realizes I am listening to her.

fact that many horses learn to not complain, "speak out," or act even slightly annoyed while they try and understand and "guess" what people want them to do shows the level of cooperation horses are willing to offer while trying to please and not create conflict. Horses want to make friends and are willing to reach beyond their comfort level in order to gain our approval.

The key here is to always listen to your horse, even if it means stepping outside the "training paradigm" (fig. 8.11). Leadership established purely through dominance will eventually break down. You can tell your relationship and training is based on friendship when your horse chooses to stay with *you* over other horses. Building a relationship on friendship may take longer, but the rewards are far greater.

To Lead or Not to Lead

Having leadership skills, whether working with horses or people, is an advantage. Leaders who help build emotional bonds and harmony in a group also build trust, respect, and loyalty. These skills create a "social facilitator." In this book, I use the term "leadership" to refer to the person's ability to be emotionally connected to the welfare of her horse—and thus gain her horse's trust, respect, and loyalty.

Horses like to be with people and other horses who communicate clearly, make good decisions for their welfare, and can empathize with their feelings. Since the horse naturally "follows," getting a horse to follow you everywhere is fairly easy. What is often missed in training is teaching the horse to stand still and keep his attention on you. This, too, is a leadership skill and requires awareness of your horse.

When you have a horse with leadership skills who knows more about horses than you do, then you are wise to collaborate with him. This is particularly true when working with mares who are going to look out for your safety and for their own safety because they are "wired" to be alert and sensitive to potential danger. So you will have to convince your mare that *your* awareness is equal to or more attuned to her safety than hers is to gain her confidence and cooperation.

★ **TIP:** / "Spooky" horses are often just sensitive horses who are rightly reacting to danger. It is the human's job to think and feel like a horse to develop the horse's respect and trust. /

★ **TIP:** / Awareness is a critical skill for leadership with horses. Yours must be equal to or more attuned than your horse's to prove you can be trusted. /

Domestication has altered the behavior of stallions tremendously by not allowing them to fulfill their natural roles as leaders protecting mares, reproducing, and assisting in rearing young. There are few if any domestic facilities that allow horses to form family bands and the larger social structure they would inhabit in nature. But horses are resilient and adaptable.

Geldings in domestic life are usually allowed to have the most "natural" existence, often living in co-ed social groups with mares and being able to participate in the guidance of younger individuals. In nature, they would most closely relate to a bachelor band of nonbreeding males. Since they are not a threat to anyone, they can form social relationships with both males and females in domestic situations. Because of their generally more docile natures, it is geldings who have contributed partially to influencing people's beliefs that horses like to have a leader and be told what to do. As we learned in Part One and just touched on again, horses *do* like to have social facilitators (roles usually held by mares) who make good decisions for their welfare, which translates into leadership. But this also means many mares want to be the decision-makers and leaders, so trying to establish a training method based on a leadership model used for geldings often will fail with mares—and smarter horses in general.

Humans can become the social facilitator with their horses using similar models as in nature to develop their own leadership skills. Specifically, pay attention to the gender and personality of your horse as you develop a closer relationship. Most training models have evolved around male horses and do not take into consideration the differences in gender. It is important to understand the ways male versus female horses establish "leadership" when determining our own actions in their company.

★ **TIP:** / "Leadership-based" training models that succeed with geldings will often fail with mares. /

Horses who are leaders like to think things are their idea because their temperament is wired for decision-making. You have to earn these horses' trust and friendship and prove to them you understand how horses communicate. Learn horse etiquette, social greetings, spatial awareness, and how to create safe spaces for you and your horse. In order to gain the interest of a smart horse, you will have to have a motivator. *Curiosity* is often the best motivator as horses like to learn (see p. 221). While you can use food for positive reinforcement to gain the interest of a horse who may be afraid, it is best to use horse-appropriate rewards such as interesting smells, a gentle nose touch, and a good buddy scratch. These reinforce those things horses want: friendship, safety, and comfort.

★ **TIP:** / To earn a horse's loyalty and love, you must provide safety and comfort. /

Here are four ways to safely offer leadership to a horse or develop your leadership skills and take your "friendship" to a deeper level:

1. **Create safe space** for you and your horse where you both can get to know each other on equal terms by simply spending time "being together" and not asking the horse to do anything. Practice the social greeting described earlier in "How to Make Friends" (see p. 140). Sleep in the barn with your horse, or have a meal together. Leave objects that smell like you in his stall or paddock. Bring novel objects for you and your horse to explore together (see more on this on p. 240). Help your horse learn how to stand in a safe space with you.

2. **Offer comfort** by making sure your horse physically is not in pain. Ulcers, for example, can make a quiet horse spooky and worried as he has no idea what is causing the discomfort. Get your hands on your horse and find the places he likes to be scratched, rubbed, and massaged.

3. **Deepen friendship, and develop awareness, confidence, and collaboration** by doing things together on the ground. Forage together

> **Curiosity is often the best motivator, as horses like to learn.**

for tasty new plants, providing motivation for your horse to want to be with you. Play awareness and spatial respect games in a paddock (see p. 142). Teach your horse how to focus by gaining his attention with an object or a food motivator.

4. **Establish the right kind of leadership.** Cultivate a positive relationship for the age, gender, and personality of your horse. Horses and humans are resilient social creatures, and there is no one right way to form positive relationships with horses. Keep in mind these guideposts: consistency, clarity, safety, and fun—for you *and* your horse.

Once some horses have developed a friendly relationship with you, it will be in their nature to "test" you to see how aware you are and whether you can help keep them alive or not. "Saving your horse's life" a few times, or being able to establish a large space without the horse walking into your space, earns respect and leadership status for you. If you are unaware of these "horse games" of "one-upping" or social dominance, then you may find yourself getting injured or having a horse who walks all over you and chooses not to cooperate. Know that since social structures in herds are somewhat fluid, you need to continue reinforcing your social position with your horse on the ground through spatial awareness and respect techniques.

Remember: If you already have a good relationship with your horse, meaning your horse enjoys being with you and you with him, and you work well together, then there is no reason to change what is working. People often create problems when they did not exist before by adopting new training methods for no reason and thus confusing their horses.

Interspecies, Trans-Species, and Equestrian Psychology

The fields of interspecies, trans-species, and cross-species communication, cognition, and psychology are rapidly developing and feed into the study of *equestrian psychology*. Because the fields have developed from various avenues of psychology, sociology, cognitive science, and consciousness (among others), the terms used to describe these fields have varied. But basically, all explore the relationships across various species of animals, including

→ [**8.12**] This pony mare demonstrates *interspecies interest* as she smells the young human before her. Prior to this photo, she had raised an orphan Thoroughbred foal twice her size, and her nurturing instincts were now directed at *my* offspring, Chase.

humans. Often called the *human-animal interaction field*, hundreds of specialties continue to evolve, highlighting the emotional, cognitive, and physical relationships and communication between various species.

Equestrian psychology explores the communication and relationships between horses and humans, specifically. When doing my graduate research studying interspecies consciousness and psychology, I thought the area of focus would take off, especially in the equestrian world. It did not at that time, primarily because the field was developing too many varied directions and partly because most horse professionals were not aware or interested. Being a "whole systems thinker," it made sense to me to integrate the many approaches and study the various levels of relationships and connections between people and other animals, particularly horses. My research focused on identifying ways—such as physical parameters like heart rate—to "measure" feelings in both humans and horses.

While many studies at the time had focused on a mechanistic approach evaluating the effect of therapeutic riding on a person's physical healing, there was little research on the emotional benefits to both species based on friendship. Today it's a different story: there is growing research showing that in terms of social welfare, it is the emotional bonds that create a sense of well-being in animals who create friendships, whether within their own species or across species. In other words, in a horse-human relationship, both horse and human can benefit (fig. 8.12).

As Charles Darwin demonstrated in his 1892 book *The Expression of Emotions in Man and Other Animals*, non-human animals show many emotional expressions that people at that time thought only humans displayed.

In other words, *animals have emotions too*. Current studies are investigating tools to help humans read animal emotions. Tools such as the Facial Action Coding System (EquiFACS), which is discussed later in this book in more detail (see p. 275), have identified ways to relate facial expression to pain and overall feelings. These tools will help improve horse welfare as well as aid us in developing a better understanding of the emotional lives of our horses. (Of course, while science continues to study the degree to which animals express emotions and how we can read them, most good horsemen can instantly tell you what their horses are feeling.)

Learn more

— Similarities Between Horses and Humans

Although horse brains are wired differently than human brains, there are a number of "social drivers" we share in common, and no doubt these similarities are what have made such strong horse-human bonds throughout the last 6,000 years (fig. 8.13). While you can keep in mind the *prey-predator model* as underlying the horse's basic nature, the *friendship model* (see p. 137) is what can be applied to most horse-human relationships, and it depends on the similarities we share with horses as emotional, feeling creatures.

★ **TIP:** / **Horses and humans share a number of mental and emotional health needs.** /

For example, we both:

• Are social mammals and share a number of physiological factors.

• Desire friends and a social network.

• Have emotions.

• Want to communicate our thoughts and feelings.

• Desire freedom to move, explore, and make choices.

• Educate and raise our young.

→ [8.13] Horses and humans as social species share many similarities, such as using touch to nurture and reinforce social bonds. Here, a group of people at a clinic are reinforcing social bonds by finding places where their horses would like to be scratched.

- Tend to be *neolocal* (young move out of family unit as they mature to adults).

- Use touch and closeness to nurture each other.

- Feel emotional pain and seek comfort from our friends.

- Become stressed if we do not understand something or cannot accomplish a task.

- Store mental and emotional stress in our bodies in the form of tense muscles, ulcers, and other physical disorders.

- Can become stressed from overstimulation.

- Do not like confinement unless we are with friends.

- Are curious and want to explore our environment.

- Can become bored or depressed if we do not have enough interaction and stimulation.

- Need adequate exercise to stay healthy.

Nothing exists for itself alone, but only in relation to other forms of life.

Charles Darwin

Differences Between Horses and Humans

While horses and humans share many of the same needs, such as those I just listed, there are some notable differences to keep in mind. Having a good understanding of the horse's physiological functions and limitations can help control people's projection of human abilities and qualities onto the horse. (People often make statements regarding their horses relative to how humans might react.) Here are just a few of our differences, some of which we touched upon already in Part One (see p. 59):

- Horses are sensory beings who "react" more than they "respond." Their brains are highly evolved for quick reactions based on sensory input, unlike human brains, which have evolved to think complex thoughts and act "rationally" in many situations.

- Horses do not have a *prefrontal cortex* in the same way humans do—the area of the brain that governs complex thinking.

- Horses are wired for groups and seek other horses when frightened. (Humans may not.)

- Horses' eyes and ears can move independently, and they have almost 360-degree perception. Their eyes and ears are primarily communication and sensory input channels.

- Horses' *olfactory bulbs* (structures that receive neural input related to scents detected by the nasal cavity) are "in stereo," allowing them to identify direction of smell and bypassing the thalamus (which usually relays motor and sensory signals to the cerebral cortex, where higher-level functions originate) to cause reaction without conscious thinking.

- Horses must be able to lower their heads to clear their nasal passages and may contract respiratory infections if their heads are tied up for long periods of time (such as in shipping).

- Horses see differently. They are better at seeing movement in the distance but slow to adapt to changes in light. It can take a horse 15 to 30 minutes to adjust fully going from dark to light and vice versa. Horses do not have sophisticated depth perception and do not see color the way humans do.

- Horses have relatively small stomachs divided into two chambers—*glandular* and *non-glandular*. Horses can't vomit, so food must pass through the digestive system. This usually takes about 15 minutes. (Comparatively, humans take two to six hours.) The horse's digestive tract is loosely packed in the abdomen cavity. (It is tightly packed in humans.)

- Horse and human lungs are similar in relative size and function, but horses can only breathe through their noses, and when cantering, exhalation is linked to the stride. They do not have the choice to increase breath and cannot increase lung capacity, but they can increase oxygenation and efficiency.

- Horses can go from normal heart rate of 32 to 45 beats per minute to in excess of 260 beats per minute when running, almost instantly—an advantage for a prey species and great for sudden bursts of speed. (Humans cannot do this as quickly.)

- Horses can increase red blood cell levels by secreting adrenaline, which causes the spleen to contract and increase concentration from 30 to 40 percent to 60 to 70 percent, thus increasing oxygenation potential.

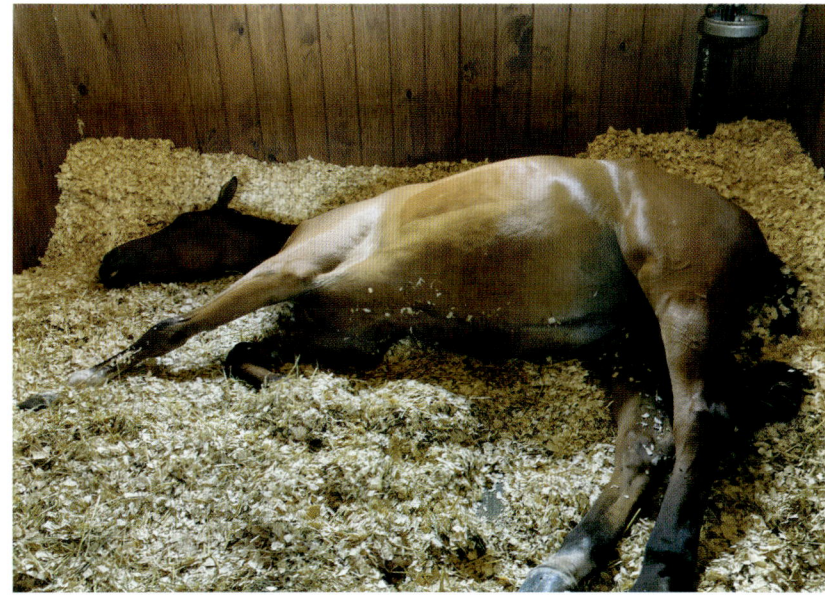

→ [8.14] Horses often only sleep three hours out of 24 hours and can sleep standing up for some of that sleep cycle. However, they need to get 30 to 40 minutes of rapid-eye-movement (REM) sleep daily and must feel safe and have room to lie down to do so. Here, Coco is taking her morning nap, catching up on her much-needed REM sleep cycle, as she is usually up all night "on guard" for the rest of the barn.

- Horses have four legs and are an ambulatory species that must stand and walk. They cannot lie down for very long without skin, muscle, and bone trauma.

- Horses can sleep standing up, but for deep rapid-eye-movement (REM) sleep, they must lie down (fig. 8.14).

- Horses relax by hanging their heads down; humans prefer resting with their heads up.

- Horses use their highly sensitive noses for communication, gathering food, and manipulating their environment; humans use their hands and the cognitive parts of their brains.

Obviously, there are a lot more differences, but think about how you walk through life each day as compared to your horse in the light of those I've just shared, and you will have a better understanding of your horse's world. Although horses do not have the same brains as humans, nor the capacity to analyze and use deductive reasoning as humans do, they have evolved their bodies and brains extremely well to stay alive and adapt over time to a variety of environments, and to emotionally respond to other species. This offers a foundation for varied horse-human interactions.

Perhaps humans can learn what horses have been doing "right"— for example, learning to manage mental and emotional stress in

❞
Horses are sensory beings who "react" more than they "respond."

[8.15] We can choose to participate in activities that both we and our horses can enjoy. Riding to the show grounds together with friends is both relaxing and social, alleviating stress in both rider and horse.

horses may also provide skills for managing human mental and emotional stress (fig. 8.15). As the fields of equestrian and trans-species psychology grow, more programs and models will be developed to assist both humans and horses.

What Kind of a Horse Person Are You?

At the end of chapter 7 I asked you to classify how you looked at horses (see p. 126). We can all acknowledge that people have horses in their lives for various reasons. Understanding why horses are in your life can not only help you make the most of the information in this book, it can help you better understand yourself and develop the best possible relationship with your horse. Because there is a need to define the "use" of horses in our lives, people often seek to learn how to "train" horses instead of learning how to be in a functional relationship with another species. It should be noted that people who have good relationship skills with other humans and understand the nature of social creatures in general, often have very good relationships with horses, as well as other social species. So being honest about your own self-assessment and skillset is, therefore, critical in determining the type of relationship you will have with horses.

Growing up, loving horses, I assumed everyone who had horses wanted to know everything about horses. But I realized early in my teenage years while competing at shows that not all people were involved with horses for the same reasons. Luckily, most of the kids I grew up with loved horses, too. We spent hours, sitting around, talking about the personalities of our horses, as each individual horse came from a different background and had a different story. But there were others I grew up with who later became more interested in how much money they could make buying and selling horses with little regard for the horse's welfare. Struck by this difference, I wanted to know what was driving people to "be with" horses. Exploring the horse-human relationship, and starting with knowing each individual's nature, was the place to start.

★ **TIP: / Understand your own nature and you will build a better foundation for any relationship with horses. /**

Being able to explore and understand your own nature will help you better relate to horses. As social creatures, we and horses may share similar attributes related to "personality type," but people need to be careful not to "project" their feelings and personality traits directly onto horses. For example, people who feel that they have been abused will frequently "feel sorry for" a horse who is not responding to what a human wants him to do and assume the horse has been "abused" also. At the other extreme, a "controlling" person who may have been brought up in a strict and rigid household may feel that all horses like being told what to do.

Over the years, I have given numerous clinics that teach you how to assess your own human nature, as well as your horse's nature. Running my sales business Equines LTD Horse Brokerage and Sales for many years also allowed me to observe common traits in customers, and I learned how to connect those traits to the types of horses that would be the best match. In graduate school, I spent many hours reviewing surveys, combined with the data I'd gathered through experience, and tried to develop simple tools to help people be better for horses by knowing who they are, first.

While this book does not go into depth on the subject, I have outlined the five general categories that I find most people can identify with, to some degree, below. Realize that most people do not fall into just one category but rather have a strong tendency toward one while exhibiting traits in others, as well. Balance, in the end, is the goal. All the traits I mention here can integrate into positive "leadership skills" with the right focus.

The Five Major Personality Types of Horse People

1. CONTROLLER—You seek and value "obedient" horses who quickly learn what to do. You have a strong work ethic and expect your horses to work just as hard. You have a low tolerance for horses trying to think on their own. You feel happy when things are in order and you are in charge.

- **Best Horses for You**: Those who like consistent and regular work and either seek to please or will test you, looking for strong guidance.

- **Strengths**: You usually have well-trained horses. You expect people and horses to do their jobs, and they often rise to that expectation. Your barn is usually run well and horses are kept to schedules. You are goal-oriented and often perform well under stress.

- **Weaknesses**: Horses and people need to "fit your program," which may not be successful in all cases. You may rely on "shortcut" training devices to get your horses do things you want in order to be efficient. Horses are not given many choices in their relationship with you. Signs of equine stress may go unnoticed as you are results-oriented.

2. CAREGIVER—You love to nurture and care for horses and often have a "rescue" or someone else's problem horse in your barn. Often you make a wonderful groom, veterinarian, or equine bodyworker. Your joy is being around horses as you appreciate all the duties related to caring for them. You may also enjoy working in horse-human interaction and therapy fields where you can share your skills with both people and horses.

- **Best Horses for You**: Those who respond to love and nurturing, and who have sensitive natures and the ability to communicate their appreciation. You may find "rehabbing" injured horses fulfilling.

- **Strengths**: You take the time to assess and care for horses physically and mentally. You are always willing to help your horse or other's horses when they are in need. You support others and are good at getting tasks done. You are careful and stabilizing.

- **Weaknesses:** You can be prone to "compassion fatigue" from being overworked while trying to care for too many individuals or situations at once, especially those that may not be "fixable." You may end up with too many horses due to an inability to part with them or because they are horses not well suited to other living and training situations.

3. **COMPETITOR/ATHLETE**—You are looking for the competition horse to share your love of sport. Regardless of the discipline, you value a top equine athlete and enjoy working together to compete and stay in shape. Winning is important, but you value the work it takes to get there, as well. Often you are a professional rider or excel in other athletic endeavors as an amateur.

- **Best Horses for You:** Those with equally competitive natures who "love the game."

- **Strengths**: Your horses usually get regular work and are fit for their discipline. You stay in shape, and both you and your horses get top athletic training and maintenance. You are detail-oriented, logical, and usually prepared for the task before you. You want to win and will spend the money and time to get to the top if you can.

- **Weaknesses:** While you enjoy thinking and may find "difficult" horses interesting, you do not have the time or patience for horses who tend to have unsoundness issues or who are not both committed and talented. You will pass on and sell horses that do not meet your expectations. You may not be sympathetic to your horse's needs.

4. **TEACHER**—You enjoy teaching both horses and people. Spending time with young horses and educating them about life, watching a student canter around a course for the first time, or teaching a group of students about equine behavior can all bring you satisfaction. Those who identify as a "riding instructor" or "educator" usually fits this category, but many who consider themselves "trainers" are often more aligned with the Controller and Competitor categories.

- **Best Horses for You**: Young horses and those horses with little knowledge about how to "be a horse" or "be with people" are good fits, as you can help them learn essential skills.

Being able to explore and understand your own nature will help you better relate to horses.

- **Strengths**: Full of knowledge, you love to share it with others. You are a good communicator with both horses and people. You take the time to listen and develop learning exercises that fit the horse and the person. You often are the "go-to" person in the barn for information. You are frequently friendly and enthusiastic. Horses generally like you, and your natural teaching ability makes you ideal for working with both horses and people in various disciplines.

- **Weaknesses**: You think every horse and every person can benefit from learning something new from you. You may be judgmental of others when you know more than they do about what horses need, how to teach, or in general, how to do things better.

5. **LIFESTYLE ENTHUSIAST**—Horses are a "lifestyle" to you. You like the cultural aspects of the equestrian world, so owning horses, attending and supporting horse events, and socializing with family, friends, other equestrians (and sometimes celebrities) brings you pleasure. You may be a spouse or parent to an equestrian, or involved in breeding or investing in horses, own a farm in an equestrian community, or be involved in another aspects of the horse world, such as finance, real estate, or marketing. Perhaps you just enjoy trail riding and the social aspects of being around "horse people."

- **Best Horses for You**: Horses that decorate your pasture or barn "just being horses," horses in training you might ride once in a while or just watch compete with someone else, and horses you support through donations of time or money are likely to fit your personality.

- **Strengths**: You are social, outgoing, and usually have good relationship skills. You often support and contribute to horse charities, events, and equestrian lifestyle activities. You may be involved in equestrian organizations and volunteer to help at horse events, even though you may not choose to ride or own your own horses.

- **Weaknesses**: You may overlook equine welfare as you may trust and believe others are caring for horses adequately, without knowing when to ask questions. Your desire for the status of being involved with horses—whether buying and selling, entertaining, or having social connections through events and organizations—can limit your direct connection with horses themselves and involvement in the bigger picture.

⎯ So…What Does This Mean?

The better you are at assessing yourself and recognizing your strengths and weaknesses, the better you will be in developing positive relationships with horses. Horses have amazing abilities to help us adapt and grow. They can be "mirrors to our souls," reflecting back to us without judgment and helping us evolve to be not just better horsemen, but better humans.

You can have as deep a relationship as you are willing to work at with your horse. But recognize that techniques, tack, or tricks do not build strong positive relationships with horses. No clinic, new technique, training device, or behavioral modification tool will give you a meaningful, trusting, and truly natural social relationship with them. To develop trust, loyalty, and friendship, you must first *truly love horses*, then provide safety and comfort to them from a horse's perspective, *not* a human perspective. Horses respond far better to *feelings* than to *thinking*, so your motivation for being with horses must come from the heart, not the mind. Then, no matter what discipline you choose to pursue with your horse, your horse will enjoy being with you.

Most importantly, realize that the best horse-human relationships have functional, happy humans involved. If you are struggling in other areas of your life, you may struggle in your riding and in all you do with horses.

← **[9.1]** Having a conversation with a new horse at a clinic. As we get to know each other, this mare is telling me with her eyes, ears, and body language what she thinks and feels, and where she would like to be scratched.

CHAPTER

[9]

HORSE-HUMAN COMMUNICATION AND RELATIONSHIPS

> **Horses befriended humans not for food, but for friendship, and through the development of functional social bonds, humans and horses can develop meaningful, happy lives together.**

[9]

AS the fields of interspecies communication, and trans-species and equestrian psychology continue to expand, the once undefined special connection between horses and people is being unveiled. The secrets of "horse whisperers" are no longer secrets, but rather techniques that outline how to connect consciously with other species.

I have spent a lifetime studying and developing tools to help people better communicate and understand other species, and in this chapter, I will share some of my key techniques. I describe various ways horses communicate with each other that often go unnoticed by humans. From body language and smell to vocalizing and subtle energy, these communication channels can be learned and implemented not only between horses, but between horses and humans as well (fig. 9.2). I use the term "sharing awareness" often as I feel it more closely describes how humans and other species can experience information together.

When humans domesticated horses we took on the responsibility for their care. Horses befriended humans not for food, but for friendship, and through the development of functional social bonds (such as those that exist in the wild horse herds we discussed in Part One), humans and horses

← [9.2] A volunteer at a rescue sanctuary makes friends with a horse by "listening" to the horse and rubbing him where the horse "asked" him to.

↓ [9.3] A curious gelding meets a young human for the first time, communicating his interest in his expression. Here you clearly see the horse's ability to "cross species" when making friends.

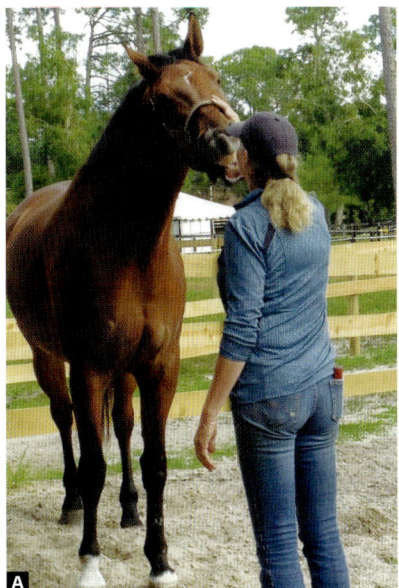

[9.4 A & B] Communicating with your horse can be fun! It can be a conversation about the two of you **(A)** or it can be a horse-centric discussion where you ask questions and see how your horse answers with his body and movement **(B)**.

can experience meaningful and happy lives together. You will see as you read through these pages that there are many ways horses and humans can benefit each other, but the foundation for all starts with a willingness to learn from each other, listen, and try to understand how both horses and humans communicate their needs and desires (fig. 9.3). Keep in mind the *whole life cycle* of your horse—from birth to death. By giving your horse the communication skills he needs, you ensure he can have healthy relationships with other horses and people, and therefore, a better life.

The knowledge presented here may challenge your beliefs a bit in some places, but the tools I give you will activate the deep-seated wisdom we all possess for communication and understanding, not just with horses and humans, but with all of life.

Feeling Safe Together—The Horse-Human Bond

Horses do not usually get to choose their humans. When humans decide to adopt, buy, or breed a horse, they become responsible for the entire life of that animal. There is no herd, family, or friends for the horse to run back to if the human fails. This is why we must think about the entire life of the horses in our care—from our horses' early experiences to how we are going to positively impact our horses' lives to how we will care for the horse if he becomes old, sick, or injured. Beginning with this mindset and commitment is core to a strong horse-human bond.

STORY FROM THE FIELD: //

A Wyoming Cowboy, His Dog, and His Horse

When I first arrived in Wyoming from California with my jumper, I was thrilled to be invited to go ride the range with a local cowboy named Paul and count "dry cows." Paul showed up at the ranch where we both kept our horses in his pickup and stock rack. I asked him where his horse trailer was—he looked puzzled and stated, "I thought you said you had a jumpin' horse." When I responded that my jumper jumped around courses of fences in an arena, he asked me, "What good is that?"

I pondered the fact that he had a good point.

Then Paul whistled for his horse to come out from the pasture, which he did. I noticed his horse had dried dirty sweat marks on him and asked him if he washed off his horse after working him. Paul looked at me in that puzzled way again and said, "Would you want to get messed with if you just worked hard for 12 hours straight? Or would you rather go roll in the dirt and eat with your friends?"

I made note of another good point.

Paul brushed his horse off a bit and threw the saddle on him, checked his feet, and put the bridle over the horn of the saddle. Then he whistled again, his dog jumped into the cab of the truck, and I watched his horse jump into the back of his pickup, ducking as he went under the horizontal bar of the stock rack. I asked Paul how he taught his horse to jump into the bed of a pickup truck, as I clearly did not think my jumper would do that. He took a deep breath as if trying to be patient with me and said, "I didn't, he just wants to go to work with me."

I realized that Paul, through his hours of silence with his horse and dog out on the range, was completely tuned in to his animal friends. He was horse- and dog-centric in his point of view. And his connection with them was so much a part of his life that he did not see it as unusual at all. //

Science may argue that the bond is simply a "needed association" for horses, because science has only begun to accept and explore the deeper abilities and implications of interspecies communication. But true horsemen do not need science to tell them how to bond with their horses. There is an inherent understanding, a silent communication of empathy and knowing. They know that establishing trust and open communication is the foundation for friendship and leadership with horses, and the most important priority (figs. 9.4 A & B).

Horses have built their entire social order on being able to trust their herdmates for their safety as well as being able to choose their own friends. So when you become your horse's herdmate and he chooses you to be a friend, he will do anything for you. This is the "special connection" people seek, and there are ways to help develop it through fair and clear communication (fig. 9.5).

★ **TIP:** / Try to think and sense like a horse, becoming aware of how your horse communicates with you and seeing the world through his eyes. /

It is extremely important when you are going to get on the back of another creature that you have made friends with him. My policy has always been, no matter what horse I have been about to ride—my own or another's—to make soft eye contact and "ask" the horse how he is feeling. A horse with distant or dull eyes who will not look back into your eyes does not want to communicate. It is then *your* responsibility to find out why. When a horse does not want

→ **[9.5]** After I acknowledged this mare's concerns and indicated she was safe with me, she turned her head around for a proper horse greeting.

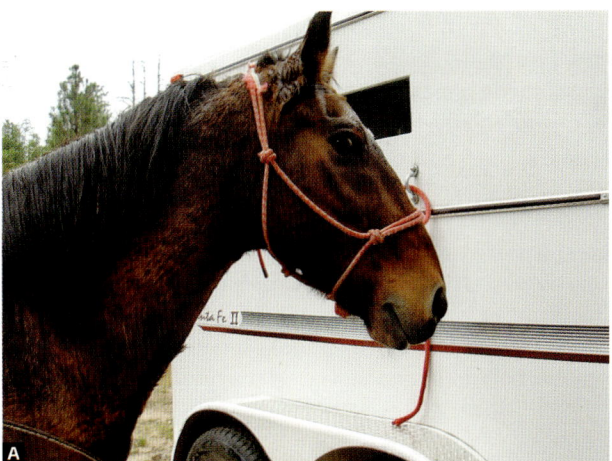

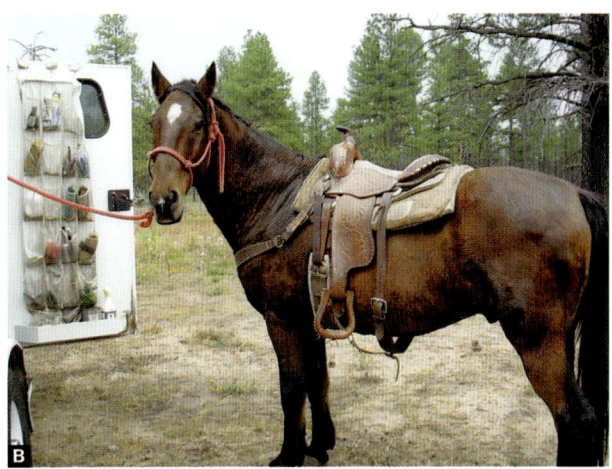

↑ **[9.6 A & B]** This horse was loaned to me from the local rental stables to ride up into the White Mountains and conduct a wild horse investigation. He would not look at me at first, and his eye was worried, his body tense, and his lips tight **(A)**.

He was used to being shut down with strange people on his back and just staying in line with other horses. We were about to embark on a private journey into the mountains, and I wanted the two of us to know each other. After talking, bribing with carrots (that he was not interested in at first), and finding a place where he finally released body tension after scratches from me, he relaxed and let me get to know him. Eventually, he looked at me with a relaxed facial expression as if to say, "Okay! Now I am ready to go" **(B)**.

to "talk with" humans, there usually is a good reason. The horse often has not been "heard" when he *did* try to communicate, or he is "shut down." Regardless of your goal with horses, a successful partnership always starts with "being friends" and that necessitates good communication (figs. 9.6 A & B).

The same communication skills apply across species when it comes to establishing friendly, open-hearted, and honest relationships. They all start with being "present" and mindful—opening the heart to an emotional connection, and being a good listener, nonjudgmental, and empathetic and clear in your communication. In the past, science has been quick to deny that horses and other nonhuman animals have consciousness and emotions, claiming they are too complex for the nonhuman brain. But leaving semantics aside, research

STORY FROM THE FIELD:

Wild Horse to Trusting Friend (In Just a Few Minutes!)

Fred Wyatt was head of the BLM Palomino Valley Wild Horse Center in Nevada when I was doing research in the area. He and I would frequently do demonstrations with recently captured wild and feral horses to show how easy it could be to communicate with horses when you let go of old beliefs and "think like a horse." We showed audiences that a wild horse directly off the range and put in a small space with two humans will soon relax and even approach—if the humans are relaxed and not chasing him.

Fred and I used relaxing signals like yawning, stretching, walking around while quietly making soft eye glances and talking in a low gentle voice, and the horse would soon let go of any fear and become curious, often wandering up and standing behind us. Usually within 15 to 20 minutes we would be petting the horse. We would offer a rope for him to investigate, gently put the rope around him and allow him to put his own head through a rope loop. To deepen our point, if the horse seemed relaxed and eager to engage, we would even get on bareback to show that when you are friends first and both horse and human feel safe, you can do just about anything in no time at all. //

→ [9.7] Making friends with Chevez, a highly sensitive and spooky wild horse at Wild Horse Rescue Center.

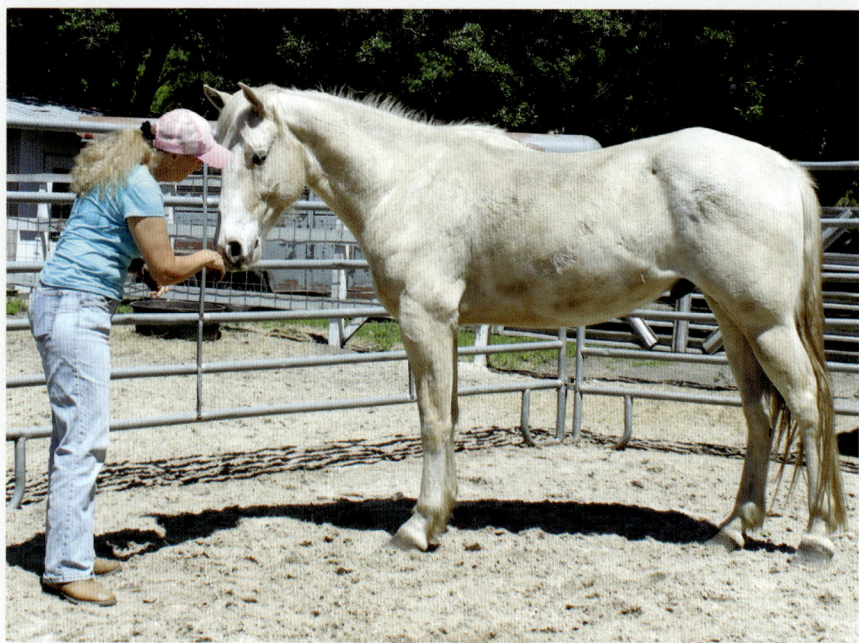

Learn more

↑ [9.8] Gabriel and Peter are communicating and connecting with each other through multiple channels. Note the horse's split ears and eyes listening to the rider, who also is using his voice as well as his energy.

continues to demonstrate the conscious and complex emotional lives of other animals. In 2012, at a conference bringing a cross-section of neuroscientists together to address consciousness in humans and non-humans, the evidence was so overwhelming to support animal sentience that the members wrote a declaration known as the "Cambridge Declaration on Consciousness" (scan code for more information).

★ **TIP:** / **Clear communication and good relationships are the foundation for functional social interactions, regardless of species.** /

While one does not have to be a scientist to understand interspecies consciousness, there has been a hesitancy to incorporate any form of science into horse-human relationship models, especially by horse training "experts." These historical interactions with horses have created a "rut" in how humans think about horses. Training and communication methods may vary, but at the core of any positive horse-human relationship is establishing a conscious connection that allows both the horse and the human to freely communicate their thoughts and feelings and have each individual understand what the other one is "saying" (fig. 9.8).

★ **TIP:** / **The more mindful you are of your own thoughts and feelings, the more aware you will become of your horse's thoughts and feelings.** /

So...How Should We Communicate with Horses?

Social creatures communicate their emotional state through their facial expressions, body language, voice, and overall energy. Horses are keenly aware of all these signals for emotional communication, which we

[9.9 A & B] Smile and have fun with your horse—learning together should be enjoyable for both of you **(A)**. Soft eye contact and gentle words keep this horse connected and interested in communicating with me and my assistant **(B)**.

discussed in Part One of this book (see p. 65). Mares are usually the best teachers for equine "language skills," as foals easily mimic their mothers' conversations, from ear positions to head and facial expressions—much like humans and primates have been shown to do in studies. But with domestic horses, forms of communication may also be learned from humans.

★ **TIP:** / Horses can read your facial expressions, so smile. /

A recent study conducted by the University of Sussex in Brighton, England, demonstrated that horses (ages 4 to 23) shown unfamiliar video images of men either smiling or angry could "read" and respond to the facial expressions (scan code for more

Learn more

Learn more

> **When subtle equine communication goes unnoticed by people, it can lead to mental, emotional and physical issues in domesticated horses.**

information). The team measured the horses' heart rates (as a correlation to stress) when seeing the images and found they were elevated when viewing faces that were considered "angry." If horses are stressed just seeing unfamiliar conflict in a still image, think how they might feel if a live person or horse is angry. Happy people usually have happy horses (figs. 9.9 A & B).

★ **TIP:** / Horses are eager to communicate with humans, if we are listening. /

It is relevant to recognize that horses can read human facial expressions because humans seem obsessed with teaching other animals human language, instead of trying to learn the other species' language. This is why communication between humans and horses is often difficult to clarify. Both species are operating with different assumptions and expectations.

★ **TIP:** / Horses can learn many languages, but the easiest is their own. /

Humans also typically want to communicate with horses to "make them do something"—groom them, ride them, train them, move them. Horses, on the other hand, usually want to communicate with people because it is simply in their nature to try to "make friends" and talk. The horse who sticks his head out of the stall, eyes bright and ears forward, when you walk by is initiating a greeting. But, if enough people walk by and ignore this request for a conversation, then the horse learns not to bother with people. Yes, the horse who bangs the stall door at feeding time makes his feelings heard loud and clear, but other communication can be much more subtle (figs. 9.10 A–M). When this subtle communication goes unnoticed by people, it can lead to mental, emotional, and physical issues in our domesticated horses.

Horses who are interested in communicating with us are talking all the time. They study humans and try and determine how to get our attention. They often learn that pawing, banging, and making noises in their stalls is what works, and they determine that any attention is better than no attention, particularly when they have something to say.

HORSE CHAT

[9.10 A–M] Observe each of the horses' expressions in these photos. Their eyes, ears, mouth, and head position are all telling us what they are thinking or how they are feeling. You can learn to recognize these kinds of messages; your horse will appreciate that you are listening.

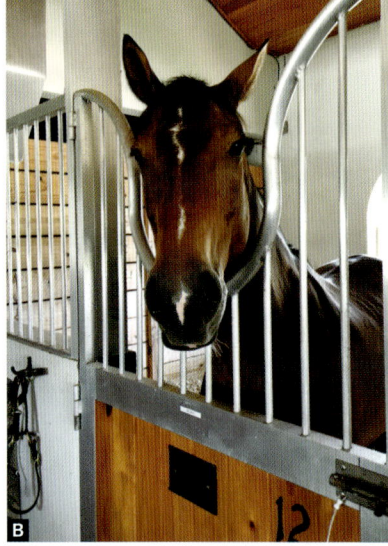
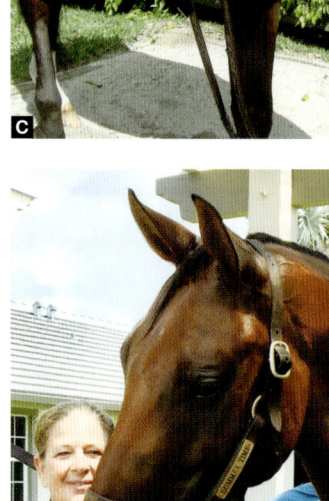

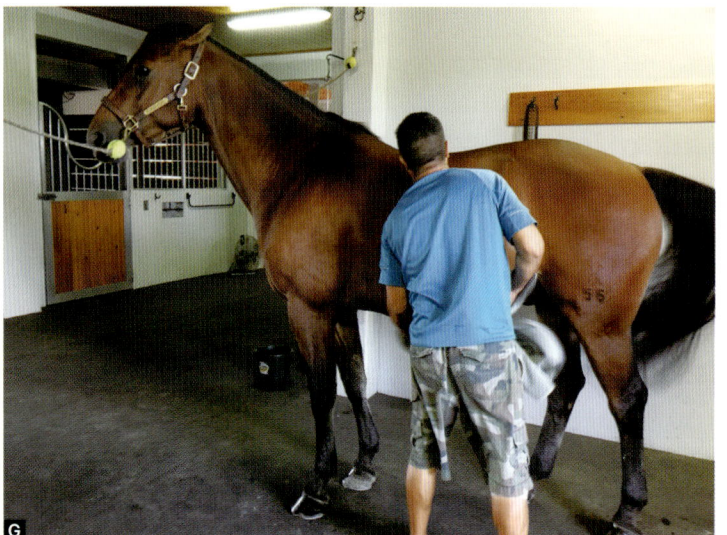

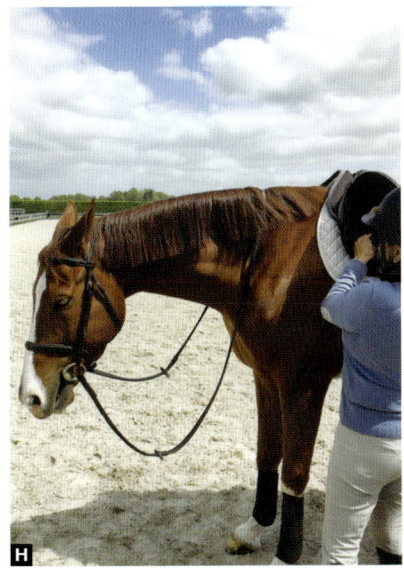
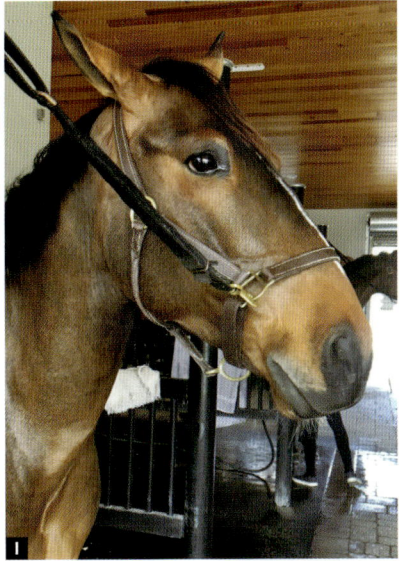

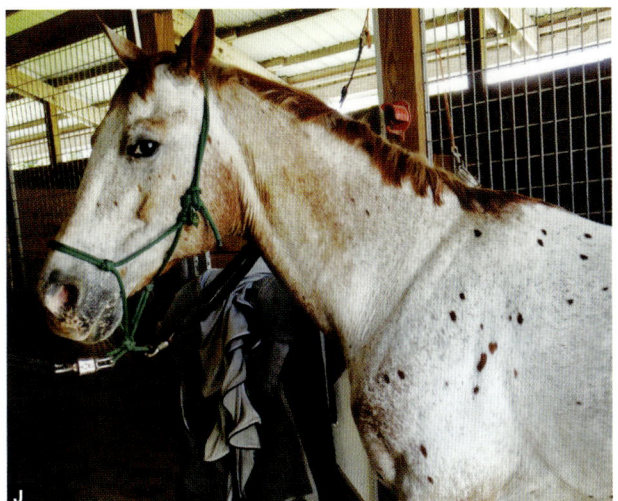

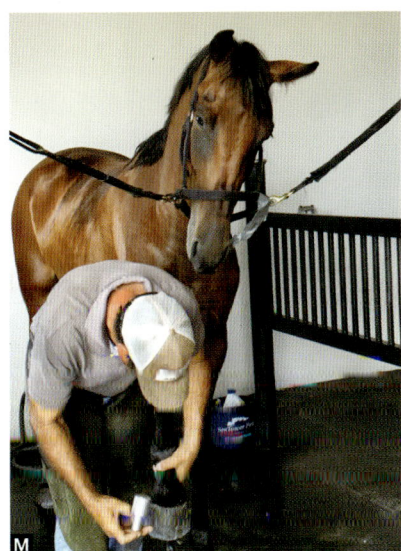

[A] "Are you coming to see me?"
[B] "Come over here and feed me."
[C] "I don't like this bit or being handled by people right now."
[D] "Wait, come back!"
[E] "Do you have something interesting for me to see?"
[F] "You smell different."
[G] "I hurt there. Do not rub my stomach."
[H] "It hurts and I worry when I get girthed up."
[I] "I'm a little worried about what is going on behind me…"
[J] "I'm not sure what is going on or if I trust you."
[K] "Those blue things were not there yesterday! What are they?"
[L] "Please don't make me move, I am sore."
[M] "You are so interesting! What are you doing?"

PART TWO / CHAPTER 9 / Horse-Human Communication and Relationships

[173]

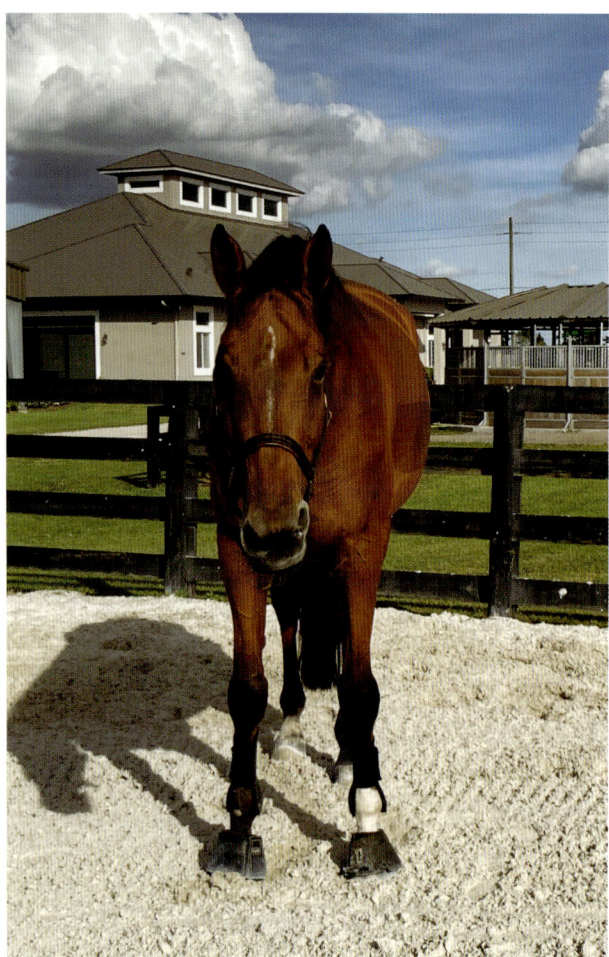

← [9.11] Coco—using her eyes, ears, and mouth, along with her confident stance—is clearly communicating that she is paying attention to me, although she is voicing her slight annoyance (using her wrinkled lips) with having to give me authority and halting outside my space.

↓ [9.12] Coco leans her head on Bill, asking politely for more "goodies."

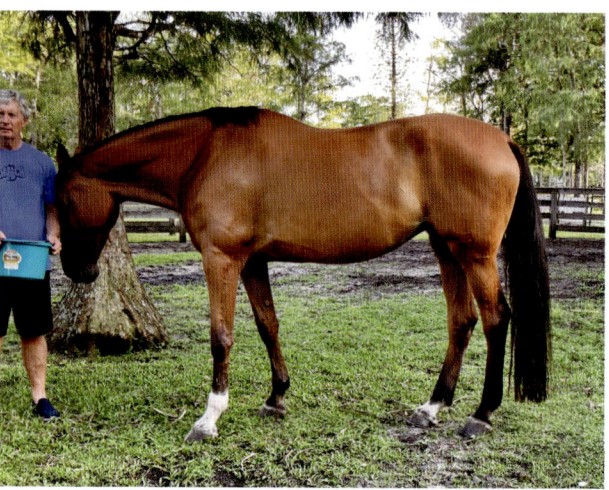

Learning and understanding your horse's rhythms can give you the subject matter and timing to start a two-way conversation with him (figs. 9.11, 9.12, and 9.13 A–C). For example, horses know when it is feeding time, so having a conversation about whether your horse likes his food is a good topic at that time, while checking to see if his tack fits will be less interesting to him. When he is moving into a sleep cycle, it is a good time to slow down your own energy and thoughts and just "be" with him, and when you are getting to ready to ride, that is a good time to discuss how his body feels. Synchronizing your communication with your horse's natural rhythms will naturally allow your horse to be more interested in what you are saying and more focused on information important to both of you.

On the following pages are some examples of different horse expressions and what they are most likely trying to communicate. Since horses learn from watching other horses (and humans), some horses may have developed "odd" language skills. For example, the young Thoroughbred whose mother pinned her ears back when he came to nurse because it hurt her, taught him that pinning your ears is a social greeting. Obviously, he gave the wrong communication to both horses and humans when he went off to the track to race and pinned his ears at everyone. In cases like this, you must pay attention to more than one signal (are his eyes bright and happy?) as well as the overall energy of the horse.

★ **TIP: / Horses use "nose bumps" to greet each other or ask for something. Mares use the gentle nose bump as a polite greeting, while geldings often use a nose bump when they are worried or asking you for something. /**

Two-Way Conversation

The first thing to consider when talking with your horse is to allow a *two-way conversation*. If all the communication is one-way and the horse tries to talk back, but he is not allowed to say anything or the person does not understand what he is saying, then the horse can shut down and stop trying

↓ **[9.13 A–C]** Sergio is worried and timid **(A)**, then he discovers he can "talk" to people—note the change in his energy, eyes, and overall expression **(B)**, plus his wide stance and high head as he attentively tracks people, trying to get attention **(C)**.

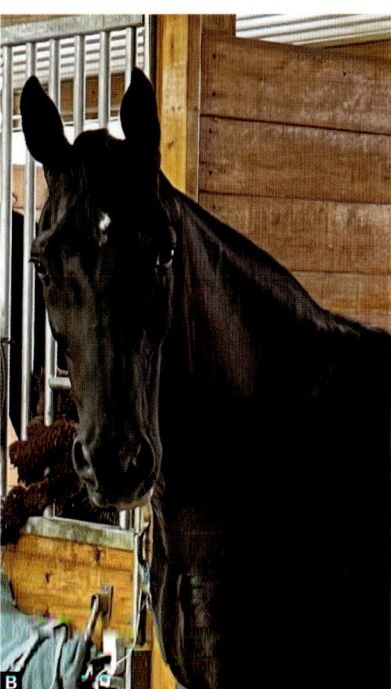
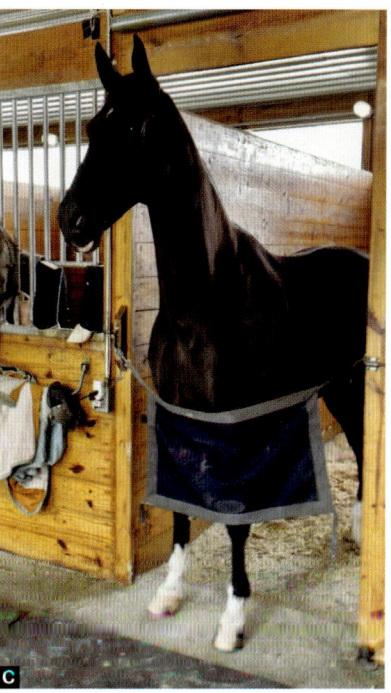

PART TWO / CHAPTER 9 / Horse-Human Communication and Relationships

STORY FROM THE FIELD:

A Horse with a Headache

When I was eleven years old at a horse show in Santa Barbara, California. I had a pivotal moment that made me realize not everyone could "hear" or "see" what horses were trying to communicate.

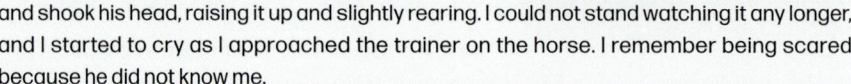

There were lots of horses going around in the warm-up ring, but one horse stood out to me. His eyes had a worried look as if he was in pain. As I would trot by on my horse, I could see his expression. The horse kept flipping his head up; the trainer had put draw reins on him to try and stop the behavior.

The horse continued trotting around, trying to communicate his discomfort. Eventually he stopped and shook his head, raising it up and slightly rearing. I could not stand watching it any longer, and I started to cry as I approached the trainer on the horse. I remember being scared because he did not know me.

As I got closer, I watched the trainer take a crop and smack the horse over the head when his head came up. It wasn't hard or done in a mean way—I knew the trainer was trying to prevent the horse from rearing again—but the horse fell to his knees.

When I got next to the trainer and the horse, I asked, "Why would you hit a horse with a headache?" I saw by the look in the trainer's eyes that it had not crossed his mind that the behavior was just the horse trying to communicate pain, not disobedience. The trainer did not say anything to me, but turned and walked away, dismounted, and told a groom to put the horse away.

In that moment, I realized I needed to learn how I knew what I knew, and I needed to learn the skills to be able to teach others how to understand what other species were communicating. The trainer that day was not a bad person, but he was ignorant, and I recognized, even at my young age, that causing undue discomfort and pain in horses, whether physical or emotional, is often not intentional, but due to a lack of knowledge and understanding. //

↗ **[9.14]** Horses are trying to tell us things all the time, but if they have no freedom to move their mouths, jaws, tongues, or necks—like this horse—then we have severely limited his ability to communicate.

to have a conversation. So be open to what your horse might be trying to tell you, and if you don't hear anything from him, ask him a question. You can even verbalize it out loud! Often, when you do this, the horse will look at you, not unlike a child or dog, as if he is wondering, "What are you saying?" This simple back-and-forth is the start of connection.

— Motivation

The second thing to consider is that there has to be motivation for your horse to communicate with you. While food might provide initial motivation to get a horse interested in you when he is shut down, it is not a true "horse motivator." *Social interaction* should be the motivator. For example, you can use your horse's natural curiosity by having new smells or objects to explore together, or you can initiate a grooming-scratching session (fig. 9.15). When horses know you make them feel safe and feel good, they are interested in being your friend, and therefore, talking to you.

★ **TIP:** / Become a good listener and encourage your horse to communicate with you. /

Communication Channels

Humans are generally "verbal thinkers" and communicators, while most other animals we communicate with are "visual thinkers" and communicators. So, as you read through the various channels for communication that follow, keep in mind that communication is simply the exchange of information from one being to another. Recognize that your thoughts and feelings carry information to your horse, whether you are aware of it or not. And your horse may be sending information related to what he smells, tastes, or feels about a situation. While horses do not speak words, they recognize human languages by

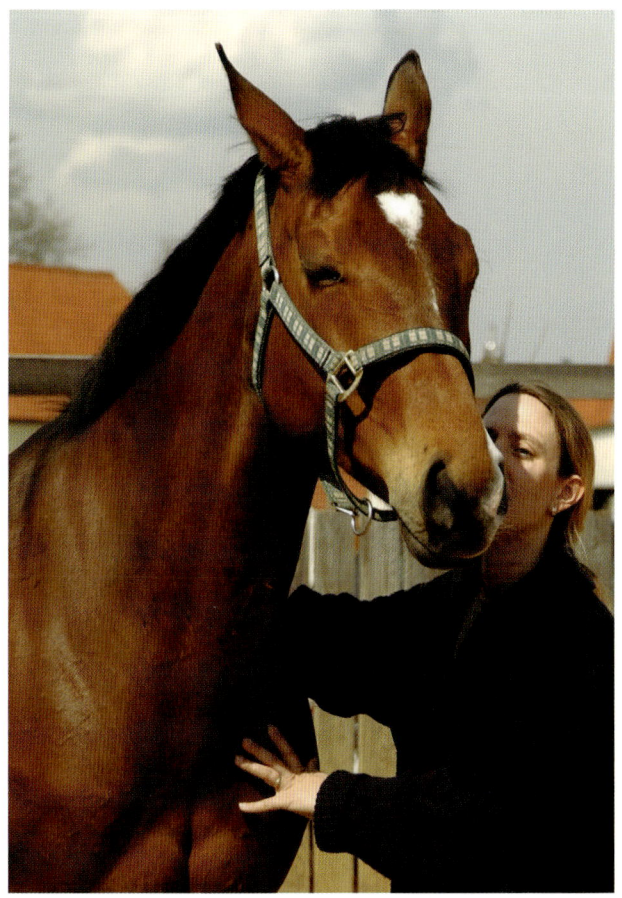

↑ [9.15] Two-way communication can start with "motivating" your horse to communicate with you about how he "feels" (assuming your horse feels like "talking"). Note the body language of this horse and person as they engage in a conversation about how the horse's body feels. The woman is using touch and voice to ask her horse if what she is doing feels good or not. With eyes starting to close and ears softly back, paying attention to the woman's touch, the horse is starting to relax, enabling the two of them to enter a deeper state of communication.

sound, and many will respond appropriately to what we ask verbally. However, you will have far greater success in communicating with your horse if you clear your mind of "verbal chatter" and try instead to be open to what your horse experiences through all his senses—sight, smell, taste, touch, hearing, and even his extrasensory perception (ESP).

There are different levels or "channels" of communication between horses and humans (fig. 9.16). Traditional training approaches tend to look at communication strictly on how to get horses to do what humans want them to do, with little opportunity for horses to "talk." More progressive training approaches allow horses to "speak up" a bit more by increasing the human's awareness of equine behavior, but still, the expectations related to the communication are driven by the humans. I like to think that there are five "channels" of communication when it is considered from a "horse-centric" point of view. These are directly related to the ways that horses communicate with each other in the wild, which we discussed in Part One (see p. 64). Recognize that no one way is used when communicating with horses—both horses and humans integrate various channels. Find which ones work best for you and your horse.

- **Channel 1: Visual** This includes all body language in the horse—facial expression, eyes, lips, ears, muscle tension or relaxation, body carriage and movement, tail position—as well as in the person. It can also include any visual stimulus.

- **Channel 2: Sound** Sound and vocalizations carry images and feelings. Soothing vocalizations indicate reassurance and security, while loud or stern ones can mean someone is unhappy or demanding. Music can also be used in communication as both horse and human can respond to the same frequencies (I discuss this in more detail on p. 184).

It's Simpler

Temple Grandin's book, *Thinking in Pictures* (Vintage, 2006) is a good place to start when trying to understand the sensory world of both humans with autism and other social mammals, such as horses. Horse thinking is simpler than our own, and their needs reflect the simpler desires of moving *toward what they want* and *away from what they do not want*. As you read this section further ponder what a more "horse-centric" approach to communication might be like.

→ **[9.16]** Coco and I maintain eye contact as I use my voice to tell her she is a "good girl" by doing what I asked her to do—"be aware, pay attention, and stand still." Note her eye on me and her right ear slightly cocked, listening to me. I walk over to reward her with a gentle nose bump and will allow her to smell the back of my hand, and then I will also give her a scratch on her neck or withers where I know she enjoys it. These are the social rewards of engaging multiple channels of communication.

- **Channel 3: Smell** Familiar smells may indicate social support, comfort, and security, while other smells may indicate danger, food, water, health status, or reproductive readiness. Every horse and human has a unique smell. If your horse likes you, he will associate your smell with a pleasant and safe relationship. A fast track to the horse's brain, smell can be associated with learning and other actions. Aromatherapy is an effective communication tool as both horses and humans may respond similarly (see more about this on p. 186).

- **Channel 4: Tactile** This is touch—a gentle nose bump, a hand on the withers, finding the scratching spots that offer social reassurance, security, and make your horse feel good. When riding, this can be a shift in your weight, body, or balance, or a touch of the rein.

- **Channel 5: Energy** This includes your emotions, thoughts, and heart coherence (see p. 59). It means being mindful and aware, sensing and breathing together, using intuition, and slowing your brainwaves down to match your horse's communication frequencies.

★ **TIP:** / Horses share facial expressions related to their emotions, similar to other social mammals. /

Be "mindful" and aware of how you communicate with your horse, as well as how your horse tries to communicate with you. As I've already mentioned, horses often "give up" or have little interest in communicating with humans when they have tried and no one has listened or responded. Engage all your senses and attention when communicating with your horse.

[9.17] Teaching a student how to use her energy, body language and voice so her horse understands "Stand still," or "Move." Her horse was very sensitive and understood "Move," but we had to soften and open our energy so he understood "Stand still" while we approached.

Channel 1: Visual Communication and Body Language

Remember what we learned in Part One: Horses see differently than people. They do not adjust to changes from light to dark or vice versa as quickly as we do. And while horses are excellent at spotting movement off in the distance, they are not as good as humans when it comes to depth perception and close objects. However, horses are talking all the time with their eyes, mouths, ears, and bodies (body language), and they *do* learn visual signals from each other and from humans (fig. 9.17). In fact, they will actually respond similarly to dogs when it comes to hand signals.

While horses are pretty good at recognizing visual signals, it usually is the energy behind the visual signal that they pick up on. Consider the young horse who learned to put his ears back when greeting others. His eyes might express eagerness, while his ears are laid back, which might be confusing to another young horse (or to us!), but older horses, particularly mares, see right through this misuse of body language and read the eager energy because they often have the ability to look into the other horse's eyes to read deeper expression.

We can mimic body language in horses to begin a conversation with them—in particular, their "calming signals," such as yawning, licking your lips, sighing, lowering your head, gentle pawing of the ground or rubbing of a leg, and eye blinking (figs. 9.18 A–E). As social mammals when one animal starts to relax and sleep, other individuals in the group may also. But, as

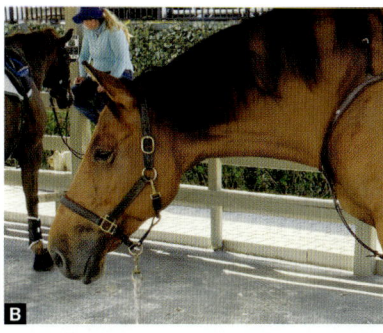

→ **[9.18 A–E]** Horse shows are good place to practice calming signals. I start yawning and slowing my energy down to a "sleepytime feeling," and Coco looks at me as if asking, "What are you doing?" **(A)**. More yawning and eye-blinking from me, and Coco's head goes lower as she tries to stay alert—it is a little like hypnotizing your horse **(B)**! As she gets sleepy, her lower lip droops, her ears get relaxed, and her head drops a little more **(C)**. Then she starts yawning as she tries to fight off sleep, but she cannot help herself **(D)**. Coco gives in and rests. Note her relaxed ears, eyes, facial muscles, and lips **(E)**.

mentioned previously, horses use integrated communication channels, so any body language needs to be reinforced with the right thoughts, emotions, and energy (see p. 208). Giving calming signals while thinking about a negative experience will not gain the same reaction from your horse as when sending out relaxing thoughts.

★ **TIP.** / As social mammals, calming signals such as yawning and blinking can relax you and your horse. /

[9.19 A–D] Chase and Coco greet each other **(A)**, and Coco nuzzles and smells Chase, enjoying the encounter, but she is aware of another horse walking by **(B)**.

Coco communicates to the other horse to "stay out of our space" with her ears going back **(C)**. She flattens her ears more strongly and gives the other horse an eye glance, showing the whites of her eyes, clearly saying, "Do not get any closer to us." This is a protective communication, as she and Chase are having their own special conversation, and she does not want it interrupted.

If you were not paying attention to all the interaction going on in the barn with other horses walking by, you might mis-read this communication as aggressive toward Chase, but it was directed toward and meant to warn the other horse **(D)**. (Because mares are usually more aware than geldings, they often will have more broad-spanning and sophisticated communication.)

_ A Note About the Eyes

As I discuss elsewhere in this book, when you learn to look into the eyes of your horse, you will be able to read what he is thinking and feeling more accurately than all his other body language. This is because his eyes reflect joy, excitement, sadness, depression, pain, worry, curiosity, and all the other emotions, and what the eyes tell us is not learned (from a horse's mother or pasture-mates or trainer) but rather a "window into his soul," as the saying goes.

There are some training methods that advocate that we should *not* look into the eyes of a horse, and this is very confusing to the horse. Now, it is important to understand a horse does not like being looked *at*, but a meeting of the eyes allows him the chance to judge what you are thinking, as well. In my field work I first noted how mares would often use a "soft glance" at each other, just to maintain social eye contact and connection.

_ It's Not Necessarily By the Book

There are a number of books written on body language in horses, but this gets a little tricky as you must take into account the situation, your horse's temperament, and what your horse has learned about visual signals from

 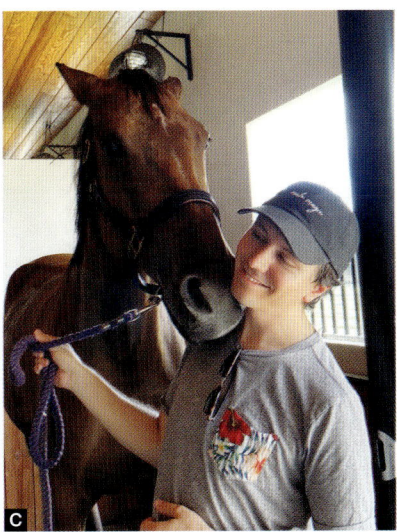 

others before you can determine what he is saying (figs. 9.19 A–D). For example, a gregarious gelding who likes to use his mouth may grab or bite things to communicate interest in interacting with you, while a timid horse may use the same body language to show he is worried or hurts. Another example is the often-referenced "licking and chewing"—this can mean a horse is relaxed *if* the horse also has a soft eye and relaxed facial and body muscles, but licking and chewing is also a sign of stress because a stressed horse may have a dry mouth and will make the same motions as he tries to secrete saliva. Thus, you must look at the whole body, energy, and history of the horse to accurately interpret communication and put it together, as we will discuss further (see p. 219).

★ **TIP: / When reading a horse's body language, you need to be aware of how your horse is using visual communication. The same body language can mean different things, depending upon the situation and gender of the horse. /**

Channel 2: Voice, Sounds, and Music

While horses may not understand every word you say, they do seem to enjoy hearing a soothing voice. Through the years, studies have shown that languages such as French appear to be preferred by horses when it comes to rewarding statements, while languages similar to German are understood as commands to obey. Obviously, vocal tone can be interpreted by horses, so whatever your language, using even tones and singing or whistling can help communicate relaxation, happiness, and appreciation, while strong, excited vocalizations may create alarm or imply you are displeased (fig. 9.20).

When humans speak, we often have an image or feeling in our minds. Talking to horses keeps us focused on the communication, and thus horses may be able to interpret what we are saying. While there is little research regarding whether horses can learn the complexities of human language, there is plenty of clinical

[9.20] Coco tried to challenge my space. My body language, voice, and energy (see p. 180) stopped her. Note her expression, ear position, and tight lips as she thinks about it.

evidence to demonstrate horses do understand simple words such as, "No," or "Come." Since horses are good at reading human facial expressions and energy (see pp. 178–9), they may be able to put the whole picture together.

★ **TIP:** / Use soothing sounds to encourage relaxation and to reward your horse as it has been shown horses respond to verbal praise. /

Music can also be used in communication. It has been shown that horses learn various notes and rhythms mean different things. When you use certain music for grooming and different music for riding or for when the farrier comes, horses can easily learn what to expect and most often will relax to the familiar sounds. Music is a great communication channel to help people and horses "synchronize."

★ **TIP:** / Music is a wonderful language to share rhythm and feeling between horses and humans. Many horses have music preferences and respond to different frequencies. /

Sound is often used as a "bridge" in communicating to teach horses desired behaviors. For example, "clicker training" uses an audible *click* to gain the horse's attention prior to performing a requested task. Rewards are used when the animal performs the correct task. Although it is more of a one-way conversation, I do find that appropriate use of voice, sound, and music can be helpful in creating a happy environment for horses.

STORY FROM THE FIELD: //

Music Preferences and Effects on Horses

The response of other animals to music has always interested me. Having grown up in a barn and at horse shows where music was always playing, I decided to test and see if all horses liked music. While most horses in my study did respond favorably by showing behavioral signs of relaxation or calm curiosity and interest to appropriately selected music, a few horses either were disinterested (I recorded no behavioral changes) or actively walked out of their stalls (into run-in paddocks) when the music was playing, as if to escape the sound. In an attempt to eliminate variables, I conducted a number of music preference trials, and it seemed horses had various music tastes.

For example, Neptune, my Thoroughbred, loved Celtic music. When it would play, he would come and stand right next to the speaker. Shamber, a captured wild horse, was by nature shy and sensitive. She seemed to enjoy Native American music and would walk in from her paddock to stand close to the speaker when drums and flute music were playing. Cedar, another adopted wild horse (from a different herd than Shamber), seemed indifferent to music, as she never changed any behavior regardless of what was playing. I have used music in my clinics since the 1980s and so have been able to observe hundreds of horses' and riders' reactions to it. The evidence is clear that certain types of music instill relaxation in horses and riders and no doubt allows them to more easily connect with each other. One funny example was when I was teaching a three-day clinic at an eventing farm in the United Kingdom. I arranged the music to be synchronized to walk, trot, and canter, and although the horses were fresh and a number of new horses had hauled into the clinic, the horses all responded calmly and rhythmically, seeming to have fun and enjoy the interaction. However, when it was time to go home, none of the horses who had hauled in wanted to get into their horse trailers. Even food was not motivating the horses to load to go home. Obviously, the horses had formed new friendships.

I asked the students to "think like a horse" and suggest a solution, hoping they had learned what to do from the clinic. While there were a number of horse-human interaction techniques that would work, one trainer said, "Let's put the music back on." As soon as the sounds we had used for the clinic were audible, the horses peacefully and willingly loaded in their trailers. The music relaxed the minds, opened the hearts, and synchronized horse and human energy. //

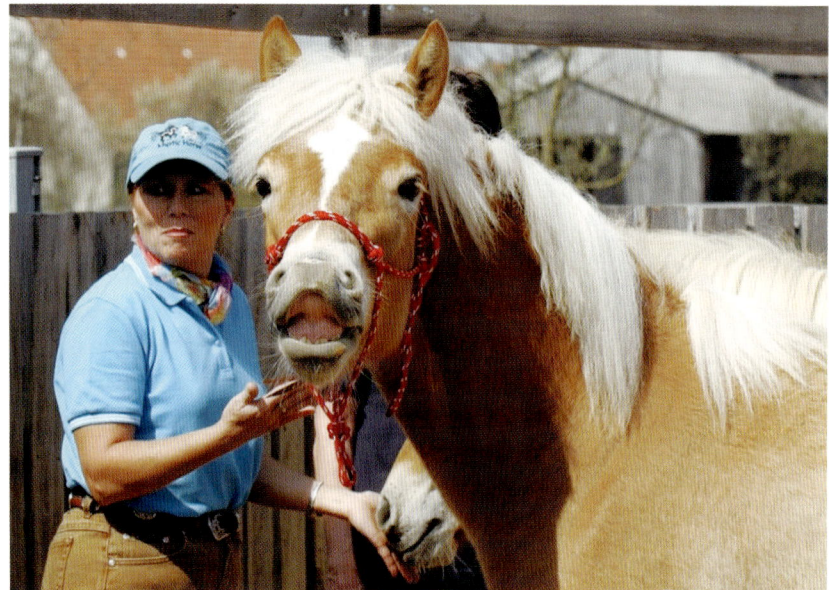

[**9.21**] The young horse with me makes a flehmen response to a particular smell.

★ **TIP:** / Speak to your horse in first person—*to* him, not *about* him. Horses respond well to our voices, thoughts, and feelings, even if they do not understand all our words. /

Channel 3: Olfactory Communication—from Manure to Aromatherapy

Probably the most underutilized communication channel between horses and humans is smell. Horses use smell constantly to assess their environment and each other's well-being (see where I discuss this in detail in Part One—p. 50). Smell to horses is like reading the newspaper or social media. They can interpret all kinds of communication from a simple sniff (fig. 9.21). They even take stock of themselves, as horses will often pass manure and then turn around and smell it just to make sure "they are who they think they are."

★ **TIP:** / Smell may be the most under-recognized channel of communication between humans and horses. /

★ **TIP:** / Aromatherapy can be used to encourage horses to relax, focus, or even learn a new lesson. /

Horses can smell other horses trotting around the arena with them, smell when a particular plant is in the pasture, smell when a storm is coming,

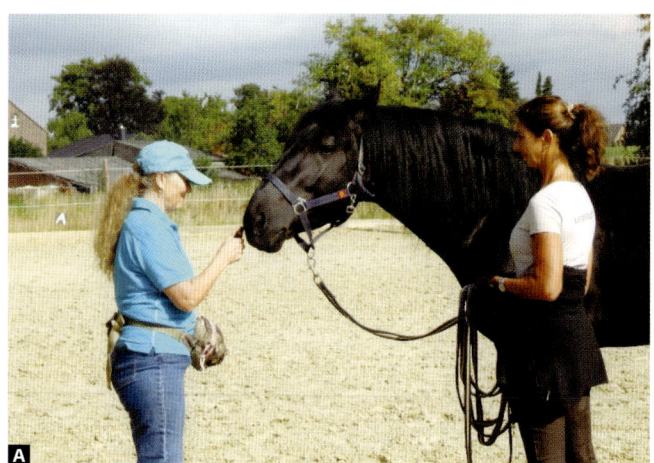

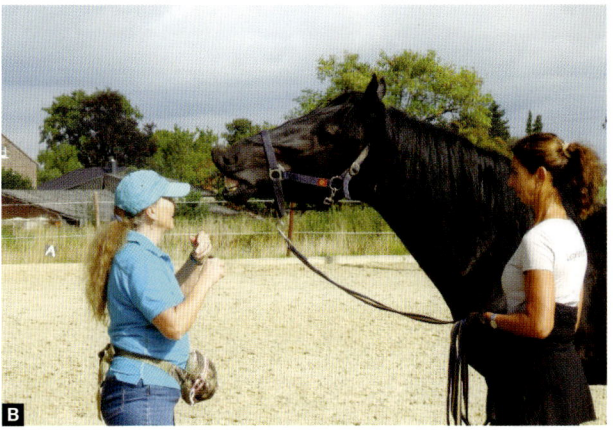
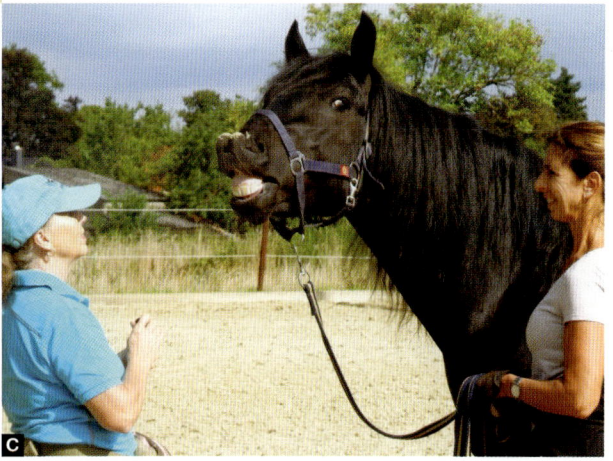

↑ **[9.22 A–C]** I offer this horse a new scent as I attempt to find a "signature scent" that he enjoys **(A)**. I try several different-smelling essential oils. Smell can "wake up" the brain, providing interesting stimuli to horses. Here the horse makes a flehmen response, reacting to a new smell or information as he processes the various scents I present **(B & C)**.

smell when another horse is sick, smell water, smell danger, smell fear—you get the point! Because horses are good smell communicators, humans can take advantage of this by having a "signature smell" your horse recognizes and enjoys (figs. 9.22 A–C and 9.23 A & B). Your own personal body odor might be enough, although meat-eaters can smell unpleasant to a horse. Try fragrances like lemongrass or peppermint, which have been shown to appeal to horses. Other scents—like lavender, chamomile, ylang ylang, clary sage, and tangerine, for example—may have a relaxing effect on your horse (and you).

★ **TIP:** / Strategic placement of your horse's own manure or his equine friends' manure can help convince your horse to enter a trailer or feel confident in a new situation. /

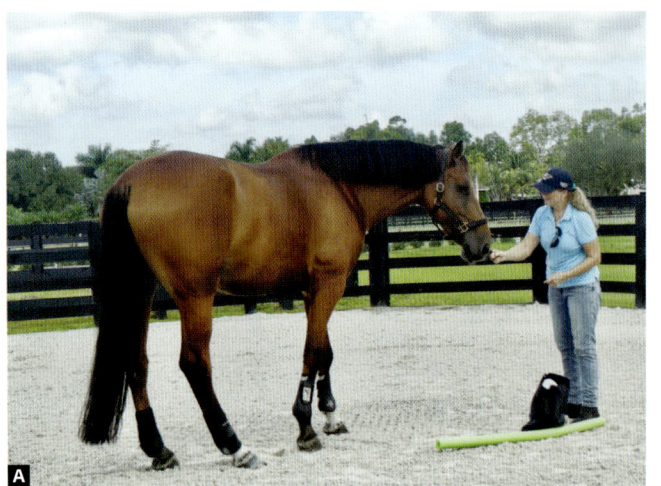

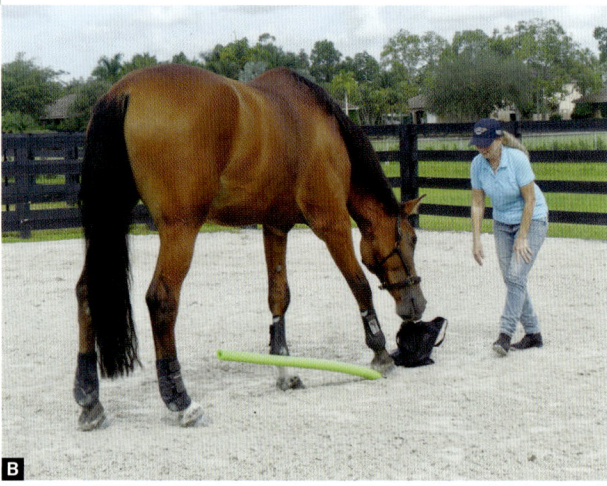

↙ **[9.23 A & B]** I offer Coco various scents to explore as I test her sense of smell and watch her reaction **(A)**. Which nostril does she prefer to use? What are her eyes and ears saying? Coco recognizes scents I used earlier and steps forward to investigate the bag that holds the different scents to see if the smell she is looking for is in there **(B)**. Some horses are actually used like "scent dogs" to track smells they are given.

Because horses rely so heavily on smell, if it smells safe and looks unsafe, they will often choose to trust their sense of smell over what they see—*unless* they have negative associations with the visual. You can convince horses to do almost anything—perhaps even against their better judgment—if it smells like a good idea. Smell is perhaps the largest sense organ in their brains so it can override thinking and initiate a response. Scent can also leave a lasting impression in a horse's mind.

★ **TIP:** / Because horses use their sense of smell much more than sight for investigating close surroundings, allow your horse to smell a strange environment before asking him to enter it. /

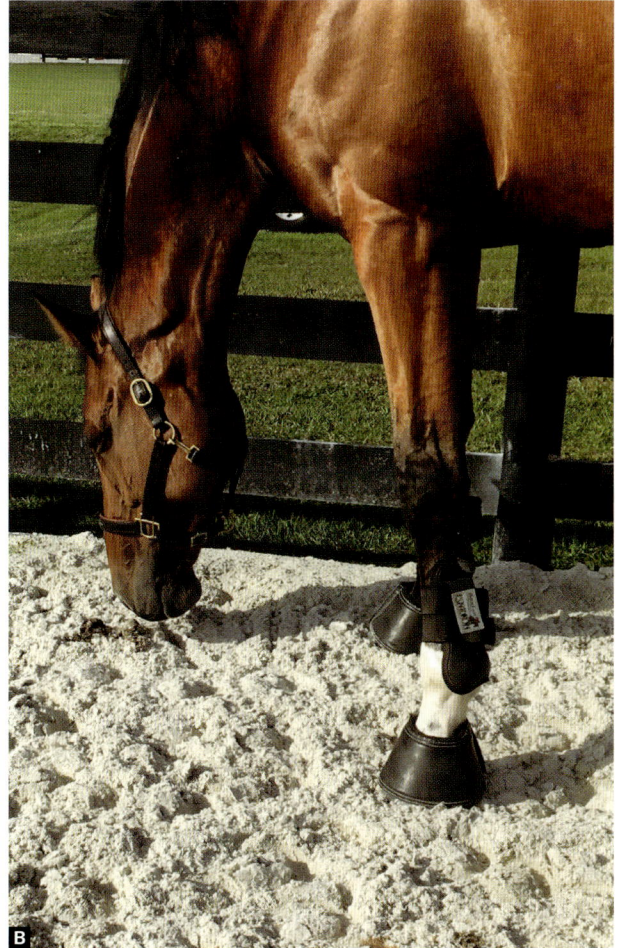

↑ **[9.24 A–C]** The jumper mare in **(A)** was afraid of poles, so she was given time to investigate one on her own, using smell to determine safety and better understand what it is. In **(B)**, Coco smells another horse's manure to determine if she knows this horse, what the horse has been eating, whether the horse is male or female, and other interesting information. In **(C)**, Coco smells Bill's hat. Note her ears and eyes, and how she is processing the information to see if she recognizes any smells.

You can use the horse's highly sensitive sense of smell to override other stimuli (such as a stallion smelling nearby mares). Specific scents can help your horse focus, concentrate, remain calm, and feel happy. While not all horses respond the same to aromatherapy, many horses enjoy the process of choosing the smells they like, and this exercise gives you insight into what your horse needs.

_ Channel 4: Tactile Communication— Hands On Your Horse

One of the best ways to start communicating with your horse is to get your hands on him. As part of a proper "horse greeting" (see p. 209 where I explain all the

[**9.25**] Gentle touch between human and horse can lead to friendship. Horses ask to be scratched by their friends and are very good at directing you where they want "grooming" when they are allowed to do so.

steps), horses often start a buddy scratch, usually at or just in front of the withers. Offer gentle scratching until you find a place your horse enjoys touch—since the withers, shoulders, and neck are common "horse conversation" spots, they are a good place to start (fig. 9.25). You will know when you find your horse's "sweet spot" when he extends his neck and lip or acts like he wants to scratch you back in the same place. You do not have to be a pro to determine whether the touch feels good or not—your horse will tell you. If it appears to hurt in any way, back off and move to another spot. (Horses who do not like to be touched in the withers area may have neck or back pain.) If your horse tells you a spot feels good, scratch deeper and in different motions to see what he enjoys the most.

★ **TIP:** / Mares often do not like their faces touched by strangers. /

Horses will often use their mouths and heads to touch humans when they want to "talk." Males, particularly young males, will use their mouths frequently, grabbing or biting. While this is a normal behavior, humans need to establish acceptable safe touch communication behaviors. Mares, on the other hand, normally are not as "mouthy" as male horses. However, mares will use a gentle nose bump as a way of greeting or asking you for something. (They are also generally very picky about who touches their mouths and noses and prefer to touch you first.) Male horses use the nose bump more when asking for reassurance when they are insecure about something.

Notice how much your horse uses his mouth. Because of the tremendous sensory ability of the nose, horses will engage in oral behaviors often when they want a distraction from a source of discomfort or worry.

★ **TIP:** / Male horses generally are much more "mouthy" than mares until they mature. /

★ **TIP:** / Mature geldings who constantly grab things in their mouths may have discomfort somewhere in their bodies. Being "mouthy" takes their minds off their pain. /

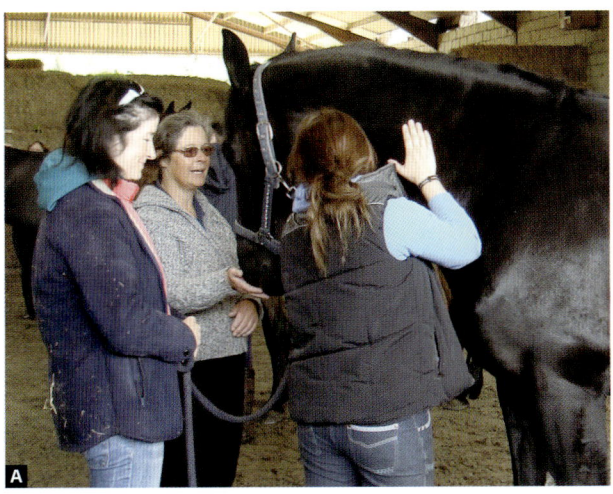 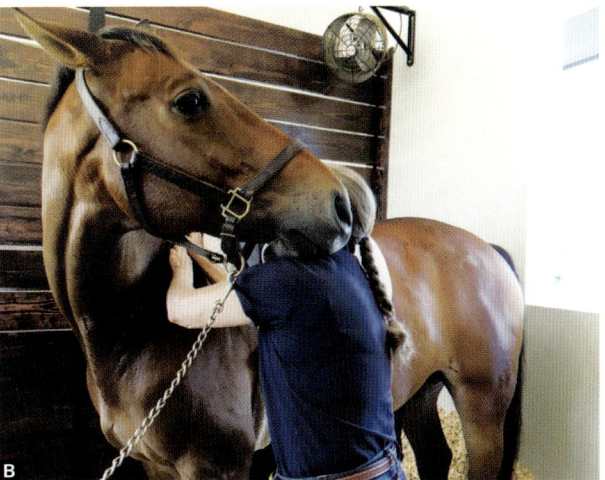

↑ **[9.26 A & B]** Students in a clinic check for side-to-side tension in a horse's neck **(A)**. Robin finds a tight spot on Summer and helps release his tension **(B)**. Note that he responds with touch, wanting to make Robin feel good in return.

_ Using Touch to Locate Pain

Tactile communication isn't just about pats and scratches and nose bumps. In our relationship with horses, we also use it to identify and solve physical issues they are trying to tell us about. Because horses naturally tend to be on their forehands, they have to learn to carry us more from the hind end to balance our weight on their backs. Soreness, "blockages," and imbalances are not just from too little exercise, poor management, or poorly fitting saddles—they also stem from improper use of tack and training devices, as well as asking horses to complete tasks their bodies are not balanced to perform.

We can communicate through touch to help ease discomfort and free tension (figs. 9.26 A & B). Jim Masterson's Masterson Method® in his book *Beyond Horse Massage* and Dr. Renee Tucker's easy body checkups from her book *Where Does My Horse Hurt?* are great places to start.

Learn more

Learn more

★ **TIP:** / Most horses like to be scratched and touched. If they do not, then they are likely experiencing pain or worry. /

★ **TIP:** / Emotional stress can be stored in the horse's body and present as physical issues, such as ulcers or tension throughout the body. /

★ **TIP:** / The most common areas horses store emotional and physical stress are the poll, neck, shoulders, withers, back, and sacroiliac (SI) joint. /

_ Major Areas for Storing Stress and Tension

- **Poll/Head/Jaw** Tension and discomfort here are often related to worry from a badly fitting bit or bridle, a tight noseband, or being asked to do something the horse cannot do.

- **Neck** Look for balance in the feel of the neck from side to side (see p. 191). Tension can be stored throughout neck muscles, particularly in the *trapezius* muscle, if the horse lacks confidence and or if the saddle does not fit comfortably.

- **Withers** This area becomes tight from compression, restriction, and compensating for poor saddle fit.

- **Shoulders and Ribs** Knots and tension can often be noted under the shoulder blades. Horses who tend to hold their breath will be sore in the ribs and possibly "girthy." Timid or worried horses can stand "base narrow," while confident horses may stand wide—both of which can cause forms of tension.

- **Back** Tension throughout the back is common in confined horses, and while emotional stress (caused by worry, fear, frustration, or sadness, for example) can be reflected in a stiff or sore back, poorly

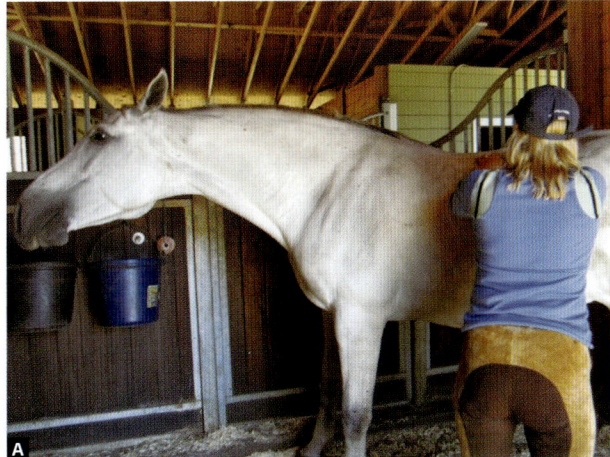

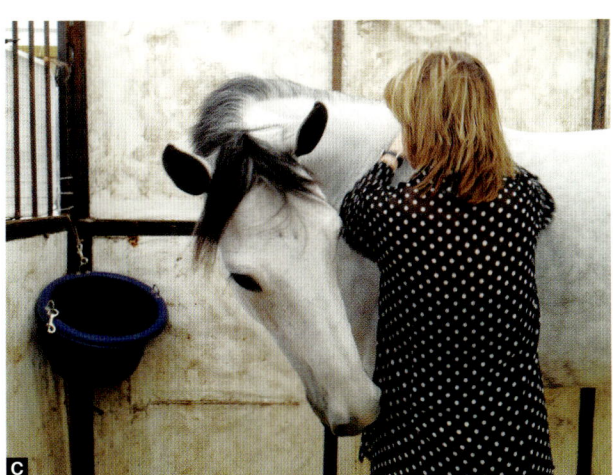

↑ **[9.27 A–D]** Conducting a regular body check before you ride allows your horse to tell you what feels good, where he is tight, and where he may hurt **(A & B)**. Horses will often turn around to offer to scratch you back or "thank you" **(C)**. Notice how Dr. Carlos Jimenez (a veterinarian and chiropractor) keeps an eye on the horse's head and ears as he works over the horse's SI area **(D)**.

fitting saddles, improper balance of the rider, limited ability to roll and buck, and general day-to-day horse-human activities can also be the cause.

- **Sacroiliac (SI) Joint** Often behavioral signs are the first indication of tension here, such as swapping leads, crow-hopping, and not wanting to pick up and hold up a hind leg. The area may be sore to the touch, and you may notice muscle imbalance in the hind end.

Getting to Know Your Horse Through Touch

Emotional stress in the horse is often stored physically, usually through soft tissue, in particular, the *fascia*—the connective tissues around the organs, muscles, joints, bones and nerve fibers (figs. 9.27 A–D). The following is a quick, hands-on, overall "body check" so you get to know what your horse's body is saying.

★ **TIP:** / Horses may also reflect the stress of their riders, so make sure you regularly get your own body balanced. /

When doing this simple body check, you are looking for:

- **Texture** (What does the muscle feel like? Can you feel the fibers? Is it smooth? Does it have knots or fatty pockets?)

- **Tension** (Can you feel tightness? Does the muscle respond when you touch it, or does it feel like a rock?)

- **Tone** (Fit muscles are firm but pliable; unfit muscles are flabby and loose.)

- **Balance** (Ideally, the muscles from side to side should feel similar.)

- **Sensitivity** (Some muscles are more sensitive than others—for example, the pectoral muscles can be quite sensitive to touch when the horse is stressed, while a tight back will show no sensitivity when you first touch it.)

- **Soreness** (Does the horse exhibit pain when you touch the surface of his body or add pressure to deeper levels?)

- **Flexibility** (Is the muscle movable? Does the muscle stretch and contract with movement?)

__ How to Do the Body Check

- **Find Your Horse's "Sweet Spot"** This is where he likes to be touched. Gently massaging or scratching

The Importance of Fascia

In all mammals, fascia works as a "spider web," connecting the systems of the body and communicating important information—from physical to emotional. Fascia lies just under the skin and covers not just the muscles but other organs, including the brain. It is the most complex and integrated communication system in the body, and touching one area of it can impact the whole being. It influences posture, musculoskeletal dynamics, connective tissue, and just about everything else! It is often the first line of "stress storage" when a horse (or person) deals with chronic worry and tension. The fascia can get "tight" and the communication system it supplements can "lock." Being able to release tension in the fascia can not only change the physical system of the horse, but his emotional well-being as well.

Learn more

→ **[9.28]** Coco relaxes with her eyes closed as her muscles let go of tension when I move the fascia on her neck and shoulder.

this area usually encourages relaxation and social bonding, and is critical with horses for both physical and emotional help (fig. 9.28).

- **Palpate the Muscles for Tension, Texture, and Tone** Muscles should all have relatively the same *texture* and *tone*. Start your palpation with the jaw muscles and poll, and work your way down the body, across all the muscle areas, looking for how your horse may be storing stress in his body as tightness and "knots" while noting differences in muscle weakness or strength. "Listen" to your horse as you move your hands over his body. When a horse is holding a lot of tension or worry, he may not enjoy being touched, so you will have to work gently and with careful attention to his reactions.

- **Check the Skin for Moveability** Fascia can get "stuck" to the muscles and feel tight. With a simple "skin roll," you can get the fascia to wiggle and often relax the muscle (figs. 9.29 A–C). Place your hands on your horse's muscle and gently push his skin. It should "roll" across the muscle. If it is "stuck," jiggle it like a bowl of Jell-O to loosen up the fascia.

- **Check for Flexibility from Side to Side, Up and Down** Your horse should be able to reach his poll and neck equally to both sides and touch his nose to his hip. He should have full range of motion, lifting his head up high like a giraffe and bringing it down between his legs, flexing the poll and neck

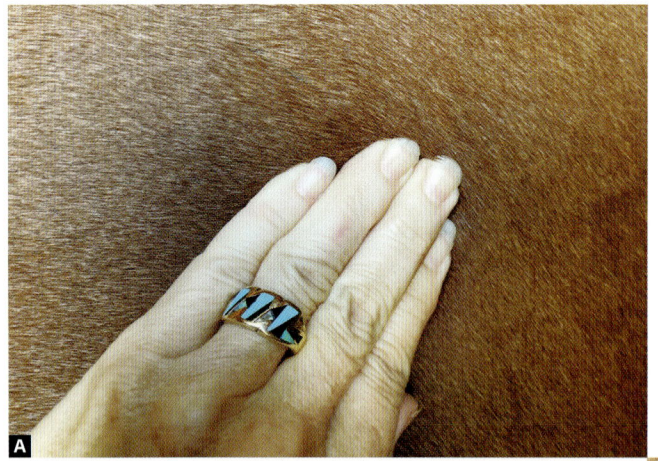

↙ **[9.29 A–C]** I place my hand on this horse's large medial gluteal muscle, and the skin does not want to slide as the fascia is tight **(A)**. By using my energy to ask the horse to let go of the tension, the skin starts to release and roll across the muscle **(B)**. When the fascia is "unstuck," the skin "rolls" **(C)**. This helps release tension in the muscle as well as the horse's entire system, as fascia is one of the body's most important communication systems.

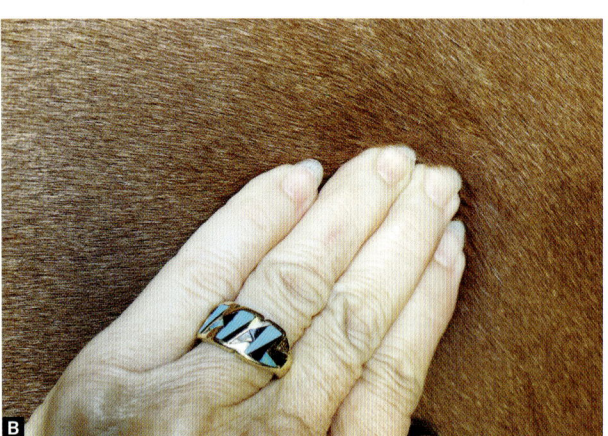

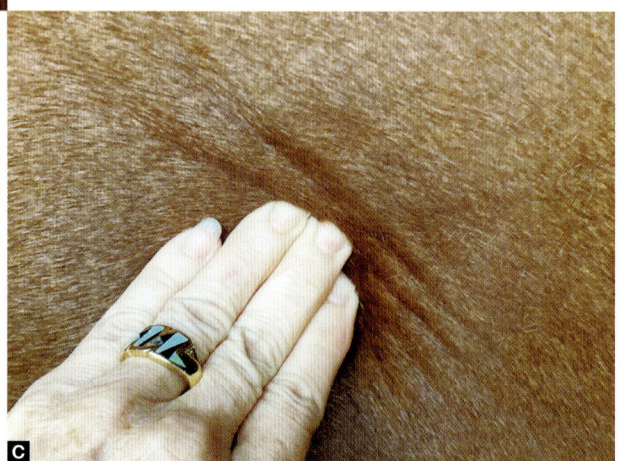

vertebrae. *Carrot stretches* are a good way to check this as they can also give you information on your horse's balance and how he might be compensating from side to side (figs. 9.30 A & B).

- **Run Your Finger Along the Topline** Using a finger with a gentle touch from your horse's poll all the way down his neck, back, and hindquarters, and down the back legs (following the *bladder meridian* used in Traditional Chinese Medicine) can bring heightened awareness to your horse of where he may be out of balance (I recommend *Beyond Horse Massage* by Jim Masterson for detailed instructions on how to use this). You may notice your horse flinch or move an ear or eye when you touch a place that needs attention. Just pause a moment, breathe, and allow your horse to adjust or release.

Through touch, you can learn a lot about your horse's body and how he stores stress. You can also help teach your horse (and perhaps yourself) to be aware of how to let go of it—both in its emotional and physical forms.

Channel 5: Subtle Energy Communication in Horses and Humans

Horses are very aware of subtle energies around their own bodies and around other beings. As I discussed in

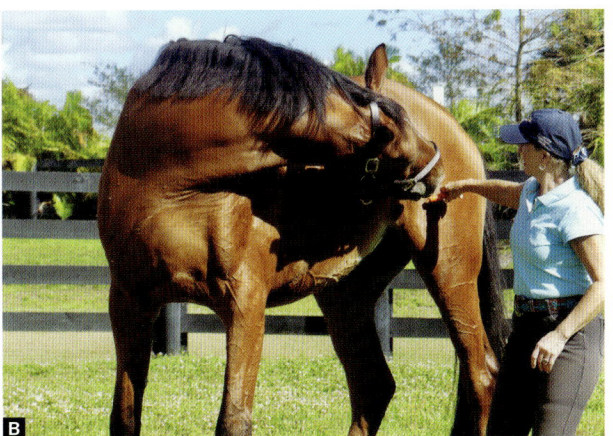

↑ [**9.30 A & B**] Coco doing a forward neck stretch **(A)** and reaching back for a carrot **(B)**. I touch the part of her body she is stretching toward first to increase her body awareness and to teach her to focus on the stretch rather than just the food.

Part One, horses have great *proprioception*, or what is also known as *kinesthesia*—the body's ability to sense its location, movements, and actions in space. Subtle energy is detected by special receptors in the muscles, tendons, and skin and communicated to the brain to tell the individual where the body parts are relative to others. Similarly, thoughts and feelings can reflect a subtle energy that can be picked up by other creatures, possibly through the proprioceptive systems, or energy "biofields," thought to surround and extend out from around all bodies of life by about 8 feet. Because biological sciences have primarily only looked at the physical and chemical relationships of life, they have missed what may be the most important influence of all—the energetic ones. Once people become aware of this energy exchange, they can communicate much more easily with animals (fig. 9.31).

← [**9.31**] Chase calmly "asks" Coco please to be nice to him as he has not ridden for a long time! Sensing his kind and quiet energy, Coco responds with her own nurturing and protective energy. This energy exchange is the foundation for a good ride.

Happily, research has begun to measure the connections between humans and other species via the fabric of energetic communication, consciousness, and emotions. For example, a number of different studies have focused on what happens to the human brain and body when we think and feel good thoughts around horses. When humans focus on appreciation and love, the horse relaxes, and both horse and human come into *heart coherence* (see p. 59), but further studies have shown there are cellular changes to the body as well. Some people call this "healing energy"; science is investigating its potential electronic and biochemical impacts.

★ **TIP: / Center yourself and become calm and open in your heart and mind before interacting with horses as they can sense your energy. /**

↑ [9.32] Consciously slowing my energy down and emptying my mind, I tune into Coco and ask her what she would like to do now, as she has done everything I have asked her to do. She drops her head and "asks" for us to relax together. We are sharing awareness and communicating on many levels.

An underlying "connective communication" allowing for an "emotional pairing" between horse and human takes place—whether we are aware of it or not—through subtle energetic exchanges. Some of these can be measured using sophisticated equipment, while others are best measured simply by monitoring the behavioral responses of horse and human. There are many different ways of describing this energetic exchange; I have always called it "sharing awareness" (fig. 9.32).

The areas of interspecies communication and human-animal interactions are continuing to grow as science gains sophistication in measuring the effects and benefits to both species. One area of research focuses on *heart rate variability* (HRV—where the amount of time between your heart beats fluctuates slightly) and heart rate coherence between humans and other life forms, including horses. Since heart rate (HR) and HRV are used to measure stress in horses, they can also be a simple tool to measure "connection" between horses and humans. Being able to observe using a simple device

← [**9.33**] The young wild horse Cedar turns to investigate my son Chase. He is the first human she has met after being rounded up and separated from her mother. Chase kneels nearby with nothing but appreciation and curiosity, sharing the same awareness as Cedar.

or app when both horse and handler or rider hearts come into a coherent pattern and the horse's heart rate is stable, you know the horse is relaxed and learning and the two are "sharing awareness" together—this is that emotional pairing and energy sharing we want. It is an easy way for horse and rider to communicate because horses are pre-programmed by nature for attunement to their environment and to each other.

★ **TIP:** / Horses can be highly sensitive to human emotions and mental states. /

The nonprofit HeartMath® Institute has conducted research that shows that when an appreciative open and relaxed state of being is established by humans, horses—given the choice—will often "pair" with the humans. Using HRV, researchers have been able to measure the effect of different interactions between horses and humans in various emotional states. Anger, frustration, and worry in humans create incoherent heart rate patterns in horses, while appreciation, love, and kindness in humans create the coherent patterns associated with mental and physical well-being. Because, as we've discussed, horses have high emotional intelligence, they often not only "read" the emotional state of humans, they respond to or mirror it (fig. 9.33).

While horses may not understand every word we are saying, it is clear they do pick up on our intentions and feelings. They are very sensitive to subtle changes in emotional and mental energy, which makes them better

Learn more

"listeners" than many humans think. As energetic and emotional creatures themselves, horses have evolved to be very sensitive to their herdmates, and to other creatures and their environment. Whether being handled on the ground or ridden, horses use subtle communication to pick up information. This adaptive ability to not only be sensitive to human and other animal emotions, but to mirror emotions, has now gained horses a slot as "therapists and teachers" in healing and facilitated-learning programs.

Part of my graduate research in Interdisciplinary Consciousness Studies explored the benefits and stressors of horse-human relationships relative to improving communication and understanding for both species. This research led to the development of various communication models for assisting equestrians and others working with horses, as well as other species. One of these was *HeartMind Speak*, which I developed before the HeartMath Institute research I've referenced above lent further scientific support to the heart connection between species.

_ HeartMind Speak Visualization

HeartMind Speak is a communication model that integrates our thoughts and feelings into a deeper connection with all life. It allows us to collaborate, create, and share awareness with other animals, nature, and people. It is based on my early research, between 1980 and 1992, investigating the subtle communication I have referred to as "shared awareness" between species and based on the rhythms and frequencies of life.

What follows is a visualization that helps develop a deeper sense of consciousness between you and your horse, allowing you to share awareness of thoughts and feelings while staying relaxed. Always make sure your horse feels comfortable with you and there are no distractions that might interrupt the process. Remember your horse has his own natural rhythms; you want to start a conversation when he is awake and interested (figs. 9.34 A & B).

1. Begin by just sitting or standing next to your horse in a safe place where you both can observe and sense each other's energy. Then the same process can be done while sitting on your horse, either at a halt or walk.

2. Take a deep breath and relax into *mindful breathing*—become aware of and focus on your breath and empty your mind. Allow your thoughts to

> **While horses may not understand every word we are saying, it is clear they do pick up on our intentions and feelings.**

[**9.34 A & B**] Fernanda slows her breathing down and opens her heart as she connects with her horse energetically, asking him to tell her if his neck is holding tension **(A)**. Her horse responds by allowing her to flex his neck and relaxing into the movement **(B)**. Breathing together is one way to synchronize your energy with your horse's energy.

drift away like clouds. *Center yourself*—that is, be really present with your breath and body. Ask your brain to slow down, then imagine everything else in your body is slowing down. Horses and most other animals function in relaxed *alpha states*—calm, yet focused and receptive—while the busy human mind stays in higher *beta frequencies*—smaller, faster brainwaves associated with mental or intellectual activity and outwardly focused concentration.

3. Focus on your heart and its rhythm, maintaining breath awareness. Try to feel your heart beating and move your awareness from your breath to your heart. Try to feel your horse's breath and heartbeat. Can you synchronize your breath with your horse's?

4. Imagine an energy field all around you and running all the way through you into the earth. Breathe in that energy. Consider how you would feel as a horse with a human on your back, and then allow your awareness to move back to feeling like a human riding your horse. Drift back and forth between being your horse and being human.

5. Visualize your heart opening and expanding with energy as you fill it with thoughts of appreciation, love, and compassion for your horse. Imagine the feeling when you and your horse are totally synchronized and moving as one.

6. Visualize your heart and your horse's heart connecting. As you again synchronize your breathing with

your horse's, imagine your breaths and heartbeats as one.

7. What do you feel and sense? Observe your horse's state of being. Are you both relaxed?

★ **TIP:** / Learn to clear your mind and be open to sensing what your horse has to say through feelings, thoughts, and energy exchange. /

★ **TIP:** / Pay attention to your horse's "space bubble" when communicating with him; you do not need to be physically in his space to "share awareness." /

Manage Your Energy

Clear nonverbal communication with horses is centered around energy management of both our thoughts and our feelings. Being able to "control our brainwaves" and regulate our energy around horses is critical for deep relationships with them to form (fig. 9.35). Some people do this naturally while others must learn it. Humans spend most of their waking time in that beta brainwave frequency mentioned on the previous page—around 15 to 40 Hertz. This does not allow us to pick up much else in or around us.

Most animals, in contrast, seem to spend more time in lower frequencies—think the alpha (9 to 14 Hertz) or theta (4 to 8 Hertz) levels. By simply slowing your thinking and thus your brain waves—think about how you feel right before falling asleep—you can relax other animals and allow them to feel more comfortable

STORY FROM THE FIELD: //

Lessons in Space

People and horses often have personal "space bubbles"—this is area of "safe space" around them that makes them feel comfortable when in new places, when near new people, or when they feel threatened.

Often when working with horses that are displaying aggressive behavior, I start from a distance where the horse feels comfortable. Mares are particularly good at using their eyes and ears to indicate where they want you positioned relative to their bodies. One mare who constantly bit people was very clear that I should remain about 6 feet away from her. She then communicated with her body and energy that I could touch her right shoulder in a specific place, as if she was testing me. I respectfully touched just her shoulder. She then turned around as if to tell me I "got it right." She proceeded to guide me to touch her body in various places to help her release discomfort, which was at the core of her aggressive biting behavior. //

around you. (This may be one reason why horses work so well with people who have learning and physical challenges.)

★ **TIP:** / **Learn to slow your brainwaves down to "sleepy time" energy when around your horse and you will relax him.** /

Emotional and mental stress in people generates negative energy that is easily picked up by horses, so making sure you are relaxed and "mindful" in the barn is key to having open, two-way conversations. Learning to turn inward, meditate, do yoga, *or* simply take a moment for a deep breath and letting go of all the thoughts in your mind before you ride can bring you into a better state for communicating with your horse (fig. 9.36). I developed what I call the *OFFER Techniques* as a process model for helping people slow down their thoughts, be present in the moment, and open up channels of intuitive communication with other species.

↓ **[9.35]** Learning how to be present in the moment and slowing your energy and brain waves opens the door for other animals to "talk" to you.

↑ **[0.36]** Coco drifting into deeper rest as we breathe and relax together in her stall.

_ The OFFER Techniques

The acronym OFFER stands for **O**pen, **F**riendly, **F**ocused, **E**mpathetic, and **R**espectful. Each word offers a technique to assist you in sharing awareness and communicating with other beings. Language is only one way to exchange information; most often it is actually much more subtle than the spoken word. Our feelings, body language, smell, and thoughts are all integrated when we communicate. (We're going to talk about each of these in more detail on the pages to come.) In addition, our belief systems, personalities, and experiences affect how we communicate. The OFFER techniques take all this into consideration in order to provide a practical path for opening your mind and consciousness to a level where you will be *most open* to receive and send intuitive information to, and exchange energy with, other animals.

_ 1 Be Open

Often your beliefs, attitudes, and judgments prevent you from being open and learning new ideas. Having an open mind is critical to creating new pathways in the brain for sending and receiving intuitive information, and for acknowledging that all forms of life are unique, but equal. Be open to honor each being for its own contribution to the world. Open the heart and the mind and ask them to work together as a team in a loving energetic vibration. Let go of old beliefs and attitudes that no longer serve your ultimate desire and purpose. Be receptive to new information.

_ 2 Be Friendly

Maintain a relaxed and friendly attitude with your horse. Overeagerness, anger, indifference, and nervousness can create stress in the relationship and prevent open communication. A light joyful attitude combined with

STORY FROM THE FIELD:

An Old Logger and His Wild Horses

"Sharing awareness" is a term rarely used in horse-human communication, but it can describe the state between two individuals who feel they are connected on some level via emotions and thoughts. The term is appropriate for many people who intuitively connect to others without analyzing exactly how they receive information.

Sharing awareness varies in horses as it does in people, but like a number of other species, they have studied our ways and learned about us. In my research, I have recorded hundreds of cognitive exchanges between people and animals that fall outside of mainstream science in their explanation. One case stands out.

There was an old timber logger who I had seen come to the BLM facility in Palomino Valley and pick up the older stallions held there. He always brought a stock trailer and took as many horses as he could fit back to his home in Southern Oregon. I was concerned, as people who took older stallions usually kept them a year and then when they had legal title, sold them for slaughter. Not many people wanted older stallions.

The next time the old man came to the facility, I went over and asked him what he did with these horses.

He was a quiet man, and I could see he was not used to talking much, especially to some young woman. He briefly said he used them for logging. I inquired about how he trained them to log and why he chose to use wild horses. At that point, I think he realized I was truly interested and not just "checking up" on the welfare of the horses he adopted. He grew more talkative, telling me his father had logged with horses before him, and that he had logged all his life. He said he had used every kind of draft team there was, and then, one day, he decided to get a pair of wild horses because they were strong and small. He had no idea how they would be to train or if he could even train them, but since they were cheap, he felt he did not have much to lose. The old man continued, explaining that this was his third load in five years, and now he would never log with any other kind of horse. I asked what so attracted him to these old stallions. (I sometimes thought I was the only one who wanted to learn their wisdom.) He looked me in the eye and said, "They're special." I urged him to explain why they were "special." And he told me what had happened to him.

It seems not long after he had acquired his first wild horse team, they were logging a very steep slope when the old man had slipped and a big log rolled on top of him, pinning him under it. He could hardly breathe, let alone call for help. One of the stallions had turned around, looked at him lying on the ground with the log on top on him, and according to the old man, "pushed his

// buddy sideways to pull the log off of him." The horses had sidestepped in unison, a movement they were not trained to do without rein aids and a whistle. He said if they had pulled forward, the log would have crushed him, and if they had backed up, they would have slipped, too. As he told me he'd never in his life seen horses "think so fast and so smart," tears welled up in his eyes. I could tell he had been "touched" by these horses and their ability to share awareness, and I knew all the old stallions he took were going to a good home.

"I won't log with any other kind of horses," the old man said to me. "I owe my life to them, and I'd trust them again with my life. They know how to think, and they know loyalty, and that's a lot more than most people." //

a sense of fun and curiosity will frequently be all you need to start sensing what other animals are feeling and thinking.

_ 3 Be Focused

Learning to empty the mind and stay focused on that emptiness is key to receiving information. The average person can only focus on one subject for approximately four seconds. (People who meditate regularly can hold on much longer.) Humans, because of our busy minds, are handicapped when it comes to simple thinking and staying "present in the moment." As we learn to still our minds, we become much better listeners.

★ **TIP: / Establishing clear communication channels allows two-way conversations between horses and humans, building trust and safety for both. /**

_ 4 Be Empathetic

Empathy implies compassionate feeling "with"—not "at" or "for"—another being. People can feel sorry for an animal, but that is not empathy. With empathy, you feel what the animal is likely feeling without projecting your own issues and emotions. The more you can expand your consciousness to include your horse's awareness with openness and clarity, the more insight about him you will gain. Having a good understanding of your horse's instincts and priorities in life is helpful.

★ **TIP: / Horses can understand our intentions, thoughts, and feelings even when they do not understand our words. /**

_ 5 Be Respectful

Being polite is not the same as being respectful. Respect implies absolutely honest feelings of dignity, worth, and esteem for another being. When a person understands the value of

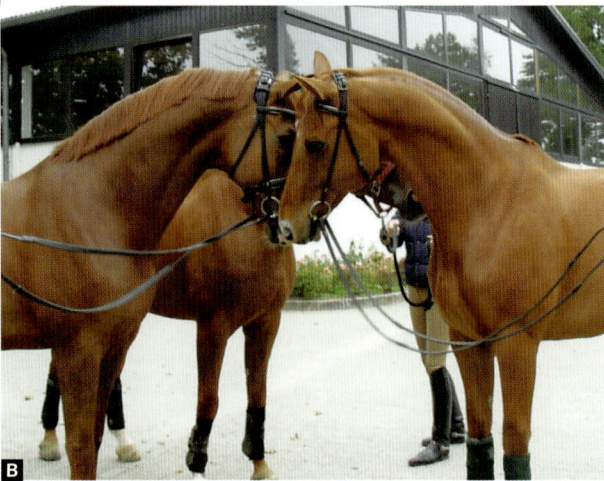

[9.36 A & B] Note the engaged and integrated channels of communication as both groups of horses exchange polite greetings. The young wild bachelor stallions **(A)** and the German-bred dressage geldings **(B)** are both having a detailed exchange of information: body language with eye contact and posturing (visual), nose blows (smell), gentle nose touches (tactile), and energy.

> **Learning how to integrate multiple forms of communication with your horse will help you better understand and manage his emotional and mental welfare.**

the presence of another being, it becomes easier to respect and communicate with him. Often people who say they "love" animals do not respect them. They want animals to do what they want them to do. Respect may mean you do not touch a horse until he requests to be touched. Honor and respect him for the beauty and value he contributes to the web of life. There is no room for "species prejudices."

Integrated Communication: Bringing the Channels Together

Communication rarely uses just one channel. We are constantly "dialed in" on many levels and integrate the information in our brains through pictures, words, feelings, and sometimes even smells and sound. As you discover your own preferred communication channels, pay attention to your horse, as they are not all the same. A horse with limited vision, may use smell as his dominant communication channel, while another may choose to vocalize his thoughts and feelings.

Learning how to integrate multiple forms of communication with your horse will help you better understand and manage his emotional and mental welfare. As I have expressed, having good social communication skills benefits your horse, and if you can teach your horse a thing or two about "being a horse," your horse will elevate you as a priority in his life.

It all starts with a polite horse greeting (figs. 9.36 A & B). Whether from horse to horse or human to horse, learning the nuances of eye contact, a "nose bump," and a comforting touch or scratch on the neck and withers can go a long way to establishing open channels for further communication. As you read in Part One, this is an essential skill horses learn when they are young, and it is the basis for starting off a friendship, as I described on page 78. So I encourage you to practice it daily with your horse.

_ Polite Horse Greeting

Normal equine social greetings include all of the communication channels: calm approach, soft friendly eye contact, blow of the nostrils, gentle nose bump, and then, if interested, a nice buddy scratch for relaxation on the withers. Starting your communication off with this "mannerly greeting" will help establish positive communication between you and your horse (fig. 9.37).

Here's how to do the greeting (figs. 9.38 A–D):

1. **Make soft eye contact** and calmly approach your horse while quietly talking to him and communicating your desire to greet him and see how he is doing.

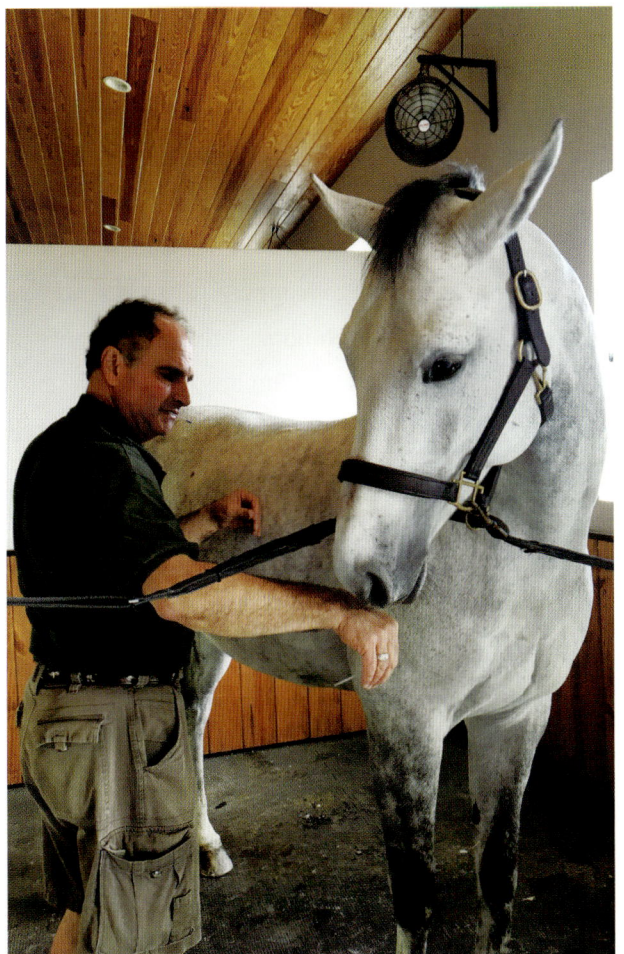

↑ [0.37] Dr. Carlos Jimenez greets a horse before examining her. Note the soft eye contact and the engaged ears as the horse processes his smell and touch.

2. **Offer the back of your hand a few inches away from the horse's nose** so he can smell you with each nostril.

3. **Gently touch the back of your hand to your horse's muzzle** while keeping soft eye contact.

4. **With one hand still in front of your horse's nose**, move back to the withers and neck and gently touch him at the withers and neck. By engaging the sensory input of the olfactory and facial nerves with friendly smells and touch while touching the withers to encourage relaxation, you quickly can connect your horse's brain to remember your smell and touch as friendly.

Note: This method assumes your horse is not afraid of humans. If your horse is or has never been touched, then start with only soft eye contact and helping the horse feel comfortable in a shared space.

Horses who were brought up with functional mothers usually know how to do this equine social greeting and respond very positively to humans who understand the social etiquette involved. But many domestic horses have not learned normal equine social greeting, and this means they have difficulty making horse friends and can be overly needy with humans. This can cause mental and emotional stress, but you can help these horses by teaching them how to greet other horses politely through your own use of the steps I've just described.

_ Riding Communication

Riding sets up a unique form of communication between humans and horses. With hundreds of years of tradition relying on obedience, many people believe that bits and reins are needed for communicating with and thus controlling horses. However, while horses can learn the system humans have developed for them using bits in their mouths, it is not the easiest way for us to communicate with them.

Using the horse's great system of *proprioception* (see p. 92), sensitive riders can "feel" and develop their own sense of proprioception to "connect" with their horses. They can sense when their horses are out of balance or not using muscles correctly. Similarly, horses can sense when riders are out of balance or squeezing with one calf more than another (as one example). They learn to react or respond to these subtle signals. Both horses and riders can connect this way. Nerve impulses from muscles and skin travel much more quickly to the brain than sensory input into the mouth (via a bit) that then needs to be processed by the brain and sent to the parts of the body that need to move.

I have developed a system to help riders better communicate with their horses from the saddle, using all the channels of communication we've already discussed in this chapter. Here is simple summary and a way to remember key points.

_ The ABCs of Horse-and-Rider Relationships

- **A = Assessment, Awareness, Appreciate** Become more aware of yourself and your horse and how you interact and affect each other. Constant **assessment** of your horse and the environment means you are looking out for your horse's safety while you ride. Become your horse's eyes and ears. How do you perceive differently from your horse? How aware is your horse? Close your eyes while sitting on your horse and allow your senses to take in everything you can hear, smell, and feel. Increase your **awareness** to feel and think like your horse. Open your heart and emotions to **appreciate** your horse. Smile and relax. Now open your eyes and

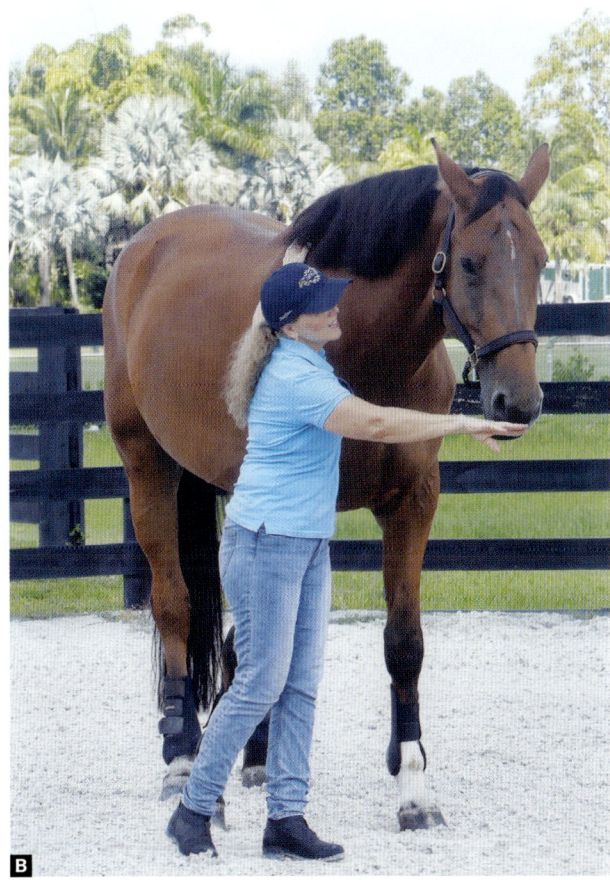
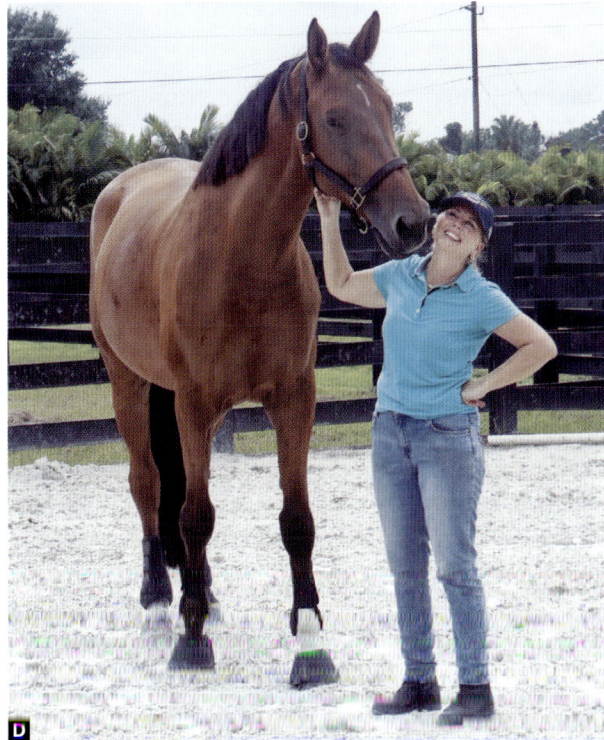

↑ **[9.38 A–D]** Maintaining soft eye contact and smiling, I approach Coco with my hand out for a polite greeting **(A)**. Even though we know each other, she appreciates this. I move one hand to her withers to communicate both with smell and touch **(B)**. Then she asks if I want a "buddy scratch," too **(C)**. We hang out, smelling, touching, and talking a bit longer as Coco had some things to "discuss" **(D)**!

↙ [**9.39 A & B**] Use of a bit should be to enhance communication between you and your horse, not to assert control or inhibit expression, as seen in these two examples from the jumper ring. In **(A)**, the horse's open mouth and neck bent behind the bit shows his discomfort with the too-tight noseband that is also adjusted too low. The horse in **(B)** is wearing so much tack he is being very careful to find relief.

take in all that you see. Notice everything from shadows, to dips in the dirt. Your horse does.

- **B = Beliefs, Breath, and Balance** Recognize that your **beliefs** about horses drive your actions (see What Kind of Horse Person Are You on p. 154). Realign your beliefs to be more "horse-centric." Take a deep **breath** and feel it move throughout your body. Try and feel the sides of your horse move as he breathes in and out and synchronize your breathing with your horse. Breathing together is often observed in synchronized herds. Bring your awareness to your body, and **balance** your energy from side to side, aligning your spine with your horse's spine. Imagine you are suspended like a carousel horse. Remember, your horse can sense the slightest difference in your weight and balance from side to side. You are proprioceptively aware of each other.

- **C = Concentrate, Communicate, and Connect** Bring your awareness and focus into the present moment. **Concentrate** on "being"—sensing and feeling your own body. Then allow your awareness to sense your horse and what he is experiencing. What information do you receive from him? Open up the channels for **communication** and expect collaboration. It is a two-way street. Stay present with your horse and do not be distracted by thoughts or external influences. This assures your horse you are there for him. When your awareness is distracted while you are riding, your horse senses a lack of **connection**.

Communicating Through the Body, Bridles, and "Feel"

Bridles, bits, whips, and spurs have long been considered "essential" tack in equestrian disciplines. But for them to be safely used as communication tools, it is important to have a least a basic understanding of equine neurophysiology to know what and how they are communicating. So, before you use any bit, bridle, or training aid on your horse, determine what you want to communicate, if the tack in question will communicate that in a fair manner, and make sure the tack fits your horse comfortably.

★ **TIP:** / Horse mouths did not evolve to hold bits or wear bridles. If you use a bit, make sure it is fitted properly and is comfortable for your horse. If your horse keeps his mouth open when the bridle is on, then the bit may not be comfortable. /

The variations of bits used on horses—from leather pieces to large metal spades—were all designed primarily for "control" of the horse, not for open communication (figs. 9.39 A & B).

Because horses were designed to use their mouths for eating, their noses for smell and tactile endeavors (remember they have prehensile noses filled with millions of nerve cells), and most of their twelve cranial nerves for inputting sensory stimuli important to their survival, they have not evolved to have bits in their mouths and their mouths tied shut. While some horses do seem to enjoy having something in their mouths (particularly young male horses who, as we've discussed, tend to be very oral), other horses do not. Mares in particular often resent it (fig. 9.40).

In addition, not all horses have mouths that can wear bits comfortably, so evaluate your horse's individual needs and your other options for communicating as may be needed. Bitless bridles, hackamores, neck ropes, halters, or bridleless riding should all be considered in order to establish the best means of communication with your horse (figs. 9.41 A–C). All tack should only be used if needed for the purpose of *enhancing* and *clarifying* communication,

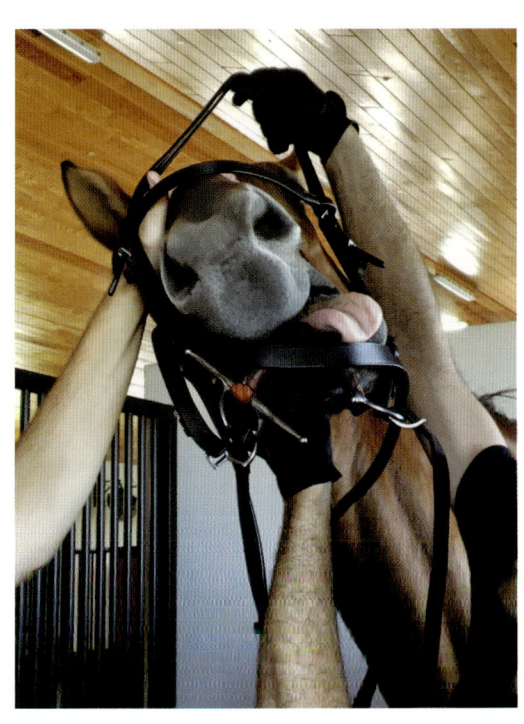

↓ [**9.40**] Mares, in particular, may not care to have a bit crammed in their mouths. Here two people struggle to put a bridle on this mare. She is telling them, "That is not how you do it!"

[**9.41 A–C**] Riding with a bitless bridle **(A)**, hackamore **(B)**, or even just a halter **(C)** can help you develop other lines of communication with your horse.

not to inhibit or cause pain or discomfort. Think about how you would like to establish communication with a human friend. Would you start a conversation assuming the friend is going to argue with you? It is always best to set up safe and open lines of communication with your horse, using all the channels we've discussed, before moving toward restriction.

★ **TIP:** / Bits and other tack should only be used to enhance communication, not shut it down. It should allow your horse to talk and be heard. /

_ Using Touch Communication from the Saddle

Using a gentle *half-halt* to the reins easily engages a horse who is listening to pay specific attention to your body in relation to his own body. The half-halt alerts him that you are going to ask him to do something. Assuming your five communication channels are open, most horses will listen. You will see their ears move backward, as well as their eyes, as they interpret the request.

★ **TIP:** / Use the half-halt to gain your horse's awareness. Use your body, touch, and energy to communicate the desired movements you would like from your horse. He will learn to adjust his own body to your request. /

Horses actually can learn very quickly to move body parts according to touch. Once a movement is "programmed" into the brain and body, the horse does not have to "think" about it. More sophisticated riders learn to first get their own bodies balanced and then use their subtle body movement (legs and seat bones), thoughts, and energy to communicate how they want their horses to move. When horse and rider proprioception and neural systems are connected and their hearts are in coherence—all the pieces we have talked about being key to communication—they can appear to move as "one" (fig. 9.42).

→ **[9.42]** This horse and rider are communicating through body, proprioception, bridle, bit, legs, voice, and energy to create a partnership in the show ring—and to move as "one."

← [**10.1**] Horses all have various temperaments and personalities, just like humans. As social mammals, humans and horses share many traits. Here, horses and humans enjoy a social outing together.

CHAPTER

[**10**]

ANALYZING THE INDIVIDUAL: EQUINE PERSONALITY, TEMPERAMENT, AND TRAINING ASSESSMENTS

> **Rather than trying to assess the individual horse or human by himself, we must recognize it is our relationships and how we fit into society that is important to both species.**

[10]

Horses, like other social species, have various temperaments and personalities. Assessing your horse's general temperament can help you prevent stress and develop a training program to fit your horse's learning style. Generally, it is easiest for people to think about how we interact and relate to others, rather than trying to assess the individual, horse or human, by himself. It is the relationship and how we fit into society that is important to both species.

While many people have an intuitive connection with their horses and seem to understand their horses' unique personalities, it can still be helpful to think about various ways to assess equine temperaments and personalities. This is particularly important before starting any training with or working on horses in a clinic situation. By understanding a horse's particular needs, you ensure the safety and welfare of both the horse and any humans involved. For example, separating a horse who is a group learner and feels safer with others, and who is also sensitive and lacks confidence and cooperation, could lead to injury in a new situation. But if you know the horse's temperament ahead of time you can appropriately assist the horse's ability to learn by including another horse in the training session, which may also help him feel safe.

Today there are a number of ways and systems offered to identify your horse's personality: using scientific questionnaires, observing body language, considering psychological models, having your horse's genetics analyzed, even assigning him elements from Traditional Chinese Medicine. Whatever method or model you choose to assess your horse, the purpose is simply to help you better understand the psychology of your equine friend and how to best limit stress in his life. Think of yourself as your horse's teacher, therapist, and friend, and you will certainly make better decisions in his education and welfare.

Since horses in domestic situations need to learn to work with humans and other horses, it is helpful to think about their temperaments according to how they learn and interact with people and others. Keep in mind that your horse may act differently with other horses than he does with you, so observing both his interaction with other horses *and* with you will give you a better overall assessment.

On the following pages I offer a number of ways to assess your horse's nature relative to the things we humans do with horses.

★ **TIP:** / **Horses have as many different personalities as people. By understanding your horse's temperament and personality, you can design a training program to best suit his learning style and reduce his overall stress.** /

In general, the better you get to know your horse as an individual being, the more your relationship will grow in positive ways. By determining your horse's learning style and temperament, you can select the appropriate training techniques that will help your horse best learn. Horses are social creatures, like us, and seem to enjoy interacting with us for companionship and exercise. But

Temperament or Personality?

Temperament and *personality* can be interchangeable and are often how behavioral traits are expressed. *Temperament* refers to behavioral style, usually related to the biological and genetic influences of the animal and how they dictate his reactions to the world. *Personality* includes temperament, but reflects its development and expression based on individual influences.

Behavioral science is far behind when it comes to accepting that animals other than humans have "personalities," as it implies individual choice and development of a more sentient and conscious creature than previously thought. I purposefully use both in this book.

they require motivation in order to learn and look to their "social facilitator" often for guidance. When they express their feelings to us, it is our responsibility to listen and respond as caretakers and friends. In doing so, both of us can benefit from a fun and fulfilling relationship.

General Observation Assessment

Good horsemen can intuitively assess a horse's temperament and personality, but they will not overanalyze a horse because they feel what their horses feel. And while there is no one way or right way to assess a horse's (or human's) personality, the following should be helpful in establishing a baseline of understanding "who your horse is."

Horses may have preferences based on genetics and life experiences, but realize these can change as they learn and develop. So, this assessment simply helps you evaluate *some* key traits and behaviors to observe when you are evaluating a horse, and multiple may apply. Keep in mind that as a social herd species, horses are best evaluated in how they interact with other horses and people, not always by themselves.

⎯ Group Learner vs. Individual Learner

Group learners become very stressed when separated from friends but learn very well with others. *Individual learners* often are distracted by their friends and want to disrupt and play all the time. They learn best in one-on-one situations with few distractions.

★ **TIP:** / Young horses tend to learn faster when around older horses performing the same tasks. /

Note: Learning to be alone with humans and separated from other horses is not natural, but horses learn to transfer their needs for safety, leadership, and social interactions to people. Horses who are *developmentally slow* (see p. 231) often are extremely social when young (two to four years old) and can be fast learners, but as they age, they develop worry and anxiety when alone without other horses. This type of horse needs to have an aware person who knows how to keep him feeling safe or he can become a problem, as he does not comprehend and learn in the same manner as normally developed horses.

⎯ Eager vs. Cautious or Curious vs. Vigilant

Horses become curious when fear is low. However, some horses are naturally more curious than others, and have an eager temperament. Often these horses will seem brave because they are curious, but they will frighten themselves because their curiosity is higher than their confidence. Vigilant, cautious horses usually are highly sensitive and some have negative associations related to stimuli. They usually do better with strong confident friends around to help them learn.

Eager horses typically have high curiosity and cautious horses will tend to spook or run away from something they do not understand, which is also called being "vigilant." Recent research has identified a genetic mutation that can identify behavior to be more curious versus vigilant (see p. 237 where I discuss this in more detail). This may be related to how domestically adaptive a horse may be and how stress will affect them.

★ **TIP:** / Eager, curious horses tend to enjoy learning with people, while cautious, vigilant horses may be stressed. Thus, training programs need to be adjusted appropriately. /

Demonstrative vs. Passive or Externalizer vs. Internalizer

Although most horses who have dominant personalities communicate well and express to you if they are upset, some have learned not to communicate. Usually this is due to bad training that does not allow the horse the freedom to express when he hurts or feels unsafe. These horses often develop resistance problems associated with anger or stress. Horses with passive temperaments often develop physical problems such as navicular disease or ulcers because they worry when they are stressed, but do not show their emotions outwardly. The healthy demonstrative horse will go out and buck and play to release stress and has no problem biting or telling you when something hurts if he does not like it. The passive internalizer wants to please and fit in, so he will often develop health issues and act like "nothing is wrong" until it is too late. You need to keep an eye on the passive internalizer as he is more likely to colic "without a reason"—or without one you can see.

★ **TIP:** / Horses who are shy and passive tend to internalize stress and are more likely to have ulcers or to colic than demonstrative, confident horses. /

Confident vs. Timid

Confident horses usually have very little stress. They are brave and take everything with a sense of adventure and interest. Confident horses who have been mistreated or badly trained will be difficult to assess because their confidence might appear low. Timid horses have the most difficult time with stress. They are generally more sensitive to what goes on around them and need support from other horses. However, even timid horses can learn to be braver horses.

Resistive vs. Cooperative

Humans expect cooperation from horses, but smart horses want a reason to be cooperative. Usually the smarter the horse, the more resistive they are to doing things that from a horse's perspective seem dangerous or not worth his time. Resistive horses require positive motivators. Cooperative horses sometimes have "I-Want-to-Please Syndrome," and you must watch them for signs of stress.

★ **TIP:** / You can build cooperation in horses by instilling confidence and using positive motivators to make learning fun. /

★ **TIP:** / Become a good observer. Know your horse's rhythms, eating habits, defecation patterns, worry level, sensitivities, and how he relates to other horses and people. /

Learned vs. Inherent

Pay attention to natural or "inherent" traits in your horse. Nature has done a good job adapting horses to their environments so they can stay safe. Understand how to separate behaviors that are learned versus those that are natural. For example, a horse that is afraid of a horse trailer but crosses over new objects like bridges or tarps without hesitation is not necessarily a vigilant horse, but instead may have had a past bad experience and has stored that memory in his brain. Or he may be engaging his "inherent" natural behavior to be cautious of a new situation that he needs to explore to see if it is safe.

Specific Horse Personalities

No person or horse should be "boxed" into a personality type, but there are some common behavioral traits you can easily learn to observe that may help you in determining how to manage and train your horse. (Note that mares, geldings, and stallions should be viewed differently because their gender and hormones present different traits. For example, mares, by nature, are more sensitive, so they normally react and use more body language than geldings.)

What follows are personality types I have observed in domestic horses when interacting with people. I developed these to help people observe and better understand their horses' nature. While these personality types can also be observed in wild herds, the parameters by which they are observed are different. We are interested in training our domestic horses to "do something," so the traits are observed relative to how horses respond to human desires and interactions. In nature, horses only interact with each other and are strongly influenced by the behavioral "culture" of the group (see p. 29).

Do not try and put your horse into just one type, but rather notice if he has behavioral traits that align with the descriptions I provide. Most horses are a combination of types. Observe the dominant and secondary traits expressed. This will help you be more understanding of how your horse perceives his world and how you can best influence both his learning and his interaction with you, leading to better performance for a particular discipline and a more positive overall relationship between horse and rider.

_ The "I-Want-to-Please" Horse

In social mammals, being able to get along with others is a key to survival, so in people and horses we see various adapted behaviors based on different temperaments to adapt. The "I-Want-to-Please Syndrome" is common in both horses and people, but only certain horses will exhibit it as part of their temperament. It may be a selective trait in domestication of humans selecting horses who want to please. Usually they are great horses, but they are also the most likely to have ulcers and other disorders as they internalize stress.

_ The Busybody

This horse always needs to be doing something. He is very curious and sometimes has a hard time focusing. He also needs more habitat enhancement as he often will start chewing wood or developing stable vices from boredom.

Understand how to separate behaviors that are learned versus those that are natural.

— The Worrier

This type of horse does not like to be alone and often needs friends around all the time for security. He is happiest when with friends and when things are explained to him before they happen. He can take on the energy of the rider, so it is important to consider that channel of communication as primary (see p. 285).

— The Sensitive One

This horse hears, sees, smells, and feels *everything*. He tends to overreact but also can be highly tuned for alertness, which can be a wonderful trait. He can take on the emotions and energy of the rider, but often makes a top performance horse.

— "Happy-Go-Lucky" Surfer Dude

This horse is into whatever looks like fun. He is often laid back and has a sense of humor but is not the hardest worker. He can be a brilliant athlete when he wants to be but is also happy just "hangin' out" with friends.

— The Take-Charge Leader

This horse is typically a mare who understands natural horse etiquette and wants the leadership role. She will always know what's going on in the barn and wants other horses to respect her. She may demonstrate body language and vocalizations to establish presence and respect more than other horses.

— The Healer/Nurturer

This horse loves to help, is affectionate, and wants to take care of others. This is the sweet horse who often makes a great equine-assisted therapy partner or lesson horse. He is not always the most talented horse, but he will take care of people and other animals.

— The "Combo"

Depending upon the day and circumstances, different sides and personality can be displayed in this character.

The SAICC Evaluation

The SAICC (Sensitivity, Awareness, Intelligence, Confidence, Cooperation) Evaluation was developed to help solve equine behavior issues by assessing horses' abilities to interact and learn with humans and thus help people better understand their own horses' nature. The SAICC Evaluation also provides a good starting point for training to determine what your horse knows about "being a horse," as well as identifying issues your horse may have developed prior to being with you.

Working with veterinarians and equine professionals—many of whom had good intentions, but limited information—made me aware early on that more clarification and knowledge was needed in evaluating horses based on how they function in social groups. Most models isolated the individual, many chasing the horse in a round pen to assess the horse's willingness to overcome fear and work with a human. Instead, the SAICC Evaluation takes into account the horse's social need to fit in and relate to other social creatures, both human and equine. It evaluates core parameters reflective of a horse's temperament and personality. Some can be tied to temperament and genetics, such as "sensitivity" or "intelligence," but others—like confidence, cooperation, and even awareness—can be trained, regardless of the temperament. In other words, even a highly sensitive horse can learn to have reduced stress under the right training model.

What is particularly interesting is that while I developed this model for horse-human interactions, it has been successfully been used in human-only organizations for evaluating "team members" and efficiency in the workplace. The reason is that most human personality assessment focus on the individual human, while this model focuses on how the individual will relate in a social group. As I've mentioned before, since humans and horses are both social species, evaluating them in their relations to others is more effective than alone.

The SAICC Evaluation uses a scale of "1" to "5" for the purpose of "scoring" the horse's overall social, sensory, and psychological fitness. I will only outline the key points and give you general guidance in these pages. (For those interested in more detailed information, please visit my website by scanning the QR code on this page.)

Learn more

_ Sensitivity

Determining your horse's sensitivity will guide how you care for and educate your horse (fig. 10.2). The highly sensitive horse, as with people, has trouble adjusting to new situations and can often feel intimidated by others, so you need to take more time to help him acclimate and learn. At the other end of the spectrum is the "dull" horse. Nothing bothers him, and he often needs a lot of stimulation to get a response. In the middle is the horse who is alert and sensitive to things he needs to pay attention to but is not overreactive. Although the SAICC Evaluation uses a scoring system, you can conduct your own relative test for sensitivity by simply introducing various stimuli to test your horse's vision, hearing, smell, taste, and touch. Start in a quiet, safe space where you have room to walk around, such as the horse's paddock or stall. Be aware that not all senses have the same sensitivity. For example, a horse with poor vision may be much more sensitive to sound or smell than a horse with all normal sensory apparatus. In addition, horses that have been *habituated*—given a horse stimulus until they have no reaction—can often produce false results. The horse may have *learned* not to react, but inside, the highly sensitive horse that has been habituated is building stress like a time bomb, waiting to go off. So, it is important to look for patterns in what is likely "normal" behavior for your horse.

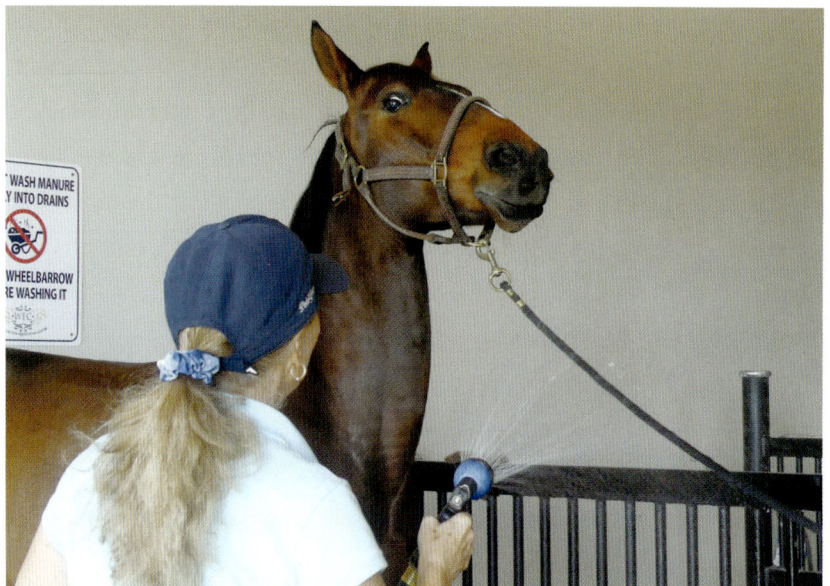

[10.2] Horses may be highly sensitive to touch, sound, smell, or visual objects when they have not learned what they are or have associated a bad experience with the sensory input. Here Radar reacts to just a little water spray as in the past he had handlers restrict his movement and spray him in his face, which he did not like. After working with him over time and giving him the ability to move away from water spray, he became much more comfortable—even playing with it.

Testing Sensitivity

- **Smell Sensitivity** Since smell is a good way to pique your horse's curiosity and increase his interest in you, you can do two things at the same time—test for sensitivity and open a conversation based on smell. Offer your horse various smells he is not familiar with to observe his reaction. An "average" response is the horse who investigates the new smell, moving from nostril to nostril. Some horses, however, have a dominant side and nostril, and will use one nostril more than another, and even turn away when you encourage the smell from their weak side. An overly sensitive horse does not want smells to get too close and may pull away, while the duller horse will appear to have little or no reaction.

- **Visual Sensitivity** An "average sensitivity" horse will often be better able to locate and understand objects far away than close up. This is the case particularly with mares, who intuitively scan the horizon, looking for movement. Not all horses see well close up, hence, they may be spooky around new objects in close proximity.

For the purpose of testing visual sensitivity, observe your horse's ability to track an object as it moves around him. A normal horse will track with ears and eyes as the object moves to the side and around from behind. A highly sensitive horse may want to move away from the object to a "safe" distance, and a duller horse may not notice the object.

- **Sound Sensitivity** Walk around your horse exposing him to various sounds, such as a squeaky dog toy, rice or rocks in a plastic tub, a crunching metal can, or popping plastic sound. Use both high- and low-frequency sounds. As you make the noises, watch your horse's eyes and body language. An average horse will look at the direction of the sound and track with his ears as you bring the sound around from one side to the other and around behind him. When a horse overreacts to one sound but not others, then he may have negative associations with it. High-pitched sounds seem to more often irritate horses' hearing—for example, the sound of clippers can generate an overreaction, while a lawn mower may not. The loudspeaker at a horse show may hurt a horse's ears and cause an abnormal behavior response, but the farrier's radio when he's at the barn is fine. You must learn to separate a "normal response" from an "abnormal reaction," and then trace back to the cause when the behavior is abnormal.

- **Touch Sensitivity** It is common that horses who have sensitivity to sound will often also be overly sensitive to touch. It can be that once these horses realize touch *can* feel good, they may happily accept massage, for example, but still be very irritated by a fly bite. Wild horses who are not used to being touched are extremely sensitive in this area because of that

STORY FROM THE FIELD: //

Odd Behavior at the Track

A Thoroughbred racehorse trainer had a winning four-year-old gelding who suddenly would not allow the trainer to come into the stall on race day. During the week, the horse trained well and got along very well with the trainer. The trainer had his vet evaluate the horse and found nothing out of the norm. I was asked to take a look at the horse, and given the odd change in behavior on such a particular schedule, I evaluated his sensitivity to visual objects and smell. I found the horse was average for visual sensitivity but highly sensitive to smell. I asked the trainer what he did differently on Sundays before races that he did not do during the week. The only difference the trainer could think of was that on race days he would wash up, put on a suit, and wear some aftershave.

It turned out the horse was so sensitive to the trainer's aftershave that the horse was afraid to be handled. As soon as the trainer ditched the aftershave, the horse was fine and allowed the trainer to walk right up to him in the stall on race days. //

highly developed proprioception we've talked about (see p. 92), while domestic horses who are used to humans and handling may not be nearly as sensitive. Taking into account your horse's history, breed, gender, level of training, and experiences is key to accurately rating your horse's sensitivity.

_ Awareness

★ **TIP:** / Horses are wired to be aware, taking in sensory input from various activities and making decisions about whether to take action or not. /

Horses by nature are usually very *aware* because it is a great survival trait and lends itself to functional social development in learning equine cultural norms. Some horses are born with more awareness, and others—at least in the wild—learn it from parents and herd members. In domestic life, horses do not learn additional awareness unless a foal's mother sees a need to teach it. Evaluating your horse's awareness level can prove to be very indicative of how your horse will fit in and deal with various circumstances. Despite this, awareness is often overlooked as an important trait because, frankly, most *people* are not very aware.

Aware horses are the ones in the barn who prefer to have their heads out of their stalls so they can watch everything going on. They keep track of the comings and goings of people and other horses. They know what everyone is doing at all times and can become alarmed if something is "out of the norm." Most often mares are these aware "gatekeepers" in the barn and will often be exhausted if they are keeping track of a lot of changes. Combine high sensitivity with high awareness in a horse, and you have a recipe for a horse who has lots to talk about and usually no one interested in hearing about all the information! (In nature, this type of horse would be rewarded for his sensitivity and awareness and would often be the social facilitator for the herd—see p. 90.)

A horse with low awareness can be affectionately referred to as "clueless." He may have average intelligence (see p. 230) and average sensitivity, but when it comes to paying attention to more than one thing at a time, he will be below average. This horse may overfocus on a single task and then be oblivious to what else is going on around him.

→ **[10.3 A–F]** I test this chestnut gelding for awareness, first asking him to halt and stand still **(A)**. Note his head tilt and his eye, indicating he wants to move away, but I continue asking him to look at me with my body language and energy.

The gelding then turns his head and looks at me and his owner and moves as if to enter our space **(B)**. I tell him with voice, body language, and energy to stand still and not move, as all I want is his attention and for him to be aware of us.

The gelding postures as if he will now move away **(C)**. Note his "thinking" ears, slightly tilted head, and eye on me. He is unsure what to do. I use body language, energy, and voice again to tell him to stay.

He relaxes his posture and listens to me, remaining still, but his body language shows he is unsure **(D)**.

Then he turns and looks at me, ears and eyes fully focused on me, waiting and watching me for my direction **(E)**. This is the look I was waiting for that indicates he understands what I am asking and that he is a horse with average or "normal" awareness.

The gelding stands still while I walk up to him and give him a positive reward with a greeting and scratch on the withers. He turns and touches my face with his nose, acknowledging that we have connected **(F)**.

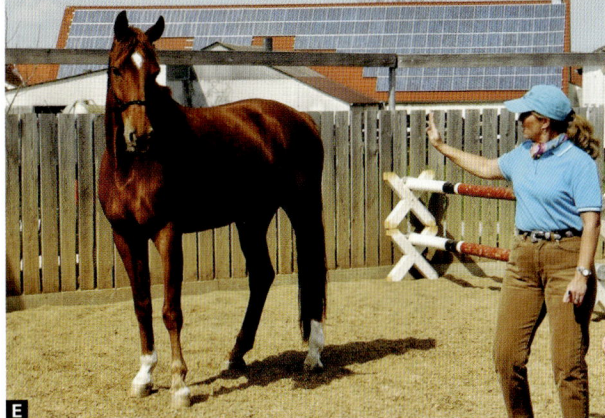
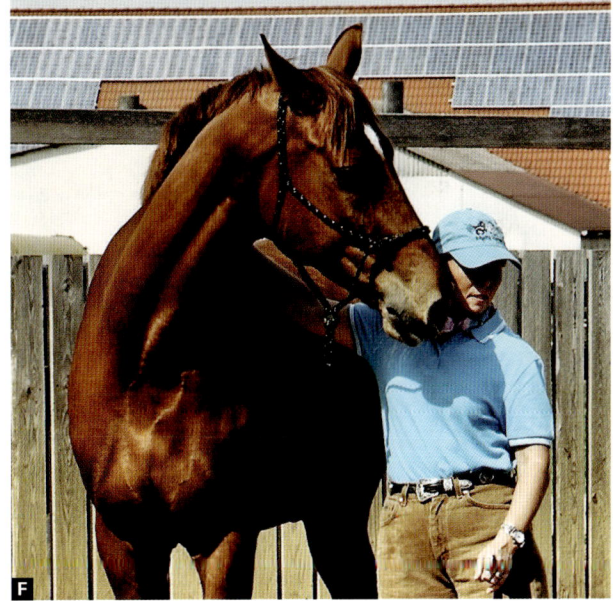

PART TWO / CHAPTER 10 / Analyzing the Individual: Equine Personality, Temperament, and Training Assessments

Testing Awareness

Since, in the wild, awareness is required to be part of a social group of horses, in domestic life, it really should be at the foundation of any training program. Understanding a horse's level of awareness is critical as it protects the human in the relationship and sets up clear communication between human and horse.

You can test your horse's awareness by doing simple exercises in his paddock, including gauging his attention to and ability to comprehend your body language, voice, and energy as you move around and "ask" him to respond in certain ways (figs. 10.3 A–F).

Another way to test awareness is to use the "carrot focus" exercise. Hold a carrot (or something else he desires) just in front of him, but do not allow him to eat it right away. This helps you measure his ability to *focus*. Focusing is an important aspect of awareness. Count out loud while asking the horse to stand still, maintaining eye contact with your horse and telling him that after "X number," he will get the carrot. Many horses give up before you can count to "10" and move away or are distracted. Highly aware horses who can focus well—usually mares—can remain aware of their surroundings and still stay focused well past a count to "20."

Intelligence

Assuming nature has produced through evolution an intelligent species able to adapt and sustain itself, we

can use modern wild horses as a baseline for intelligence. Although intelligence varies in wild horses, most horse leaders are very bright and many wild horses appear smarter than most domestic horses. This may be due to the fact that they have been raised by horses and thus had a better "horse education" than those raised by humans. Domestication may have also selected for juvenile, friendly behaviors in breeding (see p. 131), which may actually limit cognitive skills.

If a horse is highly intelligent and wants to please the human, the horse often will internalize stress and show little or no reaction on the outside, even though a stimulus may bother him. At the other end of the spectrum is the smart horse who has figured out that he is smarter than people. If this horse is mistreated, he will often turn aggressive and difficult. Intelligent horses are not always the best candidates for domestic endeavors, since many of them become bored or feel what they are being asked to do is not important to their way of thinking. But with the right person, they can be outstanding partners, assuming the human understands and recognizes her horse's intelligence.

_ Assessing "Special Learners"

Many of our domestic horses are actually "special learners" by wild horse standards. As mentioned, domestication itself has encouraged juvenile behaviors because humans have selected for friendliness over cognitive ability. Having a horse with a learning disability may be just what is needed for a top hunter to "pack" around a 3-foot course over and over again without becoming bored (fig. 10.4). So being really smart is not always a benefit in domestic horses.

★ **TIP:** / Horses with "learning disabilities" may need special education. /

★ **TIP:** / Behavior problems often present themselves with developmentally challenged horses between the ages of five and eight years old. They may have seemed like perfect horses before then, but their mental maturation slowed down. /

Using human terms in assessing other creatures is often shunned, but when there is so little data to discuss genetic mutations that lead to learning

[**10.4**] Many top hunters are "special learners," as they would be considered "cognitively slow" in a wild herd, but with human support and friends all around, they can excel.

> **Domestication may have limited genetic diversity and cognitive development in horses by breeding for human-desired traits such as talent for a specific discipline, friendliness, and body type.**

STORY FROM THE FIELD: //

Milo, a "Special Learner"

Milo was a lovely big, sweet, chestnut gelding who had won almost every hunter class he entered in his Baby Green and Pre-Green divisions and showed great promise as a three- and four-year-old. Then suddenly one day he refused to go into the ring—an arena he had shown in before. The owner was a good horsewoman and checked everything, from saddle to bridle, and even had the vet evaluate her horse to see if something physically was wrong. They could find nothing. Milo continued to school well at home and had a lovely friendly temperament, but at the next show he did the same thing. He jumped well in the warm-up ring but refused to go into the show arena. His owner tried having him led into the ring, but when the handlers left, Milo spun around to go with them.

What was up with Milo?

When I evaluated him, the gelding was overly friendly and had little awareness of his space. If you asked him to do something that he either did not understand or did not want to do, he would play with his lead rope and become very mouthy—a behavior typical of juveniles. Milo was going on six years old, and what had been fun and inspired curiosity when he was younger now was worrying him. He did not have the cognitive skills to retain lessons he learned, and so his worry increased, unless he could have other people or horses around him. Milo's safety net was to be surrounded by his friends, otherwise he was not capable of feeling protected. While he may have been classified as a *group learner* as a young horse (see p. 221), as he matured, he should have been able to learn to be separate from others. His "attachment syndrome" was beyond what would be considered "normal," as he also displayed other cognitive limitations.

Milo was able to continue showing as a Low Hunter, and he excelled in the Under Saddle classes as he was a good mover and loved being with all the other horses in the ring. As long as his horse friends went to the show with him and were somewhere he could see them when he went in the ring, Milo was relatively good—unless he forgot where his friends were. His trainer had to work with Milo on spatial awareness and respect for humans to assist in bonding, but he often forgot what he had learned, so the same exercises had to be done at least once a week. His trainer also had to help Milo gain confidence and redirect his worry. Milo required patience. Interestingly, aromatherapy was a breakthrough, as it could be used to serve as a bridge between a familiar scent and his horse friends. Eventually just smelling that scent and seeing his human friends around him made him feel safe enough to jump in a ring alone. //

and developmental disabilities in horses, it is easier to use human terms, so you better understand the type of behavior in question. As horses have been bred for conformation and ability to perform specific disciplines, little attention has been paid to what may cause mutations. Since domestic-directed genes are often associated with maintaining juvenile behavior, it is likely that some of those mutations also carry a predisposition toward abnormal learning and mental development.

Various mental abnormalities crop up more often than people realize with domestic breeding, from the *slow development learner* to the *mentally high-thinking but low social-functioning* horse. Both extremes may be due to human-influenced breeding as most wild horses, while highly alert and intelligent, will have good social skills and apply them to interactions with humans under the right circumstances.

Some horses are highly sensitive and smart, often very focused and learn things fast that they can repeat. They can become "compulsive," and they often do not socialize normally with other horses and people. In this way, their behaviors are similar to humans on the autism spectrum. Their coping skills may include avoidance of eye contact with little interest to interact with people (they do not want to smell or touch you), bolting and running toward safety if they are frightened with little or no regard for their own safety, developing repetitive behaviors such as weaving, head-bobbing or head-swinging, biting themselves, hanging their heads away from bright lights and noise to reduce stimuli, and lack of awareness and respect for human space.

It should be noted that not every horse that demonstrates such behaviors has a mental disorder; they may just be highly sensitive horses. However, having the knowledge to correctly assess horses who are potentially "special learners" benefits all involved and can help prevent mistreatment of horses due to misunderstanding.

_ Testing for Intelligence

When testing for intelligence, you must consider the priorities of the species you are testing. For horses, who are naturally emotionally intelligent, you need to think about what your horse has learned, his genetics, his temperament and personality, and how all these factors impact his priorities. Then, the intelligence test can vary, depending upon the strongest motivator for your horse. A standard test I use involves hiding food under a bucket

> **As horses have been bred for conformation and ability to perform specific disciplines, little attention has been paid to what may cause mutations.**

STORY FROM THE FIELD:

Eros at Equitana

Eros was an eight-year-old, 16.3-hand Warmblood, and a talented dressage horse. She performed well but would never make eye contact with people and avoided interacting with people if she could. I met her at Equitana in Essen, Germany, where she was living temporarily in the automatic feeder demonstration. Because of the mare's failure to respond appropriately to other training demonstrations, she was offered to me for my clinics about temperament assessment.

Eros had a young girl who was her caretaker, and the girl was disappointed that she could not lead Eros around. The mare would just pull the girl back to the feeder and ignore any efforts to convince the mare to do anything else. Was this a coping mechanism due to too much stimuli at the bustling Equitana? Was it an "addiction" to the automatic feeder? Was it stubbornness and disobedience?

I had Eros brought into my demonstration paddock with a horse friend. Eros was extremely clever at keeping her friend between me, positioned in the middle, and herself, so she did not have to make any eye contact with me. Her ears went back and forth constantly, and when she heard the main clock "tick" through all the other noise in the busy hall, she pointed her body toward the automatic feeder station and stood with her head over the paddock rail. She was keenly alert to the timing of the feeder station—and this told me what her motivation was, and what we could use as a "bridge."

I gathered up lots of feed samples from the numerous vendors at the event. Then I offered the food to Eros's horse friend, who quickly enjoyed it. Eros tried to come around her friend to see what she was eating. Once she saw the food, Eros wanted to eat, too, but it meant she had to approach me. The first time she would not make eye contact and kept her focus entirely on the food. But we had indeed found a bridge to making a connection. I had the young girl caretaker join me in the ring, and together we fed little bits to both horses and started asking Eros for a nose bump first, and then eye contact. Although brief, she began to take a glance up at us. I used an aromatherapy oil on the back of my hand when I greeted her, and she started to relax and process the smell. Now we had two bridges: the smell of humans bringing relaxation, and the association of people with food.

During the next demo, the young girl and I worked with Eros without the company of a horse friend. By then Eros was "targeted" to the food reward that came when she paid attention to the young girl. And when we buried food in the paddock and had the young girl find the food and give it to Eros, the mare became even more intensely interested in the girl. Because of Eros's intense focus and sensitivity to time, sound, smell, and food, the food was the necessary bridge to get Eros to connect with humans. Now the mare would follow the girl all around, as she associated the girl with food.

Despite our progress, Eros's eye never "went soft" in the connected way a "normal" horse's would. Instead, she was overly "attached" to the girl because of the food motivator.

Horses like Eros are often misunderstood, and because of their "autistic-type" focus and anti-social tendencies with humans, they are labeled as "stubborn" and "difficult," when actually they have a developmental disability. Eros's learning disability would require special assistance and understanding, but if she remained in good hands, she would continue to have a comfortable life, even if she never responded or behaved "normally." //

or pot (I explain how below). But if you have a horse who is strongly bonded to a friend, that bond may be a stronger motivator to test intelligence. The test you use can involve separating your horse from his friend and placing an opening in a fence that is not the usual gate so he must figure out a way to reach his buddy.

Other simple ways to test for intelligence can involve your horse's ability to engage in associative learning (learning the relationship between two separate stimuli), deductive reasoning (top-down logic), and opportunistic thinking (taking advantage of circumstances). Horses are primarily *associative learners*, and while they usually have good memories, they can find it difficult to figure out new ways of doing something once they have learned one way to do it.

Since there are no standardized intelligence tests for horses, you can be creative!

★ **TIP:** / Functional horses score high in "emotional intelligence." /

Here's how to test using food:

In a safe enclosure, hide food under a series of pots or buckets, and allow your horse to walk around and find where you placed it. A smart horse who is opportunistic will go around searching under all objects in an area for food. When you hide more in the second round of the test, he will show that he has quickly learned the game and search under all the objects again. A horse of average intelligence will learn that food is placed under a specific bucket, for example, then expect to find food there in the second round, but will not usually look beyond that particular bucket. A less-than-average horse may smell the hidden food but be unable to understand where it might be located and how to get to it. They also may not realize that you put the food under the bucket.

STORY FROM THE FIELD: //

Testing for Emotional Intelligence vs. Intellectual Intelligence

If intelligence was measured by which species has the most adaptable traits, allowing them to evolve and live the longest on earth, humans would be considered mentally challenged by nature's standards. But humans love to compare themselves to other animals, and hence most intelligence tests have been developed to favor how humans think. Intelligence tests in general not only have been human-biased, but culturally biased as well, so they have not fairly evaluated the intelligence of animals. If a person or animal has not learned about the subject matter or if they find it of no value, then it is unlikely they will score well on the typical test. Equine intelligence tests have historically primarily involved food and getting horses to learn human interests such as "how to count," or "how to push the correct symbol." While horses do seem to be able to learn these things, they do not reflect their true intelligence.

Horses have high *emotional intelligence* (the ability to understand, use, and manage emotions in positive ways to relieve stress, communicate effectively, and overcome challenges), which favors their survival by ensuring they are able to make friends and work together keeping a cohesive group. If equine intelligence is measured by how aware horses are to the needs of their herdmates or stable friends, many horses would score high. Being *associative learners* (see p. 99), horses have great memories and associate very well. And it is this that we should consider horse intelligence. Good or bad, they make connections well.

For example, let's consider a horse that spooks when going past an area of the arena he walks past every day. Humans with their deductive reasoning might think this is "dumb" because the horse cannot realize that "nothing has really changed." But when a familiar space has something new, no matter how subtle—like a scrape on the wall or a shadow on the ground—the smart horse notices it and tries to communicate to the person by spooking or resisting. The horse may, in fact, think the *human* is "dumb" for not noticing something strange that could be a threat. Again, what each species recognizes as "meaningful" to their survival has to be taken into consideration when assessing intelligence. //

Smart horses will also demonstrate their *associative memory* if they watch you put food under the buckets: they will come right to you in search of the food (figs. 10.5 A & B). This is because you are associated with the food, and therefore, you must have *more* food. This is intelligent horse thinking. Just because horses do not use deductive reasoning, like people, horses have long been considered "dumb," but nature has proven that a good associative memory brings success.

★ **TIP: / Horses are "associative learners" and have good memories. /**

Another test is similar to what we did to test for awareness (see p. 230). The difference is, when testing for intelligence, you are measuring how quickly your horse understands the concept of *space*. In an enclosed area, use your body language or an "awareness device" (such as a squirt bottle), an unusual noise (clicker, shaker), or a visual trigger (for example, a plastic bag) to define your space and prevent your horse from entering that space without your permission. An average horse catches on to the idea after three or four times (unless he is accustomed to walking all over people). A smart horse usually grasps the idea in one or two tries, particularly if he is sensitive (see p. 226). A slow learner or "below-average" horse may act offended and clearly not understand what you are asking. It may take several sessions before he understands the concept of "my space, your space." This is why so many domestic horses get injured when turned out with other horses—even after getting kicked, they do not respect

[10.5 A & B] Ava shows Radar a carrot and then puts it under a bucket. Radar returns to Ava because horses are associative learners. He understands that Ava had the carrot at first, so she will likely have more.

the idea of space. Although space is the foundation for horse social skills as well as horse language (see pp. 77 and 91), many domestic horses have spatial awareness problems because they were never taught appropriate horse behavior by their mothers or other herdmates.

By determining your horse's relative intelligence using one of these simple tests, you can better design training programs suited to his level of cognition, as well as becoming better informed about what motivates him to learn.

Confidence

Assessing your horse's level of confidence is important, because many behavioral problems originate from a lack of confidence or understanding. The genetic mutation for "vigilance" versus "curiosity" previously discussed in this book (p. 221) can directly impact confidence. While a "curious" horse may often appear more confident from birth, the "vigilant" horse who has higher sensitivity may at first

PART TWO / CHAPTER 10 / Analyzing the Individual: Equine Personality, Temperament, and Training Assessments

[**10.6 A & B**] Ty was a worried horse and spooked at many things. In **(A)** he is standing "base narrow," which is common in horses who lack confidence. After confidence-building exercises, Ty was standing "base wide" **(B)**, which tends to indicate an increase in confidence.

be "spooky" but can learn to be confident. A lack of confidence can also be caused by physical limitations or pain.

Almost every horse can benefit from being more confident as it is related to increased engagement in learning and better performance. A 2004 study conducted in Switzerland investigated heart rate variability (see p. 59) in comparison to heart rate with selected behavioral parameters. Among other things, the study found that *increased confidence could lead to better performance and reduced distress, thereby increasing welfare in horses.*

Most horses are "chicken" when it comes to scary things—until they learn not to be afraid of them. Many horses are trained "not to show fear" (rather than not to be scared at all) so they internalize their worry and eventually break down either physically or mentally.

Learn more

Less intelligent horses often are less spooky around scary things, because they have either not learned to be afraid or their reaction time is slow. In the wild, these horses would likely be dead for not reacting where it may be warranted, but in domestic situations, they are often praised by humans for being really calm. In fact, they could just be "slow learners," with below average intelligence, awareness, and sensitivity, as opposed to being highly confident.

★ **TIP:** / **Horses with high sensitivity and low confidence have a difficult time fitting into both horse and human societies. By helping them build confidence, you also build potential for cooperation and social bonding with you.** /

However, it should be noted that many "slow learners" or "mentally challenged" horses are very easy-going as youngsters, then as they reach ages five or six, they become more afraid. (Remember Milo on p. 232?) This is because as immature horses they are still in the "curious and fun" stage of life where everything is an adventure in learning. As they mature, what was once "fun" now becomes stressful, particularly if they have any pain associated with the endeavor.

★ **TIP:** / **Horses will often stand "base narrow" (legs close together) if they lack confidence and "base wide" (legs far apart) when they are confident (figs. 10.6 A & B).** /

_ Testing for Confidence

Your horse can have all the talent in the world and be very sweet, but if he lacks confidence, he is headed down a road of stress. "Externalizer" types will explode and spook, while "internalizer" types will get ulcers and store tension in their bodies. You can greatly improve your horse's future by first understanding his current level of confidence, and then assisting him in learning to be more confident. (Note that in teaching confidence, some methods are better than others. The use of *habituation*—repeated or prolonged use of a stimulus in order to decrease the innate response to it—is very common in the equine industry and can give the *appearance* of a "bombproof" horse, but as I've already mentioned, often the horse has not actually learned to be more confident, but only not to show worry or fear, which he internalizes instead.)

> **Your horse can have all the talent in the world and be very sweet, but if he lacks confidence, he is headed down a road of stress.**

STORY FROM THE FIELD: //

Bombproof or Time Bomb?

I was teaching at a clinic and one student brought what appeared to many to be a quiet "bombproof" Paint she had bought to ride on the trail. The horse did not make eye contact, and he avoided looking at objects or interacting with the people and horses all around him. This, to me, was the first indication he had learned to "disassociate" from objects or situations he was naturally worried about.

I had set up numerous novel objects in the arena, including some *on* the horses I knew well—from balloons tied to tails and tin pans hanging from manes to all forms of interesting trail obstacles. Some horses would not even come into the arena at first, but the Paint gelding walked right in. I watched his eyes; he had no reaction to most of what was going on. He failed to explore any of the objects and continued to avoid interacting with others. I pointed out his behavior to the students, explaining that it might appear he was "being confident," but it was only a matter of time before he exploded, which is the problem with teaching a horse not to react instead of helping him overcome his worry and fear and learn confidence through understanding.

The Paint *did* explode when an umbrella on the ground was caught by a breeze and lifted up in the air. He was terrified, and the incident left him on edge as he could not control his fear any longer. With a group of relaxed and supportive horses and people all around, we helped him learn to explore the objects, allowed him to show his fear and uncertainty, and reinforced that nothing was going to hurt him. He left with his eyes wide open and alert, looking at everyone and everything, and taking deep breaths and yawning—an engaged and far more confident horse. //

A basic confidence test can be done by placing "novel objects" around a familiar area like a paddock that gives your horse room to explore the objects and move away from them safely. Allow your horse to investigate on his own time plastic bags, umbrellas, squeaky toys, branches—anything that may frighten him or that he has not seen before, but is likely to come across in his life.

A horse with average confidence should walk up to the objects carefully, eyeing them, ears forward, then smelling them, touching them, and often licking or picking up the objects in his mouth, which demonstrates relaxed interest. A horse with low confidence may not want to approach the objects at all and may stand at the gate, "wanting out." The very confident horse will walk right up to the objects with a bold quick sniff, touch, taste, and move on, regardless of the variety of objects or how they look or move.

When testing for confidence, remember to keep in mind if your horse is more "curious" or "vigilant" (see p. 221). The curious horse will go right up to an object and often scare himself with the surprise of reaching it; the vigilant horse may not want to get near a new object at first and will approach cautiously. A little bit of caution in learning is a good thing, and horses who take time to understand new objects, sounds, or situations frequently retain information better than those who barge right up to everything without thinking.

Another test for confidence can be tried when you have developed a trusting and comfortable relationship with your horse, and he understands the concept of remaining still when you ask him to stand. Approach him with a novel object, such as a fluttery plastic bag or a container full of rattling pebbles—anything that your horse may not have been exposed to before (figs. 10.7 A–G). Watch for concern in his eyes and his body language. Good observation is key for safety. If your horse has any overreaction to the objects, such as spinning away, recognize his confidence limits and go back to placing the object on the ground to see if he will explore it on his own.

You can also do this test with a helper. Stand next to your horse's shoulder with your hand stroking his withers, and have another person approach you with the object. Judge at what point, if any, your horse becomes worried, and have your helper retreat with the object. Have your helper walk all around your horse with the object as you ask your horse to "hold his safe space" with you, as you relax and stroke his withers.

★ **TIP:** / Horses bond with people who make them feel confident. /

Like other tests, this assessment can be done using visual objects, various noises, various tactile stimuli, or a combination of stimuli. Know that while a horse may be fine being handled in the cross-ties where he cannot walk away, if he is not equally as comfortable with human touch in a paddock where he can escape when he is worried, then he is not truly confident or cannot trust his person. Following this test, particularly if your horse has overcome anxiety, your horse will usually signal that he is pleased with himself and understands by dropping his head, yawning, stretching his neck, licking his lips, chewing, or gently bumping his nose on you for approval. At this point you want to praise and reward the horse for his bravery. Horses have "egos," and to build a confident horse, you must help your horse develop a good sense of self. It is also good to allow your horse to have some free time to express himself after being tested for confidence.

_ Cooperation

The perfect horse may have all the right levels of sensitivity, awareness, intelligence, and confidence, but be very uncooperative. This may be fine for a wild horse, but not for a domestic one. Most horses that are uncooperative become this way because of pain or bad experiences with people. I have mentioned before that horses have good memories, and a traumatic event, bad training, or pain can all be remembered and lead to low cooperation.

It is not difficult to test your horse's level of cooperation, but you should always check for pain before passing judgment. A horse has the right to communicate to you when something is wrong by resisting. It is *your* job as the human caretaker to determine where the horse is hurting and help correct the problem (see How to Do the Body Check, p. 194). For example, many horses that refuse to go forward are telling

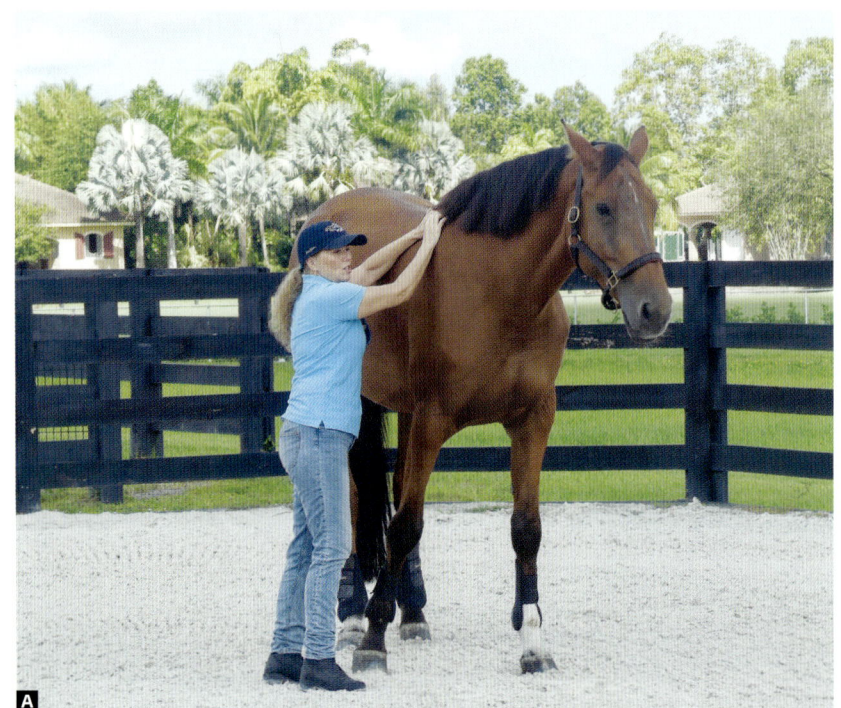

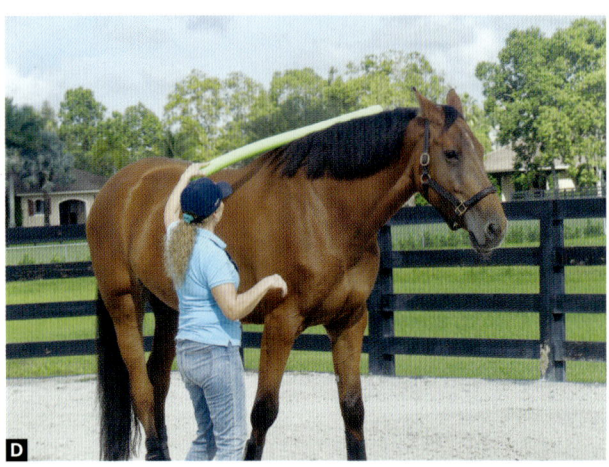

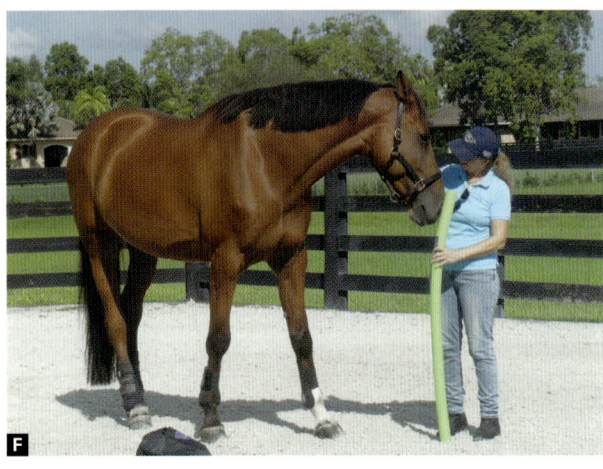
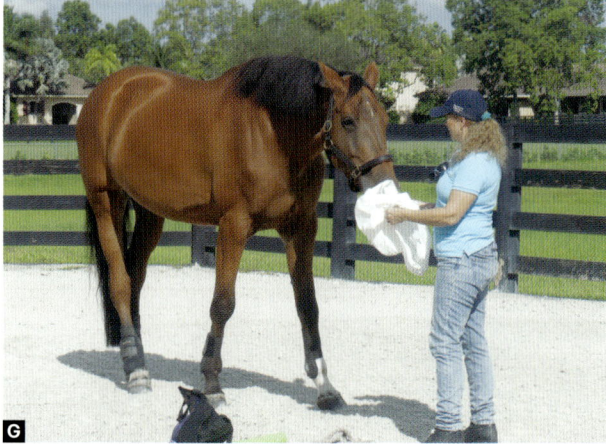

← **[10.7 A-G]** Coco has learned how to move or stand still when I ask. I reward her for paying attention and standing still with a nice scratch on the withers, which she enjoys **(A)**.

I have several novel objects lying in the paddock that I will introduce her to while asking her to stay in her space, including a foam tube **(B)**. Note Coco's wide stance and her slight lean to the left as she gets ready to move.

I wave the tube above Coco first, watching her eyes and ears to make sure she is comfortable with what I am doing. If she showed signs of worry I would take it away. Her ears tell me she is paying attention, but in a relaxed manner, with one ear tracking me as I talk to her and one ear focused on what is in front of her **(C)**. Note also her relaxed lower lip and soft aware eye.

I lower the tube into contact with Coco's neck, and she is not so sure she likes it touching her (note both ears back, tracking the object), but she trusts me enough to stand there and track the object without moving **(D)**.

I approach her left side, talking to her as she almost falls asleep **(E)**.

I tell Coco how "brave" she was for letting me touch her with the foam tube **(F)**. Her eye is soft, ears relaxed, and she reaches out with her nose to touch the tube and show me she is fine with it.

I repeat the confidence test, this time with a plastic bag that not only can change shape and size, but also has sound **(G)**.

you something is not right. Often, particularly with mares, it is because their nosebands are too tight or their saddles are pinching. With male horses, it may be they have to urinate and need the right place to do it. The bucking horse is often telling you his back, sacrum, stifle, or hocks hurt. Horses that suddenly run off for no apparent reason may have a sciatic nerve pinch that scares them, causing them to try to run away from the pain.

Horses that are good one day and uncooperative the next may have nutritional imbalances. Ulcers are a common cause of lack of cooperation because of the discomfort they cause. Carefully checking diet and making sure the horse has the right amount of nutrients and treating ulcers can make a big difference.

_ Testing for Cooperation

Most people identify an "uncooperative" horse by his behavioral issues, such as when the horse refuses to load in the horse trailer or walk through water. But while these issues can be perceived as a lack of cooperation, in many cases, the root cause is a lack of confidence, fear, or a lack of understanding.

Luckily most horses *want* to cooperate with people (unless they have a good reason not to), so testing for cooperation is rather easy. You can ask your horse to do simple tasks, like follow you, stay out of your space, and halt when you ask, all without much objection. A highly cooperative horse really wants to please and will eagerly do as you ask. An "average" cooperating horse will tell you if what you are asking makes his body hurt or that he does not really feel like doing the task, but he will still usually do it anyway, without much discussion. An uncooperative horse—often a mare who thinks she knows more than you do (and often she does)— will object louder than is normal and actually bolt, spin, rear, or generally avoid doing what is asked.

When testing for *sensitivity, awareness, intelligence*, and *confidence* you can also note how your horse responds to your requests in those evaluations. An "average" cooperating horse acts interested and wants to learn, even when he may be a bit worried about the exercises. A highly cooperative horse tracks every move you make and tries to anticipate what you will ask him to do next. An uncooperative horse does not want to pay attention, challenges your space, and in general does not want to participate in horse-human interaction.

The SAICC Evaluation provides the opportunity to not just assess your horse, but also educate and help your horse learn to be more balanced in these areas, critical for healthy interaction with people.

Benefits of Understanding Your Horse's Personality

Being able to assess your horse's temperament, in whichever way or ways you prefer, allows you to adjust your own energy and actions to form a better relationship with your horse. You can also take a look at yourself and know where you and your horse might clash and where you might connect. Remember your horse's priority, whatever his personality, is to *feel safe and find comfort*. If you can achieve this, then you will have the foundation for developing the right educational path to help your horse learn.

★ **TIP:** / Your horse's priority is to feel safe and find comfort. When you can achieve this for him, then you will have the foundation for developing a strong trusting relationship. /

Training Readiness by Age, Gender, and Experience

If your horse has not learned "basic horse skills," such as those learned at an early age in a wild horse herd, then your horse may experience added stress and increased potential for injury. Older horses who many have been trained to do human-created disciplines but missed out on early young horse development often internalize stress. They try and fit into whatever social system they are placed, often at a cost to their own welfare as they do not know how to—for example, they lack the skills to appropriately greet other horses and make friends. So it is essential to help your horse be a "functional horse" first, before training him further to do what you wish.

Female and male horses may learn slightly different things, but all horses should understand the basics of spatial awareness, respect, and status from a young age. Remember most of what horses need to know to stay in good status in a social group is learned between the ages of birth and two years old. Those who missed "Horses 101" or did not have a good upbringing will lag behind. Horses need to be aware they are horses and feel confident in how to interact and use their bodies, *before* having humans on their backs. While age is important, temperament, which we've been looking at in this chapter, is even *more* important.

★ **TIP:** / Always customize your training program to fit your horse and be ready to make changes as needed. /

Gender also does play a role. Mares may have certain days a month when they are not really in the mood to learn as their hormones are cycling. It's okay to give them a day off to let them "be mares." Same goes for stallions, and the solution for both is often the same—free-choice exercise, with or without a human directing. Hormones play a strong role in the behaviors of horses in natural circumstances, and so monitoring your horse's rhythms and adjusting your training to the horse's natural cycles can be helpful.

★ **TIP:** / Hormones and seasons can play a strong role in your horse's behavior, so adjust your training to fit your horse's rhythms and natural cycles. /

Remember, most horse behavior is learned behavior, so many domestic horses begin at a deficit, particularly if they grew up with other horses lacking the functional skills we discussed in Part One (see p. 35). Testing your horse on these skills is recommended *before* starting training and to determine his readiness for any learning. A 10-year-old horse who has done nothing but show, may have missed "K-12" in "horse school!" The good news is, it is never too late to help your horse *learn how to be a horse*.

So how can we determine what our horse knows and doesn't know when it comes to "being a horse"? I've put together the following series of questions to help you build a picture of a horse's "horse knowledge" status. These are all concepts we have already explored in earlier chapters, as related to both horses in the wild and in modern domesticated scenarios. Consider these questions as you observe your horse in the company of other horses and humans.

⏤ Basic Horse Skills Assessment—Quick Check

If your horse does not understand or know the following skills, then you need to teach him, as they are essential for both his welfare and your safety. If he does, then he is, without a doubt, ready to learn.

1. **Greeting skills:** Does the horse make eye contact, offer a nose bump or smell of you or another horse, and allow touch on his neck and withers (see p. 209)?

2. **Spatial awareness:** Does the horse understand his relative size in relationship to his environment, objects within it, and other horses and people? Does he know the parameters of his body and how to move safely in space—that is, basic proprioception (see p. 77)?

3. **Spatial respect:** Does the horse understand he should respect the space of people and other horses? Does he keep an appropriate distance from people and other horses (see p. 142)?

4. **Stand still or move:** Does the horse understand the energy dynamics of social connection and to either stand still or move when asked from the ground? Does he understand communication and how to move his body forward, sideways, and backward (see p. 142)?

As you can see, horses learn simple but important lessons in order to maintain status in a functional herd (figs. 10.8 A & B). And having a social network is key for a horse's safety and mental and emotional welfare. Not knowing "how to fit in" can cause undue stress in horses. Your horse will be better positioned in life if you teach him social etiquette and reinforce his innate

Horses learn simple but important lessons when they grow up in a functional herd, giving them skills to develop the social network that is key for the horse's safety and mental and emotional welfare.

↙ **[10.8 A & B]** Demonstrating a quick Basic Horse Skills Assessment with Chevez, a wild horse who came into the Wild Horse Rescue Center in Florida. I give a social greeting **(A)** and test his spatial awareness **(B)**. He was very sensitive and shy around people, as he did not trust humans. While he demonstrated good horse-to-horse social skills, his comfort level around humans remained too low for him to safely work with people. He lives with a group of other wild horses at the sanctuary.

equine skills. Help your horse feel safe and comfortable "being a horse" before starting any further training.

Taking a Case History

The more you know about your horse before creating a training plan, the better. Since each horse has a different temperament and personality, it is time well spent to find out as much as possible about a horse's history. While some of these questions can be answered by assessing the horse before you, others may be answered by previous owners or the breeder. In addition to the Basic Horse Skills Assessment, it is important to build out and add to a horse's "Case History" to ensure the way he is fed, housed, trained, and ridden is what is best for him as an individual.

- Was the horse raised in pasture with other horses or alone?

- Was the horse raised in a stall or with turnout?

- Did the horse grow up with opportunities to socialize?

- How old was the horse when he was weaned? Did he struggle with weaning?

- Does the know how to "make friends" with other horses and people?

- Has the horse experienced any trauma or highly stressful situations?

- Does the horse like to be touched and groomed?

- Can the horse load in a trailer and travel comfortably?

- Can the horse stand quietly for the farrier and pick up all four legs?

- Can the horse roll in both directions?

- Does the horse sleep both standing up and lying down?

Sometimes the best approach to beginning to understand your horse as an individual is just to "hang out" with him—eat together, sleep together, and generally just become your horse's pal and get to know him. Horses do not care about *what* they do with you, they just want to "be with their friends." If you qualify as their best friend, then you will find, regardless of their past experience, they will do virtually anything for you.

★ **TIP:** / **Before you start any training program, make sure your horse understands "basic horse skills" as it will make learning much easier for him.** /

Assessing Trauma

Horses do not have to have experienced "trauma" in human terms to have been "traumatized." Different aspects of temperament and personality that

[10.9] This mare was so traumatized from jumping that she could not walk over a pole on the ground. She had completely shut down. We removed all her triggers, such as the bridle and bit (she had been ridden in one that caused her pain), and just allowed her to explore ground poles on her own without forcing her to do anything until she was ready. After a couple of days, she was able to walk over the poles, and soon after that, she was free jumping without stress.

we have discussed in this chapter—sensitivity and lack of confidence, for example—as well as withstanding pain and separation and other variables can all create a "traumatic experience" for a horse. Trauma untreated in horses leads to welfare issues, as many either suffer high states of uncontrolled anxiety, creating potential safety issues for humans, or are passed through the "system," often ending up at rescue and adoption centers. Conditions related to trauma usually get labeled as "behavioral issues" or "just the way the horse is," and the horse never gets the help he needs. All people in a horse's life, the veterinarian, owner, trainer, farrier, and others, should be able to notice symptoms of trauma and then discuss what is best for the horse to ensure his well-being.

Horses and humans as social mammals share a number of emotional psychological imbalances. Despite this, while human psychology has identified issues such as PTSD (post-traumatic stress disorder) for people, little research has been done on horses who have undergone psychological stress. With over 40 years of clinical research in treating behavioral issues in horses, it is clear to me that while many issues are related to undiagnosed physical pain or discomfort, often the most dangerous behaviors are related to human ignorance of equine behavior and communication relating to the horse's mental and emotional states (fig. 10.9).

_ Common Causes of Trauma

Common causes of trauma in horses include:

• Separation from family and friends.

• Inability to make social bonds.

• Accident or event triggering the horse to feel unsafe.

• Restriction when the horse needed to flee.

• Restriction or confinement under unsafe circumstance.

• An event the horse perceived as fearful but did not understand and associated with objects, people, places, smells, or other sensory input with the experience.

• Physical abuse associated with a trigger the horse did not understand.

• Abandonment, including isolation, starvation, and no social interaction.

• Being oversensitive to sensory stimuli and lacking confidence.

_ Common Signs of Trauma

Domestic horses with the following abnormal stress behaviors often have experienced trauma (figs. 10.10 A–C):

• Lack of curiosity to interact with humans or novel objects.

Trauma untreated in horses leads to welfare issues.

- Failure to make any eye contact or show desire to interact with people.

- A "dead" or "dull eye"—no light in the eye, as if staring blankly at all times.

- A worried look in the eyes when around people.

- Looks away or hides head down in corner of stall.

- Panics in small spaces if unable to get away or move.

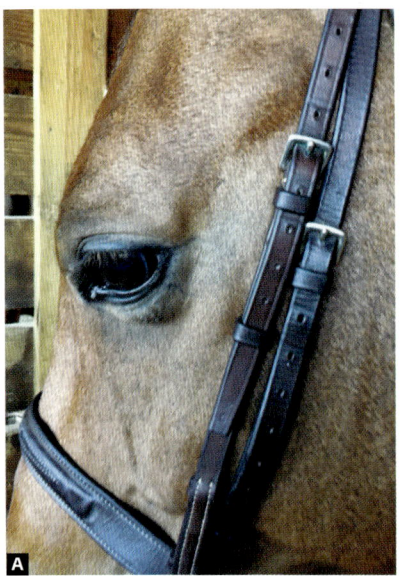 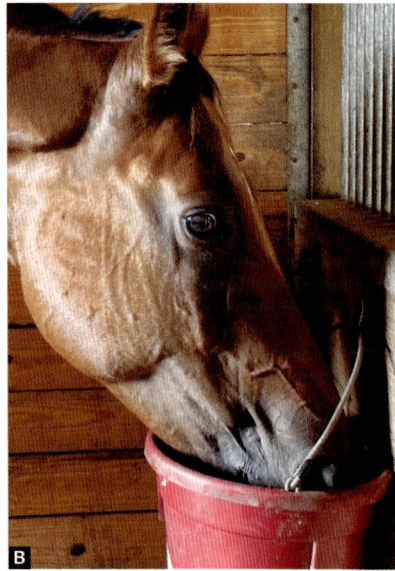 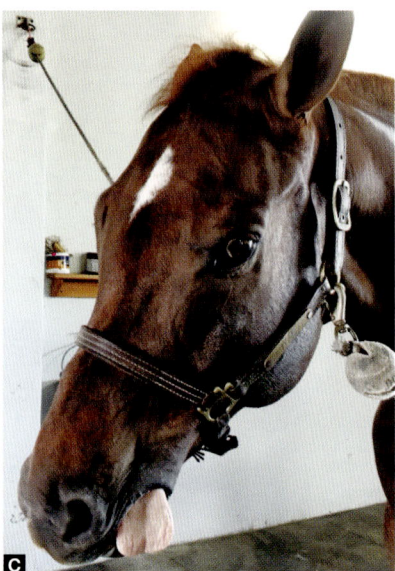

- Overreaction to stimuli (visual, tangible, audible, olfactory).

- Muscles tighten when touched; acts nervous.

- Unpredictable bolting with no regard for safety.

- Attachment to other horses, but failure to form functional social bonds.

- Bursts of anger toward humans and other horses.

- Abnormal and repetitive "vice" behaviors, such as weaving, self-inflicted biting, head-shaking, cribbing, and grinding teeth up and down stall bars.

- Over-responds to training (hyperflexes, for example).

- Holds breath or breathes shallowly when around humans; fails to relax.

- Sweats from nervousness when around people or being ridden.

- Increased heart rate when around people or being handled.

_ Consider the Context

Horses do not have to end up in rescues to have experienced trauma. They don't have to have been abused. Horses are individuals, and while we can make some generalizations about equine behavior, they are "context-specific," meaning the same circumstances can affect various horses differently. For example, the curious, very social, young male horse might not be stressed by weaning, especially if he is with friends during this period and makes new friends easily. In his case, heading off to the track or another new environment might be perceived with some enthusiasm. But a highly sensitive and timid filly, having gone through the same environmental experiences, may have ulcers and never feel relaxed enough to learn how to adjust. From a human perspective, her early life may be perceived as "normal," but from the filly's perspective, it is traumatic.

★ **TIP: / "Difficult" horses who display normal horse behaviors are usually smart horses who have learned ways to protect themselves and communicate their needs to humans. /**

Oftentimes, well-meaning trainers will have a horse who is described as "difficult." "Difficult" horses who display *normal* horse behavior are usually smart horses who have learned ways to protect themselves and communicate to humans their needs. "Difficult" horses who display *abnormal* horse behavior, such as I've just described, and are sometimes unpredictable, most often have had some form of trauma in their lives or have learned *coping behavior*. Their instinctual drivers are always on "high alert," but they have no way to find safety. Their stress levels are almost always high and will remain so until someone can help them learn new associations, let go of the trauma, and find comfort. *Stereotypic behaviors,* such as "cribbing" and weaving, are examples of what can be coping behavior (acquired due to confinement, for example). The underlying cause of stereotypic behaviors should be identified

← **[10.10 A–C]** The horse in **(A)** shows what I call a "dead" eye—staring blankly.

In photo **(B)** you see a mare who had suffered acute stress from a long trailer ride alone in a hot climate to a new facility. Note her "worried" eye as she drinks and avoids making eye contact with the humans nearby.

Photo **(C)** shows a horse (also with a worried eye) who constantly played with his tongue when in the cross-ties. When horses have no escape from discomfort or worry, they may "divert their stress" into behaviors such as cribbing, weaving, or in playing with their tongues and mouth.

STORY FROM THE FIELD:

A Traumatized Performance Horse

While conducting a demonstration on "safe space" at the Hansepferd event in Hamburg, Germany, I noticed a young horse practicing for the evening performance in the nearby warm-up ring becoming more and more stressed and worried. The gelding's eyes were scared and the rider had both sharp spurs and tight draw reins, trying to get the horse to "look fancy" but only confusing him. I found management and told them that the horse was about to explode. Management assured me the trainer "knew what he was doing."

My demonstration featured two draft horses—one a four-year-old mare and the other a more timid ten-year-gelding. I was showing the differences between mares and geldings and how spatial awareness could be achieved and learned by horses teaching other horses. I was impressed with the young mare who was an "old soul" and very comfortable holding her own space and sending the gelding into the corner with just an eye glance and ear.

Suddenly screams were heard and the warm-up ring whirled into chaos. Standing next to the makeshift arena fence, the mare and I looked to see what was happening. I had my hand on her withers, telling her she was fine and that she was strong and protective. We watched as a bolting horse and rider—the stressed gelding I had noticed earlier—galloped through the warm-up arena, knocking into other horses and riders. I knew the horse was coming toward us; he sensed the safe space the mare and I had created. The mare never flinched but instead welcomed the scared horse.

The horse and rider slid to a stop—the horse with his nose touching his chest, nostrils flared, sweating profusely, and breathing rapidly. No amount of the rider kicking the horse would get the gelding to move. I told my translator and the crowd how this horse was scared and came to be with us because we were safe. The rider eventually had to get off and lead the gelding away. I again told management this was not the type of entertainment people wanted to see and told the crowd to keep an eye on the poor horse at the performance that night.

The next day my translator told me that the horse had "jumped into the stands the night before and that management had asked the performer never to use that horse again." My heart hurt for the horse, as he was unlikely to get the mental and emotional support he would need to feel safe. His trainer was not interested in learning what he was doing wrong. The result was a traumatized horse. //

and addressed as they are often associated with poor mental, emotional, or physical welfare. Luckily, horses are very forgiving, and given a chance, will undo negative associations in exchange for a positive association, thus removing the trauma (fig. 10.11).

★ **TIP: / Horses who display abnormal behaviors and are sometimes unpredictable most often have had some form of trauma in their lives or have learned coping behaviors. /**

Stockholm Syndrome and Learned Helplessness

When assessing trauma in your horse's past, take your time. Your horse may at first appear quiet, docile, and obedient. Then, as your horse is allowed to communicate in the ways we have discussed in earlier chapters (see p. 175) and starts to feel better, he may display aggressiveness. Neither his docile previous behavior nor his aggressiveness would be accurate indicators of his temperament or personality, as the horse has not had the opportunity to develop his individuality and communicate it to you.

I've already mentioned how, often, a well-trained horse can appear quiet and obedient, when underneath he is very fearful but has simply learned how to *appear* quiet. These horses do not interact normally with people, often avoiding eye contact or walking to the back of their stalls when you approach them. It takes time to understand and accurately assess their true temperaments.

The most common "hidden" trauma in horses is seen in horses who have learned what in human terms is called *Stockholm syndrome*—a coping mechanism in a captive or abusive situation where people develop positive feelings toward their captors or abusers over time. In horses, this can occur (as an example) when a horse is chased around a round pen, roped, tied up for long periods, confined, or

↑ [**10.11**] Wild horses who have had stressful interactions with humans become hypersensitive to all their interactions. But because horses seek social friendships, most are willing to override past experiences and trauma with a person who is patient and emotionally understands their feelings. Here Diane DeLano, Director and Founder of the Wild Horse Rescue Center in Webster, Florida, approaches a horse new to the sanctuary to assess his "safe space." Note he is leaning away, but not moving as Diane is calm and talking gently to him, not making him run around, but rather asking him to accept a touch.

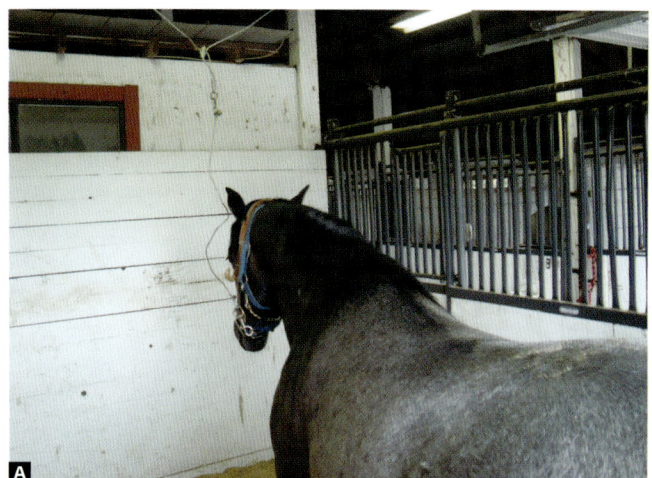

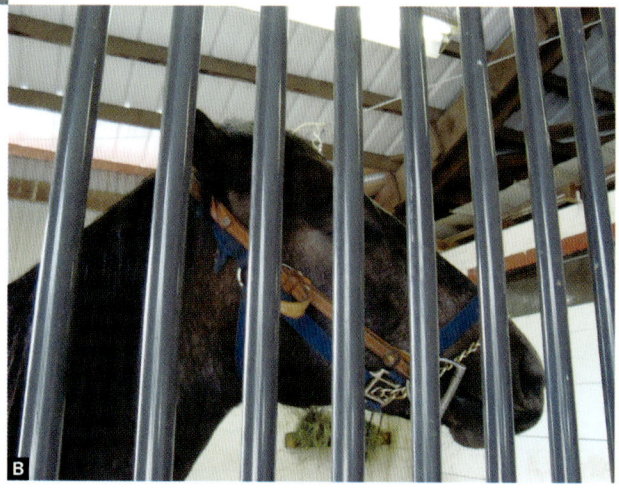

↗ [**10.12 A & B**] This four-year-old Andalusian stallion was kept in his stall 24 hours a day, secured by a cable where he could not reach food or water unless his trainer came in to release him. When I approached the owner/trainer about the horse's situation, the person was thrilled to show me how he had taught his stallion to "dance." The horse was so traumatized that he would try anything to get to food and water; when he would start prancing in place, the owner would release him. The owner also showed me how the stallion would come to him when he called. This is an example of how "Stockholm syndrome" can apply to horses. The horse had no friends, no freedoms, no access to food and water except through the owner, who was also the perpetrator of his mental and emotional abuse. The owner was reported to Animal Control, but since the horse was not "starving," they could not do anything.

otherwise stressed or made uncomfortable, then given carrots or treats from the same person who has inflicted pain or emotional discomfort. The horse "bonds" to the person who has inflicted the pain or discomfort as a way to stay safe (figs. 10.12 A & B). While this coping mechanism may seem to "work" and people may feel they have a well-trained horse, they have created a mentally unhealthy horse. These are the horses who act badly "out of the blue," suddenly taking off when their stress is so high they cannot mask it any longer. These horses can be dangerous because of their unpredictability.

The other common behavior displayed in horses with trauma is *learned helplessness*—a mental state in which the horse becomes so accustomed to stimuli that are painful or unpleasant, he becomes unable or unwilling to try and

↓ **[10.13]** This horse had no interest in socializing with people. He was in distress, and when I evaluated him, I found severe pain in his back. When I started assessing his body, he relaxed his lips as you can see here, but he remained focused inward. Note the dullness in his "dead" eye.

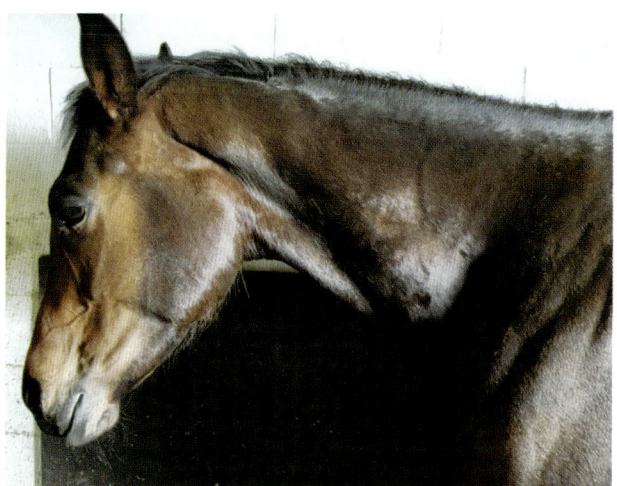

↗ **[10.14]** *Hyperflexion* is a common defense for horses who have learned to avoid pain to their sensitive mouths by curling their necks and bringing their noses toward their chests. It is frequently seen in the dressage, jumping, and reining arenas, and while it is "allowed" as a training "method," it needs to be recognized as a cause of mental and emotional stress on the horse. Western Pleasure horses learn to avoid pain by keep their heads extremely low in a similar learned behavior.

avoid those stimuli, even if it is possible to do so. The horse has learned to suppress his own behaviors and obey what the rider tells him to do because disobeying or communicating discomfort will only end in more stress—both physical and mental. So the horse continues to do what he is asked.

An example of learned helplessness often seen in sport horses is the horse with a worried eye who hides in the back of the stall when a person comes to get him. Instead of greeting the human, the horse hides his head and tries to avoid the halter. Eventually the horse gives in, but

not willingly. The horse's eye goes "blank," as if the horse has gone someplace else (fig. 10.13). This look often goes unnoticed, as the horse does not fight, but gives in to the person's desires (fig. 10.14). The horse may even perform well in an effort to try and please. But these horses will almost always have ulcers and be worried under saddle.

Most trauma in horses can be corrected by assessing the core of the issue, building trust with the horse, and gently recreating any triggers using positive rewards. Helping a horse feel safe, making him comfortable, and giving him a choice when it comes to interaction with humans are essential first steps. But we must always be aware that a deep-seated trigger can override thinking in the horse and cause panic. It is critical we create safe space, and be the eyes and ears for the horse.

← **[11.1]** Mental and emotional stress leads to physical stress in horses. This mare was stuck in a 10-foot by 10-foot stall at a horse show for weeks where she could not lie down or even stick her head out. This is typical housing at many shows, causing sleep deprivation in competing horses, as well as other stressors.

CHAPTER

[11]

UNDERSTANDING EQUINE MENTAL AND EMOTIONAL STRESS

"

The better you know your horse, the better you will be at knowing when something is 'off.'

[11]

Assessing stress is not always easy for trainers or veterinarians. This is because most have been trained to do things to horses to get the horses to perform or be "obedient." Human minds get busy with analytical complex thoughts, while horse minds are simple—they *react* to frightening things, too many stimuli, and negative emotions, but *respond* to kindness, safety, and love. Ahead I will give you tools to determine equine stress and pain, which is helpful in removing human bias (subjectivity), but there is no replacing connection and empathy between two individuals in the process, regardless of species. The better you know your horse, the better you will be at knowing when something is "off." If you become a good observer, learning to look "with" not "at" your horse, and you customarily greet your horse with soft eye contact, then you will easily observe changes in your horse's eyes, reflecting his moods and feelings.

★ **TIP:** / Horses are designed by nature to tolerate physical stress, but they are not wired to adapt to emotional or mental stress. /

★ **TIP:** / If your horse looks stressed or worried, always question what you are doing. Rethink and reassess your methods. /

Horses are designed by nature to be pretty physically tough, but they are not well-suited for handling emotional or mental stress. Hence, many horses have stress-related disorders that could be managed. It is not complicated to manage horses mental and emotional well-being when people keep in mind how horses think and feel. It is rather simple. Horses want to feel safe, be physically comfortable, and have friends around to eat, sleep, groom, and socialize with when they feel like it. Having food, water, and shelter is of course a given, and can vary from being out in pastures to living in stalls. Everything you learned in chapter 10, including how your horse was raised and what he has learned, along with his temperament and personality, are all instrumental in how well your horse will adapt to various living situations and activities, and whether or not he will experience stress.

Let's get a better understanding of the kinds of stress horses experience, how they deal with it, and how we can manage it.

Consider the Research

Horses have fallen into an odd category when it comes to welfare, and thus until recently, few research studies had been conducted to identify mental and emotional impacts on equine well-being. But the landscape is changing. As more data becomes available, people will have the ability to develop better tools and standards for the care of their horses, and they will be able to apply these standards to both competitive horse sport and recreational equestrian pursuits.

There is a disconnect, however, because academic institutions and veterinary schools publish research results primarily in academic journals not easily interpreted or available to the average horse person. Thus, the numerous studies that could benefit horses take a long time to make their way into the horse industry. Horse show and breed organizations continue to operate with little knowledge of the world of equine science, aside from nutritional and medical issues related to sponsor products.

For example, numerous studies now have been conducted that show ulcers occur in 34 to 85 percent of foals weaned too early or unnaturally, which can affect them the rest of their lives. One found that 98 percent of foals in the study developed gastric lesions within two weeks of weaning. Separation from family and friends is probably the strongest stressor in a horse's life, starting with weaning. In nature, mares decide when to wean their foals, and in some cases, foals simply lose interest and prefer to eat on their own. It is a natural cycle, and the timing is left up to the individuals.

Learn more

Separation anxiety and related ulcers do not just affect foals. They can affect mares as well as any strongly bonded horses. Mares tend to internalize stress, having ulcers more frequently than geldings. In a clinical study that I conducted between 1983 and 1993, involving the brokering and sale of horses, twice as many mares had ulcers than geldings. In one group of Thoroughbreds shipped, *all* the mares had ulcers, while only 60 percent of the geldings did. (Note that while anecdotal data, treating all the mares for ulcers eliminated behavioral and physical signs of discomfort.) In fact, ulcers in domestic sport horses are so common they are considered "normal" by many. But they are *not* normal in nature. And they are something people can better manage by reducing stress often caused by separation anxiety.

A study in New Zealand found that social isolation is a significant stressor for horses and may influence both their behavior and heart rate variability measurements, as might be associated with mild somatic pain.

Learn more

Numerous studies have looked at both the physiological and psychological effects of shipping horses and various stressors based on common shipping practices. Research is now demonstrating that *infrared thermography (IRT)* used to measure eye temperatures combined with *salivary cortisol levels* can be noninvasive ways to measure stress in show horses. A number of studies have investigated the stress caused from restrictive and painful bits, bridles, and tack, and elevated stress levels in horses with tight nosebands (figs. 11.2 A & B).

As more and more people become aware of the science behind horse-human interactions, the

paradigm related to how we interact with horses will be guided by public perception. It is therefore critical that every horse person stay current with the growing body of science related to equine stress and continue to learn how to better understand and care for their horses.

↑ **[11.2 A & B]** Many horses suffer mental and emotional stress trying to please humans. Training methods are socially accepted and being taught to each generation because they produce results in performance, but not because they are good for the horse.

★ **TIP:** / Do not mistake the horse's willingness to cooperate as comfort with performing. /

Types of Stress

In psychology, stress is divided into three main types:

STORY FROM THE FIELD: //

Separation vs. Shipping Stress

A veterinary friend of mine was conducting a study near Portland, Oregon, to measure the effect of heat stress when hauling horses in hot weather. He carefully took measurements before leaving on two horses—one to stay home as the "baseline" and one to be hauled for three hours on a hot afternoon. He loaded the "test" horse into a trailer, then drove around for the prescribed period of time in the heat, and returned to the barn where he found the "baseline" horse in a high state of stress and anxiety. After taking measurements again, the vet found that the horse who stayed home and worried about being alone had lost more weight and sweated more than the horse who was trailered in the heat for hours. In this instance it seemed separation from a friend caused more stress than shipping on a hot day! //

1. **Acute** This kind of stress is transient and usually disappears when the stressor is gone. An example is performance stress, such as jumping or racing. Acute stress is common in sensitive horses that worry and overreact as they may experience several types of stressor, triggering both physical and psychological reactions. This can be managed and usually has no long-term effect.

2. **Episodic Acute** When there are frequent triggers of stress, such as daily irregular feeding or daily handling that causes fear or discomfort, it is considered episodic acute stress. The stressors are so frequent that the horse may experience one right after another. For example, a horse who does not get fed on a regular timetable is taken out of his stall when hungry, tacked up, left to stand uncomfortably in cross-ties for 30 minutes, then is ridden with a tight noseband and draw reins and asked to do things he does not understand. This horse will first express emotional distress, but if not allowed to communicate his hunger, discomfort, and lack of understanding, he will have cognitive distress, mental fatigue, compromised learning, muscular tension, soft tissue issues, digestive disorders (ulcers, diarrhea), rapid heart rate, shallow breathing, and a compromised immune system. Therefore, episodic acute stress requires treatment on many levels, both physically and psychologically, but when corrected, it can be managed by eliminating triggers.

3. **Chronic** Horses who become habituated to stressors often become depressed and show

behavioral despair. They anticipate a negative situation because so many triggers have been associated with different kinds of stress. They are often hypersensitive horses with little confidence who have not had a functional upbringing, so they react to many stressors as they have not ever learned how, when, and where to feel safe. An example may be a horse who was abandoned, had many owners, and ended up in a horse rescue. He is a sensitive horse who has experienced long periods of episodic stress, never having stable friends or food and undergoing various forms of stressful environments and training. Both physical and psychological stress many become permanent in these horses, even when the stressors are removed. Horses with chronic stress need lots of support on all levels.

⎯ Good Stress vs. Bad Stress

Remember there is "good stress" and "bad stress." Stress helps us adapt and evolve when it is dished out in small quantities, and we can live comfortably through the experience. An example of good stress may be taking a young horse for a trailer ride for the first time with other horses. The stress level should be minimal if the young horse has been introduced to the trailer and is with friends who travel well. But, asking a sensitive young horse with little confidence to undergo this situation *alone* could tip the stress scale, and the "good stress" turns to "bad stress"—or too much, too fast—and you never get the horse in a trailer again.

Horses need to learn adaptation skills to help make their lives more comfortable living with people. Often the super-talented sport horse who can jump higher at home or run faster during training falls apart on show day or race day. Horses have to learn to handle mental and emotional stress with positive overrides of the negative worry. When they feel safe and motivated to learn, you are on your way to producing a happy horse-human partnership, whether for sports or companionship. Emotionally anxious horses may perform out of learned helplessness (p. 253) or because they "try to please," but eventually they will break down, either physically or mentally. Teaching your horse in gradual steps, one at a time, how to deal with various stressors will pay off in the long run.

★ **TIP:** / Most horses adapt well to various stressors when they feel safe and are with friends (horses and humans), which is one reason they have evolved to be our partners in so many disciplines. /

How Stress Affects Your Horse

Stress can affect a horse both physically and psychologically, regardless of origin. Pain and discomfort can cause worry and antisocial behavior. Mental and emotional stress can cause physical problems such as ulcers or muscle tension. They are all related.

Horses respond to stress in various ways, depending upon temperament, personality, and life experiences (see chapter 10, p. 222). Having trained police horses and worked with numerous wild horses, it has always amazed me how adaptable horses really are in the face of extreme stressors, once they have had time to learn. Since most domestic horses have very "sheltered" lives from birth, they do not gain the early experience needed to handle a high level of stress, and thus these horses suffer more than they need to in many performance sports. Some signs are obvious, but many more are often regarded as "bad behavior" or "oversensitivity." The following outlines how stress affects the horse's body. All three reactions can take place at the same time, depending upon the level of stress perceived by the individual horse.

1. **Behavioral** The horse moves away from something perceived as a threat or a stressor. Appears at various levels, from tensing to rearing when a saddle is uncomfortable; from stomping feet to swishing tail to remove flies. When a horse has learned not to communicate the stress he is experiencing with such behaviors (for example, being yelled at to "Quit it!" when stomping due to flies), then the stress often will be internalized and show up as a physical issue (such as ulcers).

2. **Sympathetic Nervous System Activated** A stressor causes involuntary stimulation of the horse's nervous system, creating action in the intestines, glands,

Physiological Impacts of Stress, Regardless of Origin

- Increased gastric acid, causing inflammation, digestive dysfunction, and abdominal pain, leading to weight loss, exhaustion, and colic.

- Increased sweating and diarrhea, resulting in dehydration.

- Muscle tension, leading to soreness and stiffness.

- Interruption of hormone production and distribution, causing reproductive disorders.

- Increased opportunity for injury due to malabsorption of nutrients.

- Weakened immune system, increasing likelihood of becoming sick.

- Decreased ability to focus and learn, leading to poor performance.

heart, respiration, and more. This is what triggers the *fight-or-flight response* (see p. 42).

3. **Neuroendocrine System Activated** Now there is an increase in energy flow so the horse can take further action if needed. This uses up fats, proteins, and carbohydrates, and breaks down energy rather than storing it.

Primary Stressors

Primary stressors are the starting points of the stress process. As stress increases, *secondary stressors* may occur as a result. What follows are examples of typical primary and secondary stressors for horses and their most common causes.

_ Pain or Discomfort

Learn more

- Ulcers—Studies show between 50 and 90 percent of all sport horses have ulcers.

Learn more

- Back pain—Studies show between 43 and 74 percent of riding horses may have back pain related to poorly fitted saddles. (Note that confinement and not being able to lie down, roll, buck, gallop, or engage in social mutual grooming can all lead to back discomfort, as well.)

- Mouth pain—due to bit, teeth, or gums.

Learn more

- Nose pain—due to tight nosebands.

- Inflammation/arthritis/muscle soreness—often goes unnoticed except for behavioral changes and lack of willingness to work. (Lifestyle, diet, and lack of adequate and appropriate exercise can all contribute to this.)

- Injury—soft tissue injuries often go unnoticed and the horse has limited ways to communicate.

- Disease—Lyme, EPM, viruses, bacterial, fungal, and other infections.

Environmental Stress

- Insufficient time spent in nature; lack of natural sounds, smells, sunshine.

- Inadequate habitat; poor housing quality.

- No place to lie down (dirty stall, mud, dust, hard ground).

- Confinement and inability to move at will.

- Inadequate light or constant light.

- Electromagnetic disturbances.

- Chronic noise or frequencies damaging to physical systems.

- Temperature too hot or too cold.

- Poor air quality (polluted air, presence of ammonia).

- Unfamiliar places or situations (such as horse shows).

- Insects, fungus, bacteria, pests.

Nutritional

- Food is restricted (limited availability and type).

- Poor diet (inadequate nutrients).

- Diet too rich (unable to process level of nutrients, concentrated feeds).

- Toxins in feed.

- Contaminated water.

Brambell's Five Freedoms

There is global consensus now among scientists and humane advocates that all animals should have the basic "Five Freedoms" to ensure both physical and mental welfare. Evolving out of a 1965 Humane Report in the United Kingdom by Professor Roger Brambell on farm animal welfare, the Five Freedoms are now widely accepted standards for the housing and care of all animals. Oddly, many in the horse industry often seem unaware of or overlook these Freedoms, which leads to many of the stressors I've outlined in these pages.

The Five Freedoms are:

1. **Freedom from Hunger and Thirst** by ready access to fresh water and a diet to maintain full health and vigor.

2. **Freedom from Discomfort** by providing an appropriate environment, including shelter and a comfortable resting area.

3. **Freedom from Pain, Injury, or Disease** by prevention or rapid diagnosis and treatment.

4. **Freedom to Express Normal Behavior** by providing sufficient space, proper facilities, and company of the animal's own kind.

5. **Freedom from Fear and Distress** by ensuring conditions and treatment that avoid mental suffering.

The Key to Identifying and Relieving Stress

From a "horse-centric perspective," stressors are usually related to either "safety" or "comfort." If your horse exhibits unusual behavior or misbehaves, then determine how you can make him safer or more comfortable.

- Lack of free-choice salt and water.

- Inability to forage for needed nutrients.

Social Stress

- Separation.

- Lacks ability to make friends.

- Lives alone.

- Lacks confidence to feel safe around other horses.

- Anxious around people.

- Lacks ability to understand and learn.

- Overreacts to stimuli.

- Has negative associations to specific social/training triggers such as: seeing a saddle, feeling the girth, smelling the farrier or vet, and hearing the arena gate close.

- Not allowed to communicate.

- Restricted from making decisions and choices.

- Boredom/monotony.

- Performance anxiety.

Reproductive Stress

- Hormonal swings.

- Painful heat cycles.

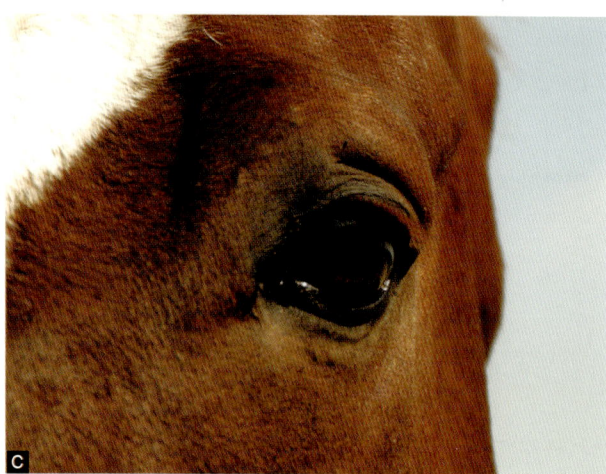 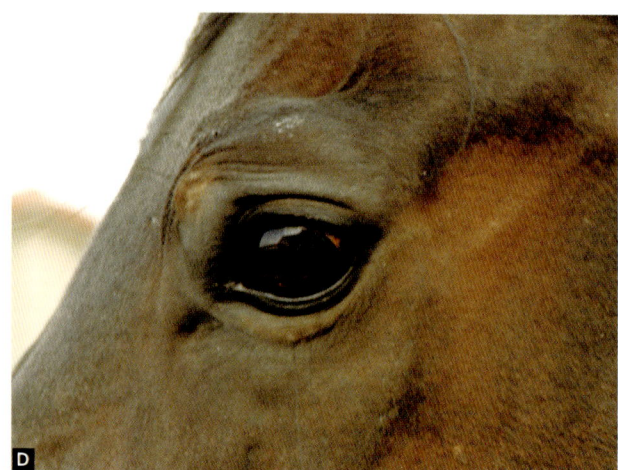

↑ **[11.4 A-D]** A horse that is healthy and engaged has a "light" in his eye **(A)**, while a stressed or depressed horse has a "dead" eye **(B)**. An eye that isn't worried is focused on what is going on around with interest **(C)**, as compared to one that shows concern when a horse is in a new environment with no friends around **(D)**. Note the wrinkles above the eyes in the horses with worry or discomfort.

- Breeding stress.

- Human-induced stress related to artificial insemination, embryo collection/transplants, hormonal regulation.

- Weaning of foals.

How Horses Show Stress
— The Eyes Tell All

You do not have to be an expert to understand when a horse is stressed. You simply need to have awareness and empathy to know if another creature is suffering. The horse's eyes show stress better than any other assessments. Even though you may have learned that a particular training method will get a desired result, you must always question whether your horse is adapting well to a program. If the eyes look stressed or worried, then rethink what you are doing with your horse. We can borrow an expression often used when evaluating shelter animals that relates to the "light" in the animal's eye. A horse who is socially engaged with you or in what he is doing should have a "light" in his eye. When there is no "light" or the eye becomes "dead"—a term I have illustrated before in this book—then there is most often pain or depression (figs. 11.4 A–D).

★ **TIP:** / The eyes show stress better than any other facial feature as they can indicate feelings of worry and pain consistent with other species. /

_ Play or Pain? When Horses "Act Out"

What we might identify as "play" or "naughtiness" can be a sign of stress due to pain. The same behavior can have different causes. A horse that is feeling good may gleefully try and nip or take a bite out of his handler in the same fashion he might nip or bite a horse friend as an invitation to run and play. Male horses spend much of their free time engaged in mock fighting and trying to "one up" each other in nature, so when they are confined to domestic life, the behaviors do not always change. Mares may not be as orally playful in the wild, but in domestic life where they do not have to worry about foals and saving energy to eat and get to water, they also will express more social play behaviors.

The best way to determine whether a horse who bites or acts out is playing or is in pain is to look at the horse's facial expression. As mentioned already, the eyes tell us a lot: A playful horse's eyes will be bright and full of energy. His ears will move back and forth, reacting to his handler's reception of the invitation to play. He will often look proud of himself if he is able to grab a "nip" or act out when a handler is not paying attention. In contrast, a horse in pain or worry will show a dull eye, worry lines, tight lips, and tense jaw muscles. His ears are often strongly back and not relaxed.

So, before you reprimand a horse, pay attention to the signs of play versus pain and try to determine the cause of the behavior (figs. 11.5 A & B).

> **❝**
>
> **If you allow horses to be happy and enjoy what they are doing, they are less likely to be injured or ill.**
>
> Dr. Scott Swerdlin, DVM,
> President of Palm Beach
> Equine Clinic

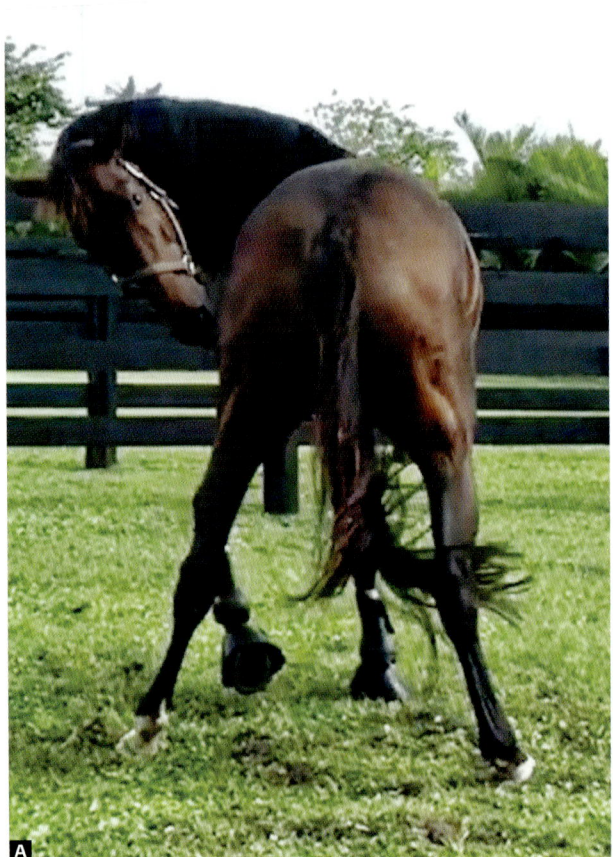

[11.5 A & B] The mare in **(A)** came to the barn labeled "dangerous." She displayed unusual behavior in the paddock. She would spin around, biting herself on her sides. Rather than being indicative of "play" behavior, this symptom, and others (like bolting or rearing for no apparent reason), are often associated with hormonal pain in mares.

In **(B)**, Coco leaps in the air and kicks out as she plays in the paddock with me. I would not let her in my space as she challenged me, and she is expressing her opinion in a joyful way. Safely playing with your horse helps develop a strong social bond as well as allowing you to observe any unusual behaviors.

Punishing a horse who is fearful or in pain will only create more worry and pain. If a horse is playing, then using your voice strongly to show disapproval is often all you need to alert your horse that his behavior is not acceptable. By far the best solution is to give your horse ample opportunity to play safely, preferably in a paddock or pasture where he can run, buck, and express himself.

Other Signs of Stress in Horses

Common signs of stress observed in horses may vary depending upon breed, temperament, discipline and situation. As we have discussed, it can be individually context-specific. While a little bit of acute (short-term) stress can help a horse learn and adapt, chronic

[11.6] This mare is sucking in and tightening her lips as a way to deal with worry. Her eye is focused on what is going on behind her, as are her ears. Her lip-sucking is her way of dealing with stress.

(long-term) or too much stress can cause harm. For example, a horse becoming slightly stressed before competition may be normal as long as the stress dissipates once the competition begins. Or a horse experiencing new stimuli might spook and run at first, but then learn the object or situation is not going to harm him and relax.

I have indicated that expressions in the eyes and "bad" behavior can be signs of stress. In addition, one or more of the following visible signs may be present:

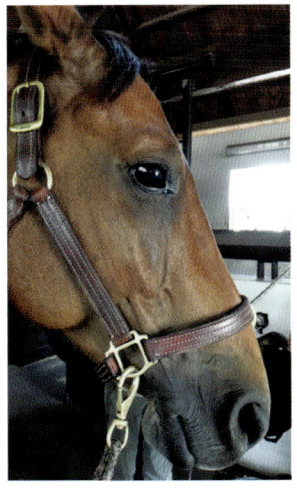

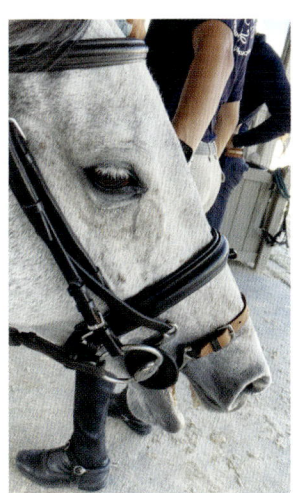

[11.7] This horse show signs of stress in his eye, tense neck and head muscles, open mouth, and flared nostril with white discharge.

[11.8] Opening the mouth and grinding the teeth can reflect discomfort or worry. Keep in mind you must look at the whole horse when considering possible stressors—male horses often will be very "mouthy" as a way to deal with insecurity or a lack of confidence, while mares who use their mouths usually are worried or in pain.

- Reduced eye blinking and increased eye flutter.

- Tight lips (bottom lip may be wrinkled—fig. 11.6).

- Open mouth, trying to avoid pressure from the bit, noseband, or reins (fig. 11.7).

- Licking mouth due to dryness caused by stress.

- Grinding teeth (fig. 11.8).

- Tense facial muscles in jaw area.

- Ear muscles tight; ears may be held tensely and slightly back, or flat back.

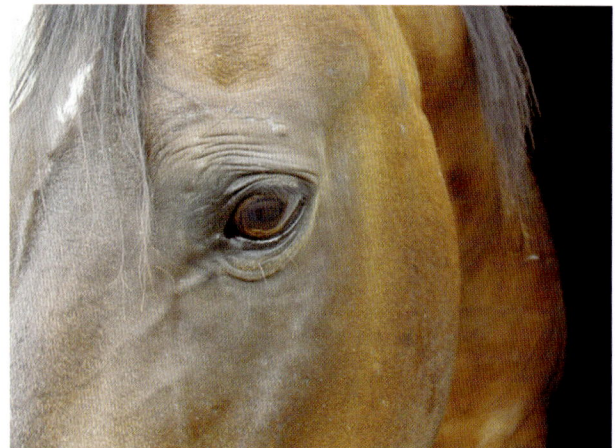

← **[11.9]** Notice the wrinkles above and below this eye. While this horse is relatively calm, his expression shows worry, which could be from pain or from anxiety. Horses who "internalize" their stress often have wrinkles in their eyelids.

- Wrinkled eyelids (fig. 11.9).

- Tense poll and neck muscles.

- Tail clamped or swishing quickly as if irritated (mares may do this more).

- Tense muscles throughout body, especially drawn up in abdomen and across ribs.

- Head-tossing or hyperflexing to avoid pressure.

- Rapid, shallow breathing or holding breath, even when not exercising hard.

↑ **[11.10]** The whites of the horse's eyes are showing along with flared nostrils and an open mouth. The neck is tight, and the muscles are braced. This horse is experiencing a high degree of stress. People often use tack and training devices in an effort to get a horse to put his head down, but a horse with discomfort and tight resistive muscles cannot physically do what he is asked. This alone can elevate stress.

- Worried sweat (suddenly and profusely in excess of normal exercise).

- Elevated heart rate (in excess of normal exercise).

- Flared nostrils; redness of nasal tissue and around eye (fig. 11.10).

- Increased, frequent defecation or diarrhea.

- Spooking, bolting, running with no regard for safety (fig. 11.11).

- Stomping, rearing, refusing to go forward (fig. 11.12).

- "Freezing" or "shutting down"; unreactive to stimuli.

- Backing up or other repetitive movement of learned behaviors in an effort to avoid.

Tools for Identifying Stress in Horses

As equine welfare gains momentum in the academic world, a number of useful tools are evolving to help identify horse emotions, feelings, and thinking, with practical uses in the area of stress and pain management. As more data is correlated, these tools will become very useful in assisting in

← [11.11] This horse was afraid of just about everything (like this dead palm leaf) because he had never been exposed to new situations or objects growing up. His level of fear caused him to get hurt frequently as he had no awareness of his size or of space. By working with him in a safe place where he was allowed to express his worry, he quickly learned to enjoy exploring new objects.

→ [11.12] This horse is rearing and refusing to go forward, his eyes, ears, mouth all saying, *I don't want to!* There is always a reason for a horse to refuse to do something, and it is most often associated with a legitimate need.

equine welfare across professions and disciplines. That said, however, there is no substitute for spending time with your horse and getting to know his idiosyncrasies, because much of horse behavior is learned and some horses misrepresent body language.

Horse Grimace Scale (HGS)

Dr. Emanuela Dalla Costa and a team of colleagues from University of Milan, Newcastle University, and the University of British Colombia set out to develop a tool to help identify pain from castration in former stallions. The *Horse Grimace Scale* was introduced in 2013 and is continually being improved. So far it has shown to be a useful tool for veterinarians and equine professionals, but seems to require training in order to accurately assess horses. The system relies on observers being able to identify the absence or presence of six *facial action units* considered to represent pain or discomfort response in horses:

1. Ears stiffly backward (not relaxed, holding tension).

2. Orbital tightening (tension and wrinkles above the eyelid).

3. Tension above the eye (often seen as wrinkled).

4. Tense chewing muscles.

5. Strained mouth and tight lips (often showing wrinkles).

6. Strained nostrils.

The scale uses pictures to help observers identify degrees of stress, allowing a coding system to be implemented (you can find it online—I've included a link via the QR code on this page). While still somewhat subjective, experienced horse people seem to be fairly consistent in how they score horses.

Learn more

★ **TIP:** / Know your horse. There is no substitute for spending time with your horse, as no system is as accurate in predicting stress and behavioral changes as a good horseman. /

Equine Facial Action Coding System (EquiFACS)

The Facial Action Coding System (FACS) was originally developed by Carl-Herman Hjortsjo in 1970 and has been applied to other animals since that time. Jen Wathan, Ann Burrows, Bridget Waller, and Karen McComb published the Equine Facial Coding System (EquiFACS) in 2015. The EquiFACS is based on equine facial movements associated with different social contexts and directed at gaining a better understanding of the emotional and cognitive states of being in horses. Using high-speed video, facial expressions can be captured and studied.

A more detailed system than the HGS, the EquiFACs uses 16 *action units (AUs)* and 11 *action descriptors (ADs)*, of which four are related to the ears. While the AUs represent the contraction of a particular muscle or muscle group, the ADs describe a general movement of either deep muscles or undetermined muscle movement or both. Those trained in reading the system show a high degree of consistency.

Learn more

Equine Comfort Assessment Scale

Colorado State University has developed a scale based on whole body behavioral observations to assess the comfort and well-being of horses primarily in a clinical practice. The system is evolving as other researchers give data and input. This scale will help veterinarians and others identify pain through abnormal behaviors and body language in horses. In addition to facial expressions, this system looks at "stalled" behaviors, such as pacing, head-shaking, biting, pawing, shifting weight, and rocking (as just a few examples).

Learn more

★ **TIP:** / As social mammals, horses and humans can experience the same physiological influences of stress. /

Equine Mental Health Assessment

Early on in my career it became clear to me that many of the behavioral issues, as well as physical issues, that horses developed stemmed from basic comfort and safety needs—not unlike children. Being able to assess the horse's mental health could give me insight into the causes. During the 1980s and 1990s, I taught a number of "Equine Behavioral Therapy" courses, primarily to veterinarians, at the same time I was brokering sport horses in California. I was also involved with a team of open-minded veterinarians (including Dr. Duncan Peters, who wrote the Introduction to Part 3 of this book—see p. 366) in developing a pre-purchase exam that would identify many factors about a horse, including his suitability for an intended use, which, of course, not only looked at the horse's physical ability but his mental ability, as well. I wanted a simple system that could help people assess horses' mental health *before* purchasing. That system has evolved over the years and no doubt will continue to do so.

After evaluating hundreds of horses from various age groups, breeds, and disciplines, one skill in horses stood out—their ability to interact socially with people. *Social interaction or lack of social interaction is a primary indicator of mental health in horses.* You could say the same of people or any other social species. Horses who have normal interactions with other horses but who choose to avoid humans have clearly demonstrated human-induced stress or have not learned how to interact with people. Curiosity and social interest are present in almost all horses and are the core of functional social relationships among and between species. Horses who do not demonstrate social interaction at any level with humans tend to maintain a high level of stress, and horses who are mentally stressed are much more likely to have physical issues.

Other associative factors (some that we have discussed in previous chapters, and some that we will look at in later pages), including temperament, genetics, environment, feed, training, history, and learned behaviors are also accounted for in this assessment. It has proven to be a simple initial evaluation for equine professionals to quickly identify human-induced stress and possible welfare issues.

The assessment uses a rating scale from "normal" behavior with little or no stress to "abnormal behavior" with moderate to high level of stress. Because this assessment, similar to the SAICC Evaluation (see p. 224), requires detailed training to complete, I have only outlined the parameters to help you determine if your horse is "socially normal" with low mental health stress or not. It is my hope that, in the future, veterinarians will be trained in mental health care for horses and will be able to help owners identify issues *before* they turn into physical problems.

Recognize that horses who have had little or no experience with humans, such as captured wild horses, many initially avoid social interaction. But once horses are acclimated to being with humans, they should demonstrate healthy social interaction skills. If not, then further assessment investigating the reasons why the horse does not want to interact needs to be conducted. (Most often, the cause is pain or trauma from training or human handling.) Note that if your horse is usually socially pleasant and you suddenly experience abnormal social interaction, you should immediately look for pain and a possible cause before doing anything else.

★ **TIP:** / Never get on a horse who will not make eye contact with you or at the very least smell you as a greeting. /

What follows is a quick assessment that assumes you are working with a domestic horse who interacts with humans on a regular basis.

1. Does the horse show normal social interactions with people?

   - **Normal Responses:** Makes eye contact; curious about smelling and finding out who you are; shows interest in getting to know you.

   - **Abnormal Stress Responses:** Avoids eye contact; walks away or stands still with head turned away; may pin ears flat back; grinds teeth; or displays another stress-related behavior.

2. Will the horse allow you to walk up to him, touch him, and put a halter on him?

   - **Normal Responses:** Stands still and allows touch or the halter put on; assists by putting his nose into the halter.

   - **Abnormal Stress Responses:** Avoids and moves away; sticks head up high; avoids eye contact; pins ears; grinds teeth; glares; tightens jaw; tries to snap or bite; tightens lips; tenses all muscles; starts to shake.

3. Does the horse want to walk with you? Does he resist following you?

   - **Normal Responses:** Willingly follows you out of stall or paddock, ears forward and curious where you are going; waits for you to walk first and

> **Curiosity and social interest are present in almost all horses and are the core of functional social relationships among and between species.**

then keeps himself next to you with his eye slightly behind your eye; nose bumps your hand to show connection; carries head up if eager and down if relaxed.

- **Abnormal Stress Responses:** Puts head up when trying to lead; will follow you but does not make eye contact and keeps ears slightly back to show worry; tension in jaw and lips and body may remain.

4. **Does the horse react to being groomed? Does he seem to enjoy touch?**

- **Normal Response:** Enjoys being touched and groomed; freely communicates likes and dislikes when touched (every horse may have some sore spots, so "normal" horses want you to find them and fix them); enjoys being scratched on the neck, withers, or other favorite spot; stands comfortably with posture that is relaxed and as "square" as is conformationally appropriate for the horse.

- **Abnormal Stressed Responses:** Remains tense even if standing quietly; tries to grab the rope or cross-ties to release worry; does not want to make eye contact; flinches when touched; tense muscles and fascia; does not release tension or seem to enjoy grooming.

5. **Does the horse accept being saddled?**

- **Normal Responses:** Stands quietly; is inquisitive about the saddle and other tack; wants to smell and interact with you.

- **Abnormal Stressed Responses:** Head comes up or turns away at sight or sound of saddle; eye becomes worried; becomes mouthy; grinds teeth; bites the

Research in the Works

While I have conducted a number of studies related to equine behavior through the years in an effort to find out what causes stress in horses and riders relative to their personality, history, breed, and training, there now are a number of other interesting projects in development. Research by Etalon Diagnostics seeks to connect temperament to genetics and a "Citizen Science" project headed up by the University of Sydney, the Equine Behavior Assessment and Research Questionnaire (E-BARQ), will assist riders, trainers, and other horse professionals in ensuring better equine welfare in management and training as more people participate and a stronger data base is developed.

Learn more

air; moves the jaw back and forth; stiffens body; pins ears flat back; stomps leg; kicks out; moves away.

6. **Does the horse accept being bridled?**

   - **Normal Responses:** Drops head; opens mouth eagerly; assists in getting the bridle on his head; shows eagerness to be interacting; may object at first, if the bridle or bit is not comfortable, but continues to interact and try and communicate with you.

   - **Abnormal Stressed Response:** Shows agitation or worry even seeing the bridle; turns head away; does not want to interact or look at you; shuts down; tightens mouth and jaw; tolerates having the bridle put on, but does not relax jaw or neck muscles.

7. **Does the horse accept a rider getting on?**

   - **Normal Responses:** Looks alert and eager; turns around to interact with you, as if to say, "Let's get going"; stands quietly; adjusts posture to support rider weight.

   - **Abnormal Stressed Responses:** Worried look in eye; stiffens body; tucks or sticks out tail to brace; raises head or hyperflexes neck; "scoots" or takes off when rider swings leg over or starts to get on; braces until told what to do.

8. **Does the horse's behavior and expression change during riding with rider movements, contact, or voice?**

   - **Normal Responses:** Looks relaxed in the eye, lips, and jaw muscles; listens to you with ears that rotate back and forth; interacts with you; breathes regularly; is responsive to ridden communication. (Note: Behavioral reactions to riding vary according to temperament, age, and degree of training.)

   - **Abnormal Stress Responses:** Holds breath; has tight lips; eyes look worried (wrinkles in brow); anticipates requests and overtries to please

> **Social interaction or lack of social interaction is a primary indicator of mental health in horses.**

STORY FROM THE FIELD: //

Giving Markley a Voice

Markley came into a sales barn where I worked. He was a fancy black hunter and had shown a lot for his age, but had been sold several times because he had become scared at horse shows. He never stuck his head out of his stall, and when approached by a person, he would turn to the corner and try and hide his head. When he was touched, he flinched and started shaking (commonly seen in horses who have been given a number of injections). He would not make eye contact, and if offered a carrot, he was unsure about even investigating it. (It took several days at the sales barn to get him to want to socialize and make eye contact with people.) He did not like to be groomed and seemed worried the whole time. When a rider was mounting, Markley looked scared. (We allowed him to move if he needed to until he felt safe standing still.)

Markley's whole body was assessed. He had several ribs out of place, and a sore back and neck. No one ever had addressed his physical discomfort, which had led to mental worry. Being a sensitive horse, he worried about the pain he felt as he had not been allowed to express his fear and was required to keep jumping. Because he bolted from pain, he had been continuously longed on a circle, which only created more pain.

Within a few weeks of behavioral and bodywork sessions, Markley was a relaxed, sweet, and very social horse. We had:

1. Allowed him to express his worry.
2. Addressed his discomfort.
3. Rewarded him for being social with people and interacting.
4. Listened to his concerns by giving him a voice.

↑ [11.13] With Markley after weeks of evaluation, bodywork, and training helped him learn to enjoy social interactions.

by overreacting to cues; resistant to moving forward; shakes head; pulls on reins; swishes tail; braces muscles; tenses back; bolts or spooks; grinds teeth; gets quick.

_ Analyzing Results

Make note of abnormal responses and follow up with further questioning of trainers or former owners with a familiarity with the horse. If a number of abnormal responses are noted, spend one-on-one time with the horse, assessing his body for pain and seeking spots he likes to have scratched. Hand-grazing or spending time hanging out with him in a paddock can reveal more clues to why he may have anti-social behavior with people.

No matter the cause, paying attention to your horse's comfort level and willingness to interact with people will give insight into his temperament, personality, history, and the best ways you can help to manage his stress going forward. I have been riding for over 60 years, and I can honestly say that I have never gotten on a horse who did not want me on him; therefore, I have never been intentionally pitched or spun off a horse. Friends want to protect their friends, regardless of species, and riders who understand and look out for their horses' level of stress are much less likely to suffer from falls and horseback riding injuries.

_ Assessment Caveat

When using any of the assessment tools I have shared, or others, remember that each horse is different—you must take into account the individual personality. Mares, in particular, use their ears to talk, and while they may pin them back when you enter their space, they are simply acknowledging you and allowing you to greet them. Many people might interpret this as avoidance behavior or a stress response, but it is not. Similarly, the horse who turns his hindquarters toward you and then looks at you, is not avoiding you, but probably instead asking for a rump scratch. Body language must always be correlated with the look in the horse's eye—as we've discussed, domestic horses can learn confusing body language if they did not grow up in a functional herd. But remember, their feelings are expressed in their eyes.

[**12.1**] We say "the horse comes first" in equestrian disciplines, but it sometimes seems otherwise. This dressage horse scored well, despite displaying discomfort and stress from the rider's spurs and hands. The open mouth, the tension in the ears, and the stiffness in the neck muscles were overlooked by judges who only saw the horse's movement and obedience.

CHAPTER

[12]

MANAGING STRESS IN HORSES

> **The primary program in a horse's 'hard drive' is to form functional relationships, grow social networks, and maintain a cohesive group for the welfare of the herd.**

[12]

First...Manage *Your* Stress

As we have already discussed, stress is necessary for survival and helps species adapt and evolve. It is the body's natural trigger for the "fight-or-flight" response in both horses and humans, flooding our systems with hormones, increasing blood pressure, heightening muscle preparedness, causing sweating, and creating a high state of alertness to help us "get to safety." But what happens when you cannot "get to safety" or appropriately process the stress and the stressors continue to build?

More and more research shows over 90 percent of all human diseases are caused by stress. This of course makes sense, as too much stress causes bodies to break down on multiple levels. People have found that being around horses can lower human stress levels. The developing field of equine-facilitated therapy has evolved from people realizing how being with horses can make them "feel better." But while there are numerous studies supporting the use of horses to help humans de-stress, only a few have investigated the effects of human stress on horses.

In relationships, whether human-human or horse-human, it is often difficult to separate the emotional feelings of the interaction. If you had an emotionally stressed creature sitting on your back, telling what to do, you would likely react in some way. Horses do. They either *internalize* and try to balance the stress, or *externalize* and try to avoid the rider's stress. On the other end of the spectrum, studies that allowed horse to have a "choice" whether to interact with people or not show that many interactions with humans can be good for and desired by horses, too. Perhaps, as with the wild foals bedding down next to me on Green Mountain, Wyoming, at night (see p. 41), when horses sense a relaxed state of being, they respond likewise.

To try to release the stress common to modern-day humans and find that relaxed state that is as good for horses as it is for you, put your cell phone away, maybe put on relaxing music you and your horse will enjoy, and try these three easy steps:

1. Take a deep breath and relax your body. Slow your breathing down and follow your breath, imagining it moving through your body, relaxing every cell and muscle. Do this three or four times.

2. Empty your mind. Let go of your thoughts—like clouds in the sky, you can watch them but do not get attached to them. Focus on your breathing.

3. Bring your awareness to how much you appreciate horses. Smile and feel the fondness you have for being able to interact with your horses. Feel your energy fill up with happiness and allow your heart's energy field to expand.

One of the variables in studies investigating whether horses become stressed when being asked to learn new things was the riders. A recent study conducted in Finland showed that horses with long-term relationships with their riders were the most confident and least stressed when faced with novel objects and experiences. This supports the importance of the horse's relationship with their human. You do not need research to tell you that horse-rider relationships vary from person to person, but with the number of horses sold based on breeding and talent alone, little, if any, attention is paid to the temperament of the horse and whether he will match well with a particular individual. Since all riders are not the same, horses react differently, depending which one is on their back. One rider might make a quiet horse worry, while another rider might relax the spookiest.

The key, then, to helping horses managing stress, is for people to manage *their* stress, and thus give their horses a feeling of safety and comfort.

Learn more

Take a moment and think about it—would you want a mentally distracted, worried person sitting on your back, restraining you from being able to think clearly or get away? What does it feel like for a horse who has evolved a sensitive system for assessing the emotional states of herdmates and being able to react instantly to changes in energy that might signify threats to learn how to balance a stressed human and still try to relax and perform?

So the next time you go to brush your horse, and certainly before you ever get on, take a moment to use the three steps I've outlined here and center yourself, relax, breathe, and become mindful of your thoughts and feelings. Because the better you manage and understand yourself, the better person you will be for your horse.

★ **TIP:** / Manage your own stress before hanging out with your horse./

Understand How Horses Think and Learn

In order to better understand how to reduce stress in horses during any human-horse activity, it is worth having a basic sense of how equine brains process and learn. In Part One, I explained how the equine brain has adapted well to movement and interpreting sensory input. Now I'd like to consider neurophysiology and how it relates to stress and training.

We must keep in mind the natural ecological and psychological evolution of cognition and awareness in horses. They have developed to have successful relationships

STORY FROM THE FIELD: //

Skills People Need to Have to Be with Horses

In my many years of study and time spent in different realms of the equine industry, there are certain qualities I have identified as being necessary if a human desires a positive relationship with a horse.

1. Willingness to slow down and spend time just "being with" horses
2. Understanding and empathy
3. Ability to manage emotions and energy
4. Kindness
5. Patience
6. Good listening skills
7. A sense of humor
8. Awareness of a horse's priorities
9. Openness to learn
10. Honesty
11. Clear communication
12. Sensitivity
13. Leadership

↓ [12.2] Kiki spends time just "hanging out" with her two-year-old stallion in the arena, teaching him how to just relax in her company. This lesson will prove beneficial as he gets older because he will recognize her quiet energy and know that it means comfort and safety. It is important to teach horses to just "be" with you and not have to "do" anything.

with other horses, and while they can transfer their social skills to people, horses' innate behaviors relate best to how they are wired for natural relationships of mother to foal, foal to foal, mare to mare, mare to stallion, stallion to foal, and stallion to stallion. While the relationships for a foal may change and develop over his life, the primary program in a horse's "hard drive" is to form functional relationships, grow social networks, and maintain a cohesive group for the welfare of the herd. Horses have developed high emotional intelligence to do this, and while they may not have deductive reasoning and think like humans, they have high awareness and good cognition. Horses do not think "individually" as much as they think with a "group mind." They have developed a high level of emotional awareness and responsiveness rather than complex thinking, like humans. Thus, horses have very well-adapted brains for what is important for horses.

★ **TIP:** / Equine welfare can be directly related equine cognition, learning and emotional well-being. /

Science does not yet fully understand how a horse's brain integrates and functions. And while we can make comparisons with humans, we need to be continually learning from the horses themselves. Currently there is research investigating the relationship between horses' emotional development and how they learn, which has a direct impact on horse welfare. The findings are similar to what we see with children—if a child or horse does not get the needed emotional support and education to succeed in life, they are likely to suffer. Horses who do not get the emotional support to learn the things they need to learn to adapt to humans and domestic life often get mistreated and "thrown away" as "dangerous," "stupid," "talentless," or "worthless." Better understanding the emotional needs of horses can help humans manage and train them better, and thus reduce welfare issues.

Learn more

★ **TIP:** / Horses who are happy learn better and are less likely to be injured or ill. /

How the Horse Brain Operates

For the purposes of thinking about how the brain and nervous system operate, let's divide it into parts. (Note that I am not a brain scientist.

Learn more

I recommend the book *Horse Brain, Human Brain* by Janet Jones, PhD, for an accurate and in-depth discussion of equine brain function as it compares to ours.) Horses have a *brain, spinal cord, spinal cord reflexes,* and *peripheral nervous system.* Their brains, as in all mammals, can be divided anatomically into three sections: *hindbrain, midbrain,* and *forebrain.* These areas all interrelate to some degree, but specific nerve functions do associate with certain parts.

- The **hindbrain** consists of the *brainstem*, in which all nerve fibers from the spinal cord pass. The brainstem is composed of the *medulla* and the *pons*, which contain groups of nerve cells that automatically control breathing rhythm, coughing, blood pressure, and heart rate. The primary part is the *cerebellum,* which accounts for about one third of the horse's brain. It coordinates movement, and receives sensory information about joints, muscles, balance, proprioception, and head and eye movements.

★ **TIP:** / Once a movement is learned and stored in the (hindbrain) cerebellum, little thought is required to initiate the movement. /

- The **midbrain** connects the *brainstem* (hindbrain) to the higher centers of the brain and is involved in major body systems, including sleep/consciousness and temperature regulation. It is the center for visual and auditory reflexes, and for voluntary movement.

- The **forebrain** is primarily made up of the *cerebrum*, which is divided into two hemispheres and is responsible for perception, emotions, voluntary movement, and most learning. There are twelve *cranial nerves* in horses that regulate smell, taste, vision, chewing, and tongue movement. As previously mentioned, the olfactory nerves "win" as they have the fastest path to the brain by passing directly to the *limbic system* of the brain without having to go through the *thalamus.* (The limbic system is associated with memory, emotions, feelings, and endocrine functions.) The olfactory nerves are also the only nerves that relate to the side of the horse on which they originate. Other cranial nerves associated with sensory input and facial muscle movement must be filtered through the thalamus and *hypothalamus*, where emotions and conscious perception, sleep, focusing, and some awareness integrate. The *cerebral cortex* is rather thin but accounts for almost half the size of the cerebrum because it is folded. Neural pathways send information from touch-sensitive receptors in the skin, through the brainstem and thalamus, providing a whole-body view to the somatosensory part of the cerebral cortex. Muscle movements are directly related to neurons in the *motor cortex,* and because the horse uses his prehensile lips and muzzle for tactile activities, these areas represent a large part of the brain. The cerebral cortex is also associated with thinking and awareness.

★ **TIP:** / The forebrain of the horse processes specific sensory data and initiates action relative to the horse's safety and survival. /

★ **TIP:** / Horses integrate sensory input quickly, allowing for fast physical reactions and decisions based on feelings. /

Unlike humans, horses keep it simple—sensory input comes in, and except for olfactory, which is even more direct, travels through the thalamus, which sends a message to the *basal ganglia* to ready the body for movement. Horse brains are designed for assessing feelings, primarily those related to fear, and quickly reacting without thinking about it. Although they do not have the capacity to analyze the way humans do, horses can think, and they can learn to override stimuli that trigger fear and panic, but we must always remember they have adapted to be sensing and feeling creatures rather than critical thinkers. Understanding this can help us reduce equine stress by acknowledging the differences between their brain function and perception and our own.

Equine Learning Styles

Understanding the various types of learning styles will also help you limit stress in training by better understanding what you are doing and what your horse can understand. Most creatures use a variety of styles of learning, and you will also see that the types of learning overlap to some degree. Think about how your horse might learn best alone, in the company of other horses, and from humans.

★ **TIP:** / Horses use a variety of ways to learn, based on their temperament, experiences, mental ability, motivation, and the social context./

Common Types of Learning

- **Associative Learning** Broad field of learning that associates a stimulus to a behavior, whether conscious or unconscious.

- **Signal Learning** A form of learning to associate sight, sound, smell, or touch with a perceived or expected action, where once learned, the horse has little thought involved in the action. Can involve using a "bridge," such as a clicker in clicker training, touching a specific point of the horse to signal an action, a half-halt to mean, "Listen for a command," or a particular aromatherapy scent to mean "Relax." This is a form of conditioning.

Learning Theory

There has been a lot of focus recently on "learning theory" in horses. First, let me say it is not "theory." Horses learned from each other for thousands of years before humans stepped into the picture, so do not confuse "learning" with "learning theory." And while there is debate over the language, ethics, and methods employed in horse training and learning theory is claimed to be a way to improve training welfare, remember to *just pay attention to your horse*. Do not get stuck in any one type of training or system. The more you observe your horse, the easier it will be to determine his learning style and the best ways to prevent stress in training. Because learning theory is becoming more popular as a way to blend some science into traditional horse training methods, I have outlined the seven basic principles below:

1. Use learning theory appropriately.

2. Train easy-to-discriminate signals.

3. Train and subsequently elicit responses singularly.

4. Train only one response per signal.

5. Train all responses to be initiated and subsequently completed within a consistent structure.

6. Train persistence of current operantly conditioned responses.

7. Avoid and disassociate flight responses.

Learn more

- **Operant Learning** Also called *operant conditioning* or *instrumental conditioning*, this is another form of associative learning that uses *positive* (addition of stimuli) or *negative reinforcement* (removal of stimuli) to punish or reward a behavior. Most riding involves some form of operant learning—pressure or lack of pressure and restraint or release, for example.

- **Perceptual Learning** This learning style relates to how experience can change perception and behavior—for example, a horse who is afraid of a novel object, but then receives a positive reward and now associates the object with relaxation. Or the horse who, once he learns how to jump, gets excited whenever he sees a jump. Can also be considered a form of *discrimination learning* (see below).

- **Discrimination Learning** Learning to differentiate sights, sounds, smells, tastes, and touch. Horses are great at sensory discrimination. The trailer that smells odd, but looks the same, may trigger a discriminatory response of refusing to go in the trailer.

- **Motor Learning** This refers to changes made in the nervous system as a result of repeated experiences. An example would be the precise muscle movements of a dressage horse that, once learned, a slight stimulus from a rider's leg can evoke. Most of riding involves motor learning.

- **Observational Learning** Learning from watching others. Can involve mimicking behavior or initiating actions similar to what has been observed.

- **Conceptual Learning** This is not a very useful learning style for horses, but humans like to "test" other animals. It involves being able to use some form of *deductive reasoning*, such as being able to select square images versus round images. There's not much motivation here for a horse, and it will probably only appeal to really smart horses who are bored in your barn.

- **Non-Associative Learning** This is when there is no apparent stimulus for the behavior, such as *desensitization* and *sensitization.*

    - *Desensitization,* also called *habituation,* is when behavior changes and an animal no longer responds to certain stimuli (see also p. 239). This is often used with scary objects or situations. Remember, though: If the

> **Understanding the various types of equine learning styles will help you limit stress in training.**

horse learns not to show stress, then he internalizes stress. Desensitized horses are obedient because their habituation and operant training has produced the behavior without allowing the horse to feel safe.

- *Sensitization* As we've discussed, horses are great associative learners, so a negative stimulus can create a heightened reaction toward even the anticipation of the stimulus trigger. For example, a horse who was beaten with a whip may overreact at even the sight of the whip, and on the opposite side of the spectrum, a horse given peppermint candies will start salivating just hearing the crinkle of a plastic wrapper.

★ **TIP:** / While horses can continue learning their entire lives, horses in nature learn almost everything they need to know to survive and be socially accepted between birth and two years of age, as their brains are very active and open to learning during that time. /

Best Practices for Applying Equine Learning Styles to Training Models

So, what does all this brain function and learning style stuff mean? In a nutshell: just keep it simple, clear, consistent, and positive. As we learned in the way their brains function, horses are not complex thinkers, and they do better learning one task in a short period of time with positive reinforcement, than being chased around a round pen or ridden for 40 minutes until they figure out what a human wants them to do. Think: "Simple-Clear-Short-Reward-Repeat."

Most training programs incorporate a variety of the types of learning I summarized earlier, so it is helpful to understand what kind of learning you are employing and how your horse is perceiving the training. This will allow you to minimize stress and enhance training experience for your horse.

Also, keep learning *fun* for your horse. Horses by nature are social creatures, and when they do not have to worry about finding food and water or being stalked by predators, then they have time for play, learning, and socializing. Being with your horse should not always be about work or training. To form fun, trusting relationships you need to do things your horse likes to do and allow him to make some decisions in his interactions with you (figs. 12.3 A–D). Horses like to have choices and they enjoy exploring.

→ **[12.3 A–D]** Keep learning fun and interesting for your horse with some "paddock play." Coco stands over a pole to learn better balance and proprioception **(A)**, then thinks I placed the poles the wrong way…I had moved them to see if she would notice **(B)**! She moves the soft pole back to where she thinks it belongs **(C & D)**.

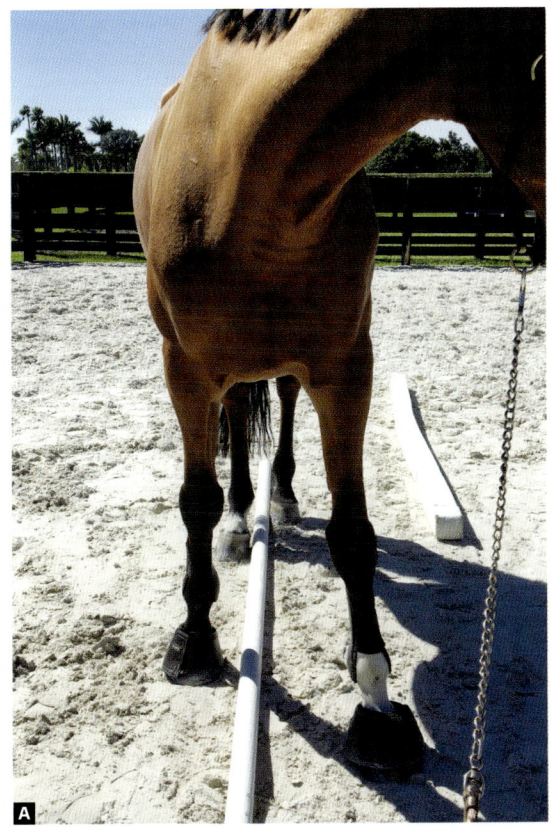

PART TWO / **CHAPTER 12** / Managing Stress in Horses

★ **TIP:** / **Horses learn by associating sensory input with experiences. When they are good experiences, horses gain confidence and cooperation. When they are perceived as fearful or uncomfortable experiences, horses may have both mental and physical stress.** /

Here are my best recommendations for educating and training horses while minimizing their stress:

- Keep it simple and short. Horses learn best when you have one thing you are trying to teach them, such as "stand still" or "move."

- Reward instantly when the horse accomplishes what you have asked.

- Be patient if the horse does not understand. Repeat the request. If your horse makes a mistake, it is either because you are not asking or presenting it in a way he can understand, or because he has pain or worry related to what you are asking him to do. Try again or try a different way and check your horse's body for discomfort.

- While fear and pain can be motivating factors to learn, and they are incorporated into a number of training models, they do not build positive relationships with humans. Positive rewards—gentle voice, pat on the withers or neck, release of pressure—can all trigger long-lasting positive learning.

- Use a three-second "timeout" for rewarding a task well done so you can praise your horse and he has time to absorb the emotional good feelings. You can also use it when your horse has "lost it" and you need to "reboot." By stopping the action and giving your horse three seconds to just relax, erase, and start over you can prevent stress building up and teaching negative associations.

- Use consistency in training so you are predictable in your expectations with your horse. Horses are creatures of habit and adapt well once they learn something.

- Allow your horse to learn new tasks from watching other horses. Horses facing arenas, watching other horses perform or jump learn faster than if they had not been exposed to the activity. Group learning, where one horse follows a leader, can be very effective for all young horses or older horses in need of retraining.

- Limit restrictive tack and training devices until your horse feels relaxed and trusts you. Anything that makes the horse feel like he cannot escape if he needs to will create stress.

- Keep training fun and enjoyable for your horse!

Reducing Stress Related to Breeding

In nature, reproduction primarily occurs in the spring once a mare has had her foal, but it can be used for social reinforcement all year long between closely bonded stallions and mares. As you learned in Part One, offspring learn about breeding from watching the social etiquette of "equine courtship and copulation," and young stallions living in bachelor groups often practice mounting behavior on each other, long before they ever get a mare of their own. But in domestic life, breeding horses has become so sophisticated with artificial insemination (AI), cloning, and embryo transplants that horses no longer have any choice in who they reproduce with and whether they raise offspring. Unnatural circumstances around the breeding process can be stressors for many horses, reducing reproductive success and causing owners increased financial investment. The best breeding farms in the world still do not come close to nature's success rate (under normal natural conditions, virtually all mares who are covered conceive and have foals), but there are many things we can adapt from nature to help improve breeding practices and reduce stress in both mares and stallions.

Whether you are a farm manager, a veterinarian, or a mare caregiver, understanding natural reproductive behavior and biology and its interconnectedness to successful reproduction is key. Stressful situations, even in the wild, cause horses to abort and "absorb" foals, so becoming more aware of what causes stress before, during, and after breeding is vital. Here are some easy ways you can change aspects of the breeding process that are commonly stressful for the mare or the stallion.

★ **TIP:** / In breeding, use smell the way nature intended, since it is the primary trigger for reproductive readiness. /

- **Use Smell and Taste** Smell and taste of the opposite sex is one of the best stimuli for both mares and stallions. It can be used in various ways, even if the mare and stallion never meet.

AI TIP: Have the stallion's manure or a little shavings with wet urine shipped in a plastic bag to the breeder or mare owner. It is best if the mare can smell it for the first time while in a relaxed and familiar situation, then smell it again during the procedure for breeding. Smell recognition is very strong in horses, and this technique has allowed many difficult mares to settle.

Live Cover TIP: Allow the stallion and the mare to have the smell of each other for at least several days before breeding. When it is time to cover the mare, both horses will have smell recognition and feel more comfortable.

Semen Collection TIP: Use mare smell by placing a little urine of a mare who is ready on a cotton ball, and allow the stallion to smell while collecting. It will keep him focused.

★ **TIP:** / Reduce stress in mares shipped for breeding by including the smell of friends. /

"

Being with your horse should not always be about work or training.

- **Avoid Separation and Isolation** One of the most stressful situations for mares is to leave their friends and herd. Often mares may abort or reabsorb foals while shipping when they have been "sent out" for breeding on their own. This is especially a concern for timid maiden mares. Mares shipped with companions, especially other mares, are less likely to worry.

- **Limit Restraining Devices** Restraint is almost unnecessary if smell is used properly along with other preparations to make breeding pleasant. The use of twitches, lip chains, hobbles, and head-restraining devices can cause severe trauma for mares and decrease success.

- **Let Them Watch** Breeding activity is a very social occasion in a wild horse herd (see fig. 5.5, p. 94). All horses watch and interact. It is a learning time for young horses and often bachelor bands of young stallions will keep their distance but observe breeding activities. For domestic maiden stallions and maiden mares, it can be very helpful to allow them to watch the breeding process before their time. Much of horse behavior is learned through observation.

- **Give Positive Rewards** Breeding should be a positive experience for both mares and stallions. Even though horses may seek to breed based on instinct and hormone levels, they often do not understand what is happening and may become frightened or aggressive. Using positive rewards, like feeding carrots the first time a horse must come into the breeding shed or stall, can reduce stress when it is time to breed or collect semen and increase success. Massaging horses can also elicit a relaxation response and breeding readiness, taking away some of the stress.

★ **TIP:** / Ensure adequate exercise for both breeding stallions and mares. /

- **Include Exercise** Nature ensures horses have plenty of energy, and there is often a lot of "running around" during breeding proceedings. Stallions who are hard to handle or aggressive are often just stallions with built-up hormones who need to "run it off." Adequate exercise before breeding can stimulate confidence, increase circulation, and prepare the horse for a pleasant breeding procedure. Exercise can also help mares be more receptive.

★ **TIP:** / Regularly check a breeding horse's diet and nutrition, and allow free-choice vitamins and minerals if possible. /

- **Good Nutrition** Although most breeders are well aware of optimum nutrition for their horses, each horse—depending upon breed, the habitat and climate in which the horse lives, and performance discipline, if any—may experience a change in nutritional requirements. Allowing your horse to have "free-choice" nutritional supplements can take the guesswork out of feeding and allow you to monitor what your horse is consuming. Presented with options, horses have an excellent ability to seek out and find what they need nutritionally. Chemical contaminants go unnoticed until something goes wrong, and they affect hormones, which can affect breeding success. There are no restrictions on growing hay in contaminated areas, and often the lovely green hay has not only been overfertilized but grown in contaminated soil. Small quantities of various chemicals can influence the highly sensitive endocrine system of the horse, affecting reproduction. Test hay and water, or only buy from non-contaminated sources.

Improving Early "Foalhood" Education and Initial Training

Foals born in domestic environments may or may not experience a "normal" horse upbringing. Various training philosophies recommend everything from "imprinting" and putting a halter on their heads when they are a few hours old to leaving foals alone out in a pasture until they are weaned. The early foalhood life of your horse can have a major influence on how adaptable and stressed your horse might become as he matures and is exposed to various disciplines and lifestyles. Early positive education of foals is critical to help horses become socially adapted to domestic life.

★ **TIP:** / Early positive education of foals is critical to help horses become socially adapted to domestic life. /

Consider the Weaning Process

The best way to avoid weaning stress, as we have already touched upon, is to allow mares and foals to stay together in their own spaces until weaning, and then slowly separate them by allowing the foals to go back to their birth stalls and placing the moms within visual distance. This process offers a more natural weaning and is less stressful than sudden separation (fig. 12.4).

Although it is common on large breeding farms to house weanling foals together in a pasture, this is best done with an old "school mom" or "uncle" horse who can supervise and teach the foals about life. Foals learn social greetings and spatial respect. Respect for space is a key element for the foundation of education in young horses. Young horses spend much of the day "space-taking" from each other. This is a game, so realize play can be a great learning language with all horses, but especially young horses.

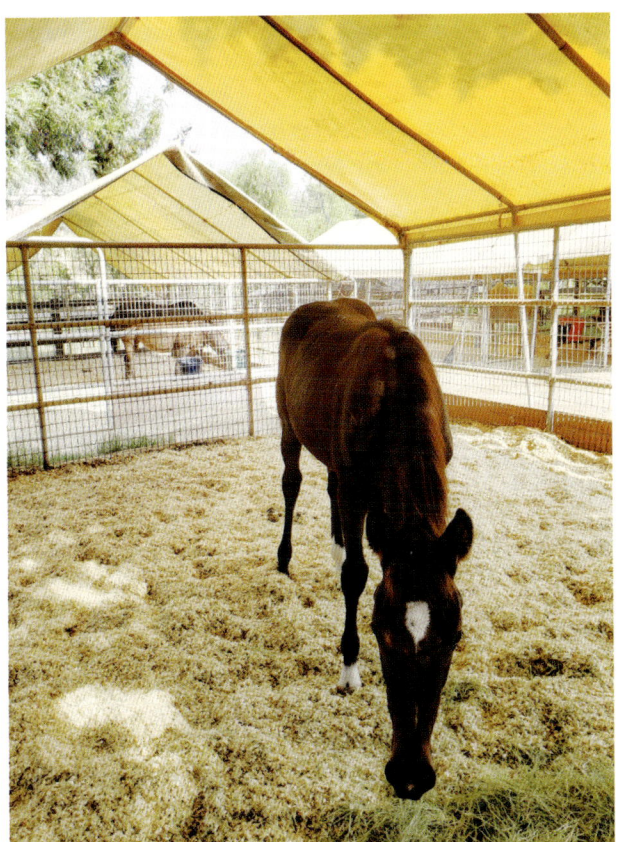

↑ [12.4] When weanlings can stay in familiar spaces and still see their moms, they experience little stress with the separation.

PART TWO / CHAPTER 12 / Managing Stress in Horses

★ **TIP:** / Foals and mares are less stressed by gradual separation that allows them to see each other. They can still eat and sleep near each other, watch each other, and socialize, but the foal cannot nurse. Foals should be old enough to naturally want to eat on their own. /

In the best situation, foals will also grow up around other functional horses that model positive interactions with people. Even wild horses learn to be social with humans when they see another "horse friend" developing a relationship with someone. There is a balance to be struck between fear and curiosity—too many negative associations and the horse will not want to be with people, while positive associations with people stimulated by high curiosity and low fear in the young horse will prove to develop long-lasting happy relationships with humans (fig. 12.5).

★ **TIP:** / When weaning foals as a group, place an adult "teacher horse" in the pasture to help supervise and show the foals needed skills. /

Consider Halter Training

Nature teaches horses to follow, so starting training using halters and lead ropes does make sense to horses. The young foal follows his mother, and his mother follows you on the lead rope. This is simple, but as foals grow, they may start challenging your space. Because temperament plays such a strong role in equine development, you have to assess the young horse's personality as he develops and design education to keep learning fun. You begin by allowing the foal's mother to help educate the youngster so he feels safe interacting with people.

To Imprint or Not to Imprint?

"Imprint training" foals has become a popular training model in order to make foals easier to handle as they grow up. But studies now show that imprint training done at birth at various times and a number of sessions did not result in better-behaved and less-reactive foals when tested at six months. While in theory, this training model might sound like a good idea, it actually can cause trauma to the foal and mother. Foals need the nurturing from their mothers to form functional social bonds, and they learn most of their early life lessons—like how to form positive relationships with people and let them do things to you—from their mothers. When the mare is comfortable with people, she teaches her foal to be comfortable with people.

With that said, imprint training may be an advantage when the mare is aggressive toward people or nervous around people, as she may pass her reactions on to her foal. Some mares do not want to be mothers and will not let their offspring nurse, in which case the human, by default, becomes the "mom." It is important not to jump to conclusions, however. Allow enough time to see how the mare and her foal bond before deciding to "imprint." The idea that a foal needs to have a halter put on his head right at birth, as some "imprint training" recommends, is interrupting the mare-foal bonding time, which is critical for the comfort and well-being of both.

Learn more

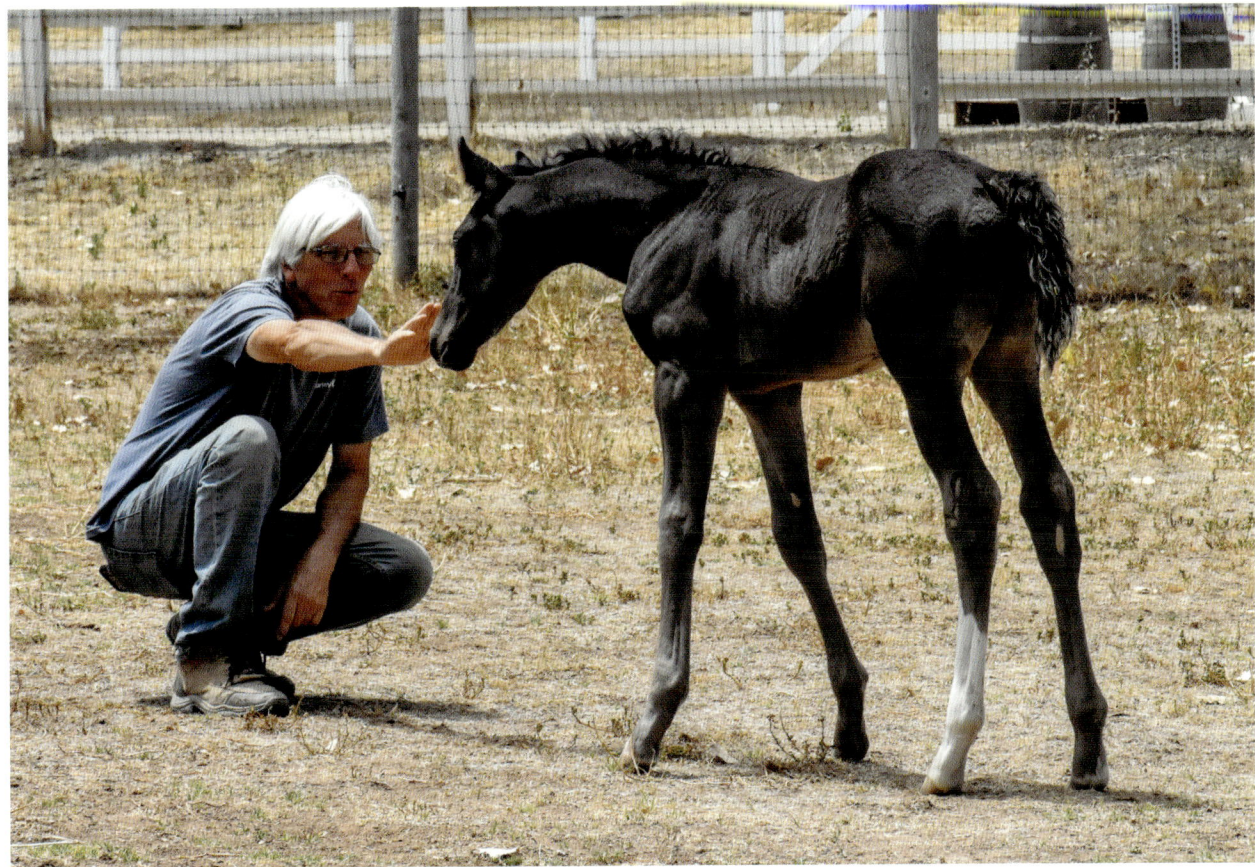

↑ **[12.5]** Foals brought up interacting with both people and their dams have very little stress when it comes time for training as they have already established social bonds with the humans in their lives. Wes Suddaby allows a foal to approach him to smell and get to know him, on the youngster's terms.

★ **TIP:** / Because temperament plays such a strong role in equine development, you must assess the young horse's personality as it develops and design education to keep learning fun (see pp. 148 and 244). /

★ **TIP:** / Older horses who did not have "normal" young horse development often will go through juvenile behaviors as they are allowed to communicate. /

Learning about ropes and halters can be a fun experience for foals—something interesting to play with. You can leave ropes lying on the ground near a foal to let him explore before you start using them. Although using soft rope halters and "butt ropes" might be needed if the environment is not suitable for educating the foal loose, note that putting a chain over a young horse's nose is surely going to cause problems. It is like starting a friendship by taping a child's mouth closed with something that hurts. The horse

[12.6] When practicing with the halter and lead rope, keep the youngster's attention with your positive thoughts and by engaging their active senses of smell and touch.

will not be able to focus on what he is learning if all he is thinking about is the discomfort on his nose—the most sensitive part of the horse's body.

Understanding how to keep a young horse's attention is critical. Tapping the shoulder with the back of your hand can trigger the hindbrain (see p. 289) response similar to a horse nipping another on the shoulder. Most horses will look up as if to say, "What?" You can also use their highly developed sense of smell (fig. 12.6).

Spend time in a paddock playing with the foal. Much like a playpen with children, the young horse will grow to associate the paddock with fun, which makes training easier as he develops. "Paddock play" (see p. 348) allows horses and humans to interact in safe ways, and it is an ideal place to teach spatial awareness.

_ Consider Future Job

Horses that learn about a sport or discipline from another more experienced horse can have a much better chance at succeeding without stress as they become competitors themselves (fig. 12.7). Ex-racing mares with foals at their sides can teach their offspring how to enter and break a starting gate and how to start down the track with confidence and calmness. From a horse's perspective, having a mother teach you from your side is far better than having some human on your back trying to convince you and a bunch of other fearful horses that you should enter a metal pen! This is far less stressful for horses than traditional methods and has the added benefit of

↓ [12.7] Foals learn how to behave and interact with people from their mothers. Doing things together with them is a great way to introduce different experiences.

producing racehorses that know their job and can run with less worry and fewer injuries.

★ **TIP:** / Young horses only have the ability to focus and learn for about 15 minutes at a time, so keep training fun and in short sessions. /

_ Consider Future Environment

Foals who will spend most of their lives in stalls appear to develop fewer stress disorders when they are born and raised in confined environments, such as stalls and paddocks, compared to those who were raised in pastures from birth to three years old. However, horses easily learn to live in both environments, so allowing your horse to learn how to be in different-sized spaces at a young age is beneficial as he gets older. Foals who grow up in a pasture with their moms and other mares and foals and are then brought

STORY FROM THE FIELD: //

Mares Are Good Teachers

In the 1990s in Washington State I was working with racehorse trainers and breeders, and we conducted a study comparing the stress levels of horses going to the track for the first time. One group was eight mares with foals who had taught their foals how to enter the starting gates and break down the track. The control group included young horses who had no exposure to the gates or the track until they were under saddle. Several of the mares with foals were very competitive, and we wanted to observe what they would do.

Two of the mares clearly used body language and facial expressions to encourage their foals to "compete" and race ahead of them. Two other mares seemed protective of their foals and wanted them at their sides, while the other four mares seemed to just enjoy getting out for a gallop and did not seem that interested in their foals being there. It was noted that the foals who had been started in gates with their mothers at home as babies did not seem stressed when brought to the track.

The trainer who brought the horses to the track their first time commented that the foals who had been exposed to the gate and racing with their mothers acted as if "they knew what they were doing" and were much calmer when they got to the track than his control group had been. He continued to monitor several of the youngsters and noted they did not sustain injuries; two of the horses that he kept track of were still racing sound as nine-year-olds. //

into a strange environment with no horse friends, experience a great amount of stress (figs. 12.8 A & B). Separation anxiety, on top of confinement and unfamiliar surroundings with no friends, is a perfect recipe for ulcers and other stress-related disorders.

★ **TIP:** / Horses may experience less stress if they are raised as young horses in similar environments to what they will experience as older horses. /

In the wild, horses under the age of three years old are rarely asked to leave their natal band unless they are causing trouble. Male horses generally are better at adapting as they would be joining a bachelor band. Since the security of the herd is crucial to horses' survival, separation from a bonded social group is a serious reason to be stressed. Horses with dominant, confident, and curious temperaments usually learn to adapt. But horses with timid, shy, or worried temperaments become very stressed and often try harder to please their human companions, internalizing their stress and worry.

★ **TIP:** / Horses living in confined spaces isolated from interaction with other horses, need to learn spatial awareness and respect from humans as they often do not have needed proprioceptive awareness of their body, size, and space. /

Consider Gender Differences

"Normal" male horses do not take life as seriously as mares, unless they are wild stallions

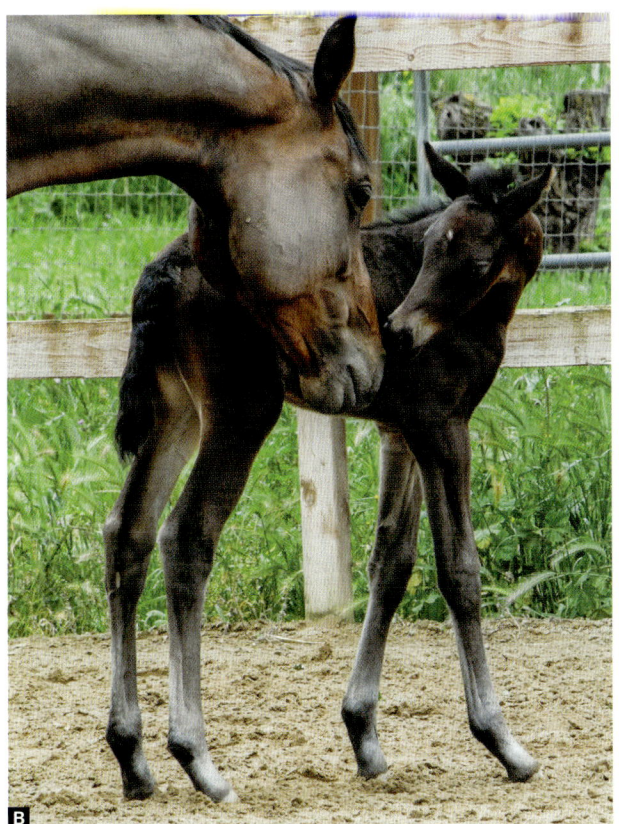

↑ [**12.8 A & B**] It is healthiest for foals to learn how to live in a variety of environments with their mothers, including both open green spaces **(A)** and dirt enclosures **(B)**.

with harems of mares. This may also explain why geldings are more common for riding disciplines, as some males in nature spend their whole life in a bachelor band just watching, playing around, and "looking for trouble." Geldings may have less stress over time than mares as it is more natural for male horses to constantly be "mixing it up" and changing places and groups. Even in domestic life male horses need more play time. They engage more in dominance behaviors, like jumping up in the air or rearing. They are following their natural instinct to practice stallion behaviors, which involves spending time on their hind legs, mock fighting, grabbing each other's necks and legs, and generally playing hard.

Wild mares do not usually leave their harem band and can form lifetime friendships with both other mares and stallions, so separation anxiety can naturally cause tremendous stress in mares. And since mares are more wired to internalize their stress (so as not to expose a weakness

STORY FROM THE FIELD: //

A Young Colt Raised by Humans Needed an Old Mare to Teach Him How to "Be a Horse"

One of the first behavior cases I had after starting my professional practice was an unruly Arabian colt named Domino. He had been hand-raised and the owners thought it was cute to let him jump on them from behind and put his legs over their shoulders. As he matured, he was shown in halter, but as he turned two, he became unmanageable for the trainer, striking and rearing in the show ring. The owners had him gelded, but that did not help, and he was starting to be dangerous to humans. They had tried whips and round pens, where he quickly learned he could be faster than humans and would strike, rear, and bite.

Domino was feisty and had normal behaviors for a young horse raised by humans, not horses. He had learned to dominate people. The best solution was to have functional horses teach Domino about spatial respect. The owners happily agreed as they had no other solutions, and I sent him to live with a group of Thoroughbred broodmares for a week. The older crippled matriarch of the herd used every communication she had to tell Domino to get out of her space while the other mares mimicked her and watched. It was the "1, 2, 3" punch.

First the mare gave Domino an eye glare and pinned her ears. He ignored her, as he had no idea what it meant. Then she snaked her head at him with her ears back, and he still approached her. On his third attempt, as if to say, "Didn't your mother teach you manners, boy?" the old mare ran Domino into the corner and kicked him three times in the chest and ribs. Domino was in shock, as he had no idea why the mare had attacked.

By the end of the week, Domino was the sweetest, most polite horse in the herd. Even the youngest mare in the group would just flash an eye at him and he would stand quietly by himself while the girls ate. When I and his owners came to pick him up, Domino trotted to the gate where we were and walked quietly and respectfully behind us with his head down, and loaded up, perfectly relaxed into the trailer. He was happy to get away from all those pushy mares!

Domino remained a very respectful horse once he learned spatial awareness and respect. //

to a predator), mares tend to get ulcers more easily than geldings. The more mature mares become, the more seriously they usually take life. Their job in nature is to be alert and guide the herd to safety. So even yearling fillies may worry about their family and friends.

★ **TIP:** / Male horses usually mature more slowly than female horses. /

★ **TIP:** / Male youngsters need more physical play time with each other than fillies. /

★ **TIP:** / Instinct still drives horses to want to be in herds, thus the herd still exists, even though stall walls or paddock fences may divide horses' space. Horses do not always view the fence as a boundary, especially mares. /

★ **TIP:** / All the skills horses need to learn to stay safe in a herd are learned between the ages of birth and two years old, regardless of gender. /

Mares are the dominant teachers in equine society. Although most mares seem to possess a natural ability for their "horse heritage," some mares do not. Some mares allow their foals to mount them and push them away from food or nurse whenever they want. This is usually only a problem with colts, not fillies. When this behavior is obvious, humans must step in and take over the horse-etiquette training and teach the foal spatial respect (see p. 77). Foals that learn spatial respect as youngsters have little difficulty with humans who understand and reinforce this lesson later in life. In the long term, these horses experience very little stress in their lives, as they understand horse social skills, spatial respect, *and* how to relate to humans.

> ★ **TIP:** / Mares are the primary teachers in equine society, so mimic what a good mare teaches her foal, and you will help create a well-adjusted horse. /

The Importance of "Preschool" for Horses

Don't forget, young horses are going to act like young horses, not domestic pets. In their natural life they would be spending most of the day playing with other horses and eating and sleeping all as a group. And while they are at a good stage of development for learning, they need breaks to play and process information, otherwise they can develop behaviors that are difficult for people to manage. Keep early training sessions to 15 minutes, and make them fun.

Young horses may nip at first, especially if they do not have any manners or respect. Recognize this is a normal behavior for a horse at this age if he is objecting to what you are asking, does not want to do it, or has too much energy to pay attention. If you punish a horse for trying to communicate like this without giving him a more appropriate way to communicate, then you are shutting down channels for connection between you and may be creating a stressed horse. Giving young horses plenty of safe "play time" when they can be outside "expressing themselves" is critical in helping them manage their energy and adapt to a more confined structure.

When the young horse does pay attention to you, it is important to give a positive reward by stroking over his withers in a way similar to his mother and using your voice to reinforce relaxation and show you are pleased with his behavior. If the young horse is particularly nippy, as some male horses tend to be, then making sure he has adequate exercise is critical. You can also use your voice to show you are *not* pleased, as well as redirect the mouthy behavior by tapping horse's shoulder and withers to bring his awareness to his hindbrain (p. 289) and body.

> ★ **TIP:** / Make sure your young horse gets adequate exercise. Young horses, particularly male horses, need lots of exercise and play time or they may develop abnormal behaviors with humans. /

"
When you punish a horse for trying to communicate without giving him a more appropriate way to communicate, you are shutting down channels for connection.

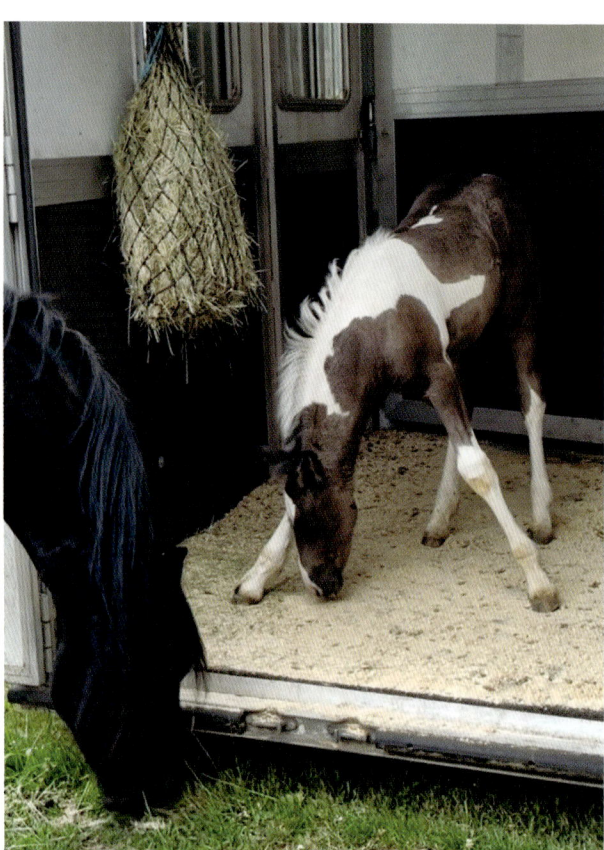

← **[12.9]** A foal explores an empty trailer while his mother grazes next to him, giving him a feeling of comfort and safety.

★ **TIP:** / Use smell, sound, visual, or touch triggers to redirect unwanted behaviors. /

Early young horse education should introduce the youngster to everything he may later need to be comfortable with. Ensuring the young horse watches other horses interact with people and exposing him to all the devices and activities he may endure in later life, such as saddles, bridles, saddle pads, brushes, loud noises, farriers, horse trailers, and so on, with his mother, is the best strategy for low-stress training (fig. 12.9). You can see foals through yearlings seem "proud" if asked to wear a pony saddle on their backs while they walk behind their saddled moms. Then, when the same horses later are introduced to new saddles and bridles, jumps, cows, trails, and horse shows, for example, there is little stress, as the horses already have a certain understanding and association for a positive experience.

★ **TIP:** / Horses have very good memories, and things learned when they are young (good or bad) stick with them their whole life. /

Preparing Your Horse to Learn (Whether Young or Not)

There are many ways to train horses (you already saw all the different ways they can learn back on p. 291), and because horses are so adaptable and want to fit in, many of these methods produce results. But not all are positive for the horse. An obedient horse who performs well may be internalizing stress, so it is key to help your horse learn to manage his stress at a young age. Coping skills, such as cribbing, and behaviors like spinning around when faced with fearful objects, can become habits and are more difficult to undo as your horse ages (see p. 249). Helping your horse learn how to be with people and understand and enjoy the various activities that humans will expect him to know—such as standing still, being tied, picking up his feet, getting a bath, trailering, and so on—will make all the difference when it comes to the horse's future level of stress. Unfortunately, few people take the time as the horse is developing to teach the horse basic adaptation skills. Luckily, it is never too late, and even older horses can go back to "kindergarten" and learn about life in a positive way, because horses are forgiving and adaptable creatures.

_ Adaptive Skills Horses Need to Learn to Be with People

The following are activities people often assume horses have either learned or should be comfortable with, but if your horse has not grown up with horses or people who have taught him about himself and how to interact with others, he may not have such basic interaction skills. Take time with assessing your horse's comfort level with each activity.

If your horse has trouble with any of the following things, then start addressing them in an open space such as a small paddock or large stall where the horse can escape if he feels unsafe. Recognize and help your horse understand you are his "safe space" with calming signals, a gentle voice, a good-feeling scratch, and a loving kind heart.

- **Allowing Humans in Their Space and Standing Still** Horses who learn this skill with other horses should transfer the knowledge easily, but "space

❝
Even older horses can go back to 'kindergarten' and learn about life in a positive way.

↑ **[12.10 A & B]** Kiki maintains eye contact with her two-year-old stallion Jet as he learns to walk with her, not in front of her **(A)**. Rather than putting a chain over his nose, she talks to him and circles him when he tries to forge ahead, keeping his attention and teaching him where he needs to be when walking with her **(B)**. If his eye gets in front of her, then he will make decisions on where he is going, rather than Kiki. It is key to stay connected and keep his eye equal to or slightly behind her eye.

games" will help your horse understand how to read your energy and body language, reinforcing his ability to stand still when you ask without restraint (see p. 91).

- **Spatial Respect** This is how to follow people and not walk into their space. It means not bumping into people or running through their space. It is part of body awareness and proprioception.

- **Being Touched and Groomed All Over, but Not Always Grooming Back** Horses love to engage in mutual grooming, and the "buddy scratch" is wired into their social brains—"You scratch my back, and I'll scratch yours." So, horses have to learn to transfer their innate desire to scratch you back to something else.

- **Go Forward, Go Backward, Go Sideways, Bend Around, Pick Up Your Legs** This is the proprioception necessary to move their bodies in small spaces with humans around and know where all four of their legs and feet are at the same time. In other words, body awareness.

- **Being Led and Following** Horses naturally follow people they trust as horses are very organized in nature. Horses often will travel single file behind each other, even if they could spread out. This is because it is easy to "follow the tail in front"—and you don't have to worry about a predator in front of or behind you when there are other horses all around. When leading a horse, think "follow me," rather than pulling on his head (figs. 12.10 A & B).

- **Remaining Calm with New Stimuli** Learning that loud noises, novel objects, odd smells, and other foreign stimuli are "normal." From fly sprays to the smell of burning hooves when getting shod, horses can learn that these things are all a part of life.

- **Allowing Halters, Bridles, and Bits** As already discussed, it is best for foals and young horses to see older horses wearing tack when they are young and curious. Then as they get older, the tack is familiar. Allowing foals to play with bits shortly after weaning gives them oral stimulation and prepares them for future endeavors with humans (fig. 12.11). Often the young foal will select the bit he likes to feel in his mouth. Since horses do not naturally like restriction, making bits and bridles seem fun at an age when they are curious can relieve stress when they are older.

- **Restraint** This includes how to stand in cross-ties, trailers, and how to be tied up. Restraint is not taught from one horse to another, but it is innate for horses to pay attention to what others want them to do and to try to please others so they can stay in the group

↑ [**12.11**] Jet carefully and gently explores the bits I am holding. Note how his eyes and body language show he is processing the experience with curiosity and interest.

[12.12] Horses can learn through observation. This horse and rider watch intently as another horse is on course in the arena.

where it is safe. Restraint is allowed by horses when horses feel safe; then there is no reason to object.

- **Confinement** If horses have friends, food, and water, then they usually are happy. Perhaps a reason horses adapted to domestication was the fact they would give up "freedom" in exchange for "friendship"—one that included food and water brought to them!

Horses started under saddle who missed the horse class on "Equine Etiquette and Leadership 101" will often have difficulties and appear stressed. Humans may have trouble with these horses, particularly if the horse has a strong or fearful temperament (see chapter 10, p. 221). The simple solution is to test your horse with the previous skills before working him under saddle and evaluate what he has learned or not learned.

Finding the Right Motivator

Starting a relationship off right is important. If your horse is already familiar with you and trusts you, then you may not need any other motivation other than being with you. However, worried or new horses easily start to relax and bond when they find there is something beneficial in the relationship for them. From a "horse-centric" point of view, ask yourself, "What is the horse getting out of this?" If every time the horse sees a new object, or is asked to do something new, there is a reassuring hand on his withers or he is fed a piece of carrot, then the horse begins to associate new objects and experiences with positive reinforcement. This becomes a motivation for the horse to learn faster (if, of course, the horse likes touch or food). Horses like to be rewarded for paying attention and doing what is asked.

★ **TIP:** / Horses learn faster when they are rewarded. /

Sometimes, motivation can come from being with friends or part of a group. Often horses seem to learn best with a friend around. Learning things with others can be much less stressful for many horses. All my jumpers start off learning to jump free in a pen with other jumpers. The seasoned jumpers like to show off to the young or inexperienced horses, and soon all the horses are running around, jumping obstacles as they play.

Many good racehorse trainers have been able to tell from watching foals in a field with their mothers who will make good racehorses. They watch the horses match up and compete with each other as they run around and play in a natural setting. The horses are motivated to run and be fastest by having other horses around. This can go for any discipline, from working cows to jumping. Horses can learn through observation and be motivated by each other (figs. 12.12).

It is important to mention that research is not in agreement on whether horses learn from observation or not. However, I will say the facts are that horses who are good at mimicking behavior, who are aware observers, and who are motivated to learn will indeed gain knowledge from watching other horses, and even people. Any barn manager knows that horses watch humans and can learn how to untie knots, open doors, and find the feed room if they are motivated (figs. 12.13 A–C). In nature, watching and learning are essential to survival. One group of wild horses I studied in Nevada had learned how to follow thunderstorms to find water, while another had learned how to knock over prickly pear cactus and suck water. And another group had learned how to jump barb wire fences to get to a stock water tank. There was clear motivation to learn (thirst), and once one smart horse figured out how to get water, the others clearly learned through observation. Thus, the primary reason I believe many studies do not agree on this point is that they do not take into account whether the horse is motivated to learn or not.

> **In nature, watching and learning are essential to survival.**

↙ **[12.13 A–C]** Fendi, the gray horse, watches riders "bob for apples" at the barn Halloween party **(A)**. He is demonstrating curiosity, interest, and motivation to interact with the people and the activity as he loves apples. Other horses who have not been watching people bob for apples try and give up, as they do not understand why the apples keep moving **(B)**. Fendi, the only horse who had the opportunity to watch the entire activity with both humans and horses was the only horse who went against the normal horse impulse to use the lips to try and grab an apple and instead used an open mouth and teeth **(C)**. His motivation, combined with his ability to focus and learn from watching others, paid off.

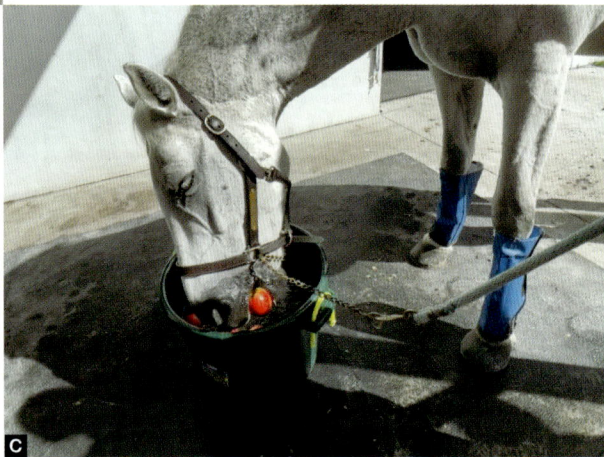

★ **TIP: / Horses need to be motivated to learn. Social interaction and positive reinforcement are often plenty of motivation for horses. /**

★ **TIP: / Go exploring with your horse to teach him about novel objects, sights, and sounds. /**

Keep Lessons Simple and Short

★ **TIP: / Keep lessons short and simple—not more than 15 minutes for young horses—and teach one new thing at a time. /**

It is easier for horses to learn new information when there is only one lesson at a time. Trying to learn how to balance a rider (who is usually not balanced), adjust to the discomfort of a saddle, and interpret the odd and often uncomfortable stimulation of a bit and bridle on nerves used for smelling and eating is a lot of lessons wrapped into one experience for most horses. Any one of the issues can override the brain with stimuli and cause worry and pain, limiting the horse's ability to learn.

★ **TIP: / Allow your horse to process what he is learning. If he does something right, take three seconds to reward and praise him. If he did not understand, then take a three-second timeout to "reboot"— take a deep breath and visualize again for your horse what you are asking him to do. /**

Break learning up into small modules. Spending 10 to 15 minutes three times a day is more productive than an hour once a day. Recent studies have found horses have various attention spans, based on discipline, age, temperament, living conditions, and overall welfare. Temperament may dictate how many lessons and how fast your horse can learn, so do not compare your speed of progress together with anyone else. Horses will learn what they can when they can, so establish a program that works for you and your horse. And remember, as mentioned before, young horses are in a high state of learning between birth and two years of age, but their attention span may also increase as they get older and find more motivation in the lessons—not unlike other social mammals.

Learn more

★ **TIP: / Horses learn at their own pace. Do not feel pressured to "make" your horse learn faster than he can learn without stress. /**

— Appropriate Use of Round Pens and Square Pens

Trust is developed between two social creatures who get to know each other *on the ground*. This can be done in a stall or paddock by just spending time with your horse. Round pens and square pens (paddocks) provide a safe confined area where you and your horse can interact on more equal terms. It is also critical that horses have some safe play time where they can roll and buck at liberty, and these spaces provide that.

★ **TIP: / It is critical for both physical and mental health that horses have a space where they can exercise freely. /**

Round pens have become popular in horse training for many as they are easy to use and do give the horse some freedom and enough space to safely move at various gaits (fig. 12.14). Round pens are beneficial for horses who have too much energy to ride and need to have a run without going anywhere, particularly stallions who need to "gallop off their testosterone" before riding. They also offer an option for free-longeing. Remember, though: Horses rarely run in circles in nature, other than foals playing around their mothers or stallions looking for attention from mares.

★ **TIP:** / Round pens need to be at least 60 feet in diameter. /

But round pens can also be confusing to horses. Horses have no place to "hide" or find "safe space." In nature, horses are more aware of other horses who demand more space, and they pay attention to those "dominant" horses. But when a dominant horse wants another horse to stand still, he doesn't chase him. He will

↓ [12.14] This 70-foot diameter round pen is ideal for liberty work, as it allows the horse ample room to move comfortably while staying connected with a handler. Here Coco has room to trot, canter, buck, and play at a safe distance while still listening to and keeping an eye on me.

communicate different energy for standing still, and he will walk up and nose bump and usually start a buddy scratch to relax the other horse. Using round pens for chasing horses around in order to habituate them to humans, while popular, is often a stressful and slow way to train horses or to teach them to trust people.

★ **TIP:** / **Square pens are best for spatial awareness training as they allow a more natural way for horse and human to interact together.** /

↑ [**12.15 A & B**] Coco walks into the corner and has one ear and her left eye on me as she is listening to me, wondering what she is supposed to do next **(A)**. In the corner she turns to look at me for direction and starts to step forward, asking me, "Are you sure you want me to keep going?" Using hand signals and my voice, I tell her not to come to me and to stay out along the fence line **(B)**.

Square pens offer a more relaxed and natural way for horse and human to interact, because they allow people to use more "horse-centric" activities to develop a friendship- and leadership-based relationship. Horses can find "safe space" in a corner and more easily understand space awareness and how to stay out of the human's space (figs. 12.15 A & B). Remember, the horse with the biggest space "bubble" is usually a decision-maker because he is able to take the space of another horse and keep other horses out of his space if he desires. You can teach the same thing horses teach: *my space, your space, stand still,* and *move*. Practicing this game of "space-taking"—walking toward the horse and asking your horse to move out of the space, or to stand still and remain in it—can only be clearly done in a more

[12.16] A round pen can become a less stressful training area if you create "corners" or "safe spaces" for the horse to rest, ask questions, show focus, and learn spatial awareness. This horse has turned to look at me, questioning if he is doing what I have asked and what is next. I have his full attention and he is learning in a calm and relaxed manner.

open training area where a horse can easily move from one delineated space to another. When this exercise is tried in a round pen, the horse will just start running in circles, unless you can designate some spots within the pen for the horse to stand and relax (fig. 12.16). Giving the horse a quiet "timeout" to sense your energy and thoughts allows the horse time to process.

Less Is Always Better: Tack and Training Devices

Just because something is made, doesn't mean it is safe, kind, or helpful to use on horses. That goes for bits, bridles, whips, spurs, and other training devices on the market. Rarely, if ever, is there any guidance or warnings on how to comfortably and appropriately use tack. It is up to the person who buys it to know how to use it. And one of the number one reasons for behavioral issues and stress in horses is poor-fitting tack. The horse who has a sensitive nose and throws his head because of a tight noseband and the horse who opens his mouth because the bit is pinching his lips are simply communicating discomfort. Sadly, many times in these cases people try tighter nosebands and stronger bits. Ironically, the cure is to remove the discomfort and have less control, not more.

_ Comfort + Communication = Control

Somewhere in the history of horsemanship, the object became to control the horse to gain his obedience. And as long as the horse was "winning" and doing what he was asked to do, then the riders felt everything was fine. Until it wasn't. When the focus of the rider's choices of tack is to have control instead of making sure the horse is comfortable, the communication between horse and rider becomes a one-way channel. And as we've discussed, when horses feel like they cannot communicate, they become stressed.

_ Bits, Bridles, and Tight Nosebands

Many of the behavioral issues I have addressed through the years had very simple fixes: loosen the noseband or find a bit that is comfortable. Having observed mouth sores, tongue sores, cheek sores, gums sores, fractured teeth, indented nasal bones, and loss of sense of smell, all due to bits and nosebands, I wondered why good people allow this to happen to horses (fig. 12.17). I realized many simply did not know it was happening, as they had been taught that bits and nosebands and other training devices had only one goal—control, not comfort or communication.

There are thousands of various bits, bridles, and other contraptions sold as "tack" to help train horses. Unfortunately, none of them come with the warning: *Do not use unless you know what you are doing to the horse.* Strong bits and tight nosebands were developed for the purpose of riding stallions into battle and keeping them from biting their tongues when they galloped, not for the recreational and competitive riding disciplines today. However, people have lots of opinions when it comes to the purpose of bits

→ [**12.17**] This sensitive horse was ridden in such a tight noseband and pulled on so hard it permanently damaged his face.

and nosebands, and manufacturers seem willing to make and sell anything humans come up with.

★ **TIP:** / Just because something is "made for horses" does not mean it is safe, kind, or helpful. /

A number of scientific studies have shown that restricting the faces and mouths of horses can increase stress by measuring increased heart rate and heart rate variability, as well as cortisol levels. Here are some common behavioral issues associated with bit or noseband discomfort:

Learn more

- **Refuses to Go Forward** This is common with mares and sensitive horses who have tight nosebands. Horses follow their heads. If their heads are restricted, they may feel they cannot move.

- **Races Off and Gets Strong** Horses run from pain and discomfort. Often the horses with the "toughest" mouths are the most sensitive, and in an attempt to avoid discomfort, they learn to lock their jaws or grab the bit and run.

- **Tosses Head** Often seen when there is an uncomfortable bit, tight noseband, or pressure points on a horse's headstall.

- **Opens Mouth** The bit is not comfortable. Many sensitive horses will open their mouths even with very soft bits, as they do not find anything in their mouths comfortable.

Remember what a horse's mouth and nose are designed to do. They are highly sophisticated sensory organs able to process multiple and specific information through touch, smell, and taste. The face and mouth of a horse has not evolved to deal with pain. When enough pain sensors are activated in these areas, horses cannot think. Instead, they try to fight or escape the discomfort in any way possible.

In her book *Horse Brain, Human Brain: The Neuroscience of Horsemanship*, Dr. Janet Jones points out the following three facts:

1. Tight nosebands compress the trigeminal cranial nerve and the infraorbital nerve. When these nerves are compressed, they cannot function properly when

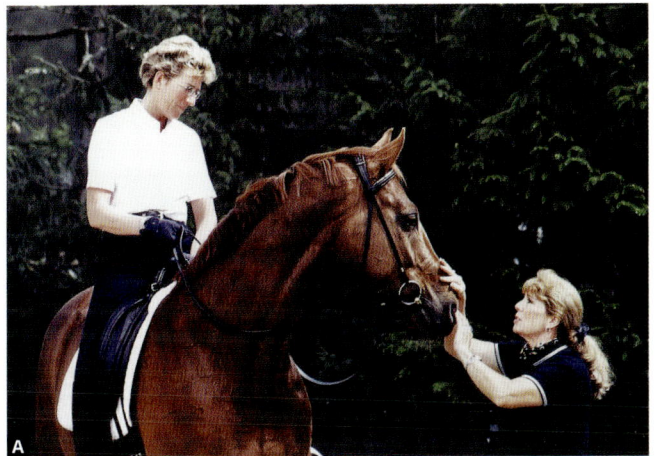

[**12.18 A–C**] Three examples of how riders can choose to connect with their horses without harsh bits and tight nosebands: In Photo **(A)**, dressage Olympian Heike Kemmer works with me without a noseband to allow her horse more comfort and freedom in communicating. In **(B)**, horseman Warwick Schiller rides bridleless to demonstrate a relaxed and comfortable horse. And in **(C)**, a mare that was very sensitive to a bridle was shown successfully with just a leather strap as a bit.

attempting to communicate with the horse's brain. They can't, for example, carry information clearly from the rider's hand aids to the horse's brain.

2. *Compressed nerves cause numbness and tingling that are likely to annoy the horse. A tight noseband worn daily in the same location bruises the nasal bones so that the horse feels pain each time it is applied.*

3. *Everything about the equine brain is designed through evolution to evade confinement. To a prey brain, close confinement means death. A tight noseband for a bridle creates stress in the horse because it is a form of confinement.*

Good riders who are connected and in communication with willing horse partners should only need very light contact while giving the horse the freedom to move his tongue, eyes, jaw, and head. Many horses prefer to communicate through their bodies, not their mouths and noses (figs. 12.18 A–C).

★ **TIP: / Bits and bridles must be comfortable for the horse to wear in order not to override his ability to think. /**

Look at Your Horse as an Individual

Understand and assess your horse before committing to a specific bit or noseband. Check your horse's palate to see its size and shape before putting a bit in his mouth. Know your horse's sensitivities and whether a bit is even an option. Some horses like things in their

[12.19 A & B] Radar prefers to have a bit in his mouth for security when jumping. His expression in Photo (A) indicates he is comfortable with this bridle. The horse in Photo (B) is wearing a full-cheek snaffle with a flash noseband and a running martingale that are correctly fitted to minimize pressure on the sensitive areas of the horse's face, and while his eye is not showing signs of discomfort, he is being very careful and attentive to avoid pain.

mouths, while others do not (particularly mares). Learn the biomechanics of your horse prior to using any training device. Some horses are just not structured to fit certain types of bridles.

★ **TIP:** / **Horses have sensitive mouths and faces. Less is always better when it comes to bits and nosebands.** /

Horses who are uncomfortable with a particular bit will open their mouths. When they are comfortable, they will close their mouths and use their tongues to feel the bit and relax (fig. 12.19 A & B). Bits can give certain horses a "feel" for the rider's hands and a line of communication. But for others, the nose and mouth are too sensitive (figs. 12.20 A & B).

★ **TIP:** /The forebrain and associated nerves of the face are wired for sensory input, while the brainstem (hindbrain) is wired more for muscle movements. /

_ Draw-Reins, Side-Reins, and Other Training Devices

The debate surrounding the use of various kinds of tack for training will continue until there is agreement on

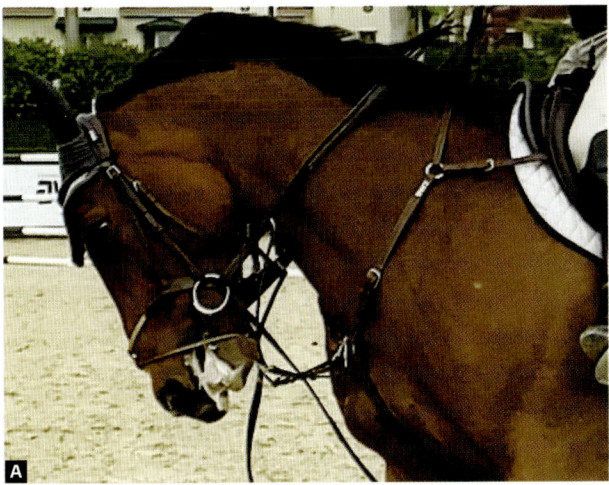

↑ **[12.20 A & B]** Remember, a horse's mouth and face are filled with nerves designed for sending sensory information to the brain, not for trying to avoid pain and guess what they are supposed to do to get away from pain. These jumpers are showing the stress their bridles are causing with open mouths and worried eyes.

a process for assessing what is best for horses. Far too many problems occur from "trainers" and riders using devices without knowing the harm they can cause. Both physical and behavioral issues arise when horses are forced to carry themselves in positions that cause them stress and discomfort.

Chambons, Pessoa Training Systems, De Gogues, side-reins, Lungie Bungies, and similar training tools are all human inventions that try to get horses to quickly do what they want them to do. If and when they are used, they should be fitted loosely enough to allow the horse to flex and bend naturally (fig. 12.21). Most importantly, they should only be used by people who understand why they are using them and how to fit them in a correct manner.

PART TWO / CHAPTER 12 / Managing Stress in Horses

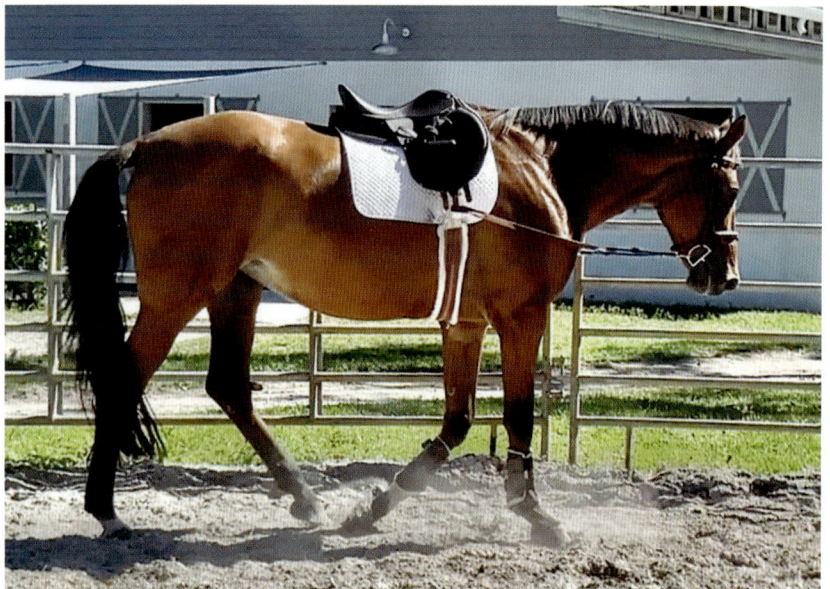

[**12.21**] Side-reins can be useful in helping guide the horse where to carry his head and feel rein contact without a rider. These side-reins are elastic and set at the correct length for comfort and light contact with the horse's mouth.

★ **TIP:** / Horses compensate in their bodies to adjust to restriction, which can cause physical and behavioral issues. Do not use training devices unless you know how to do so and have adequately assessed whether they will benefit your horse. /

I often hear people say that they ride in draw-reins to have more control. This may give them the *illusion* of control, but when you pull the horse's head down, he will often brace his back and stiffen and shorten his neck. While *inviting* a horse to drop his head is relaxing, forcing it down can make him spooky and worried horse, since he cannot see or move naturally (figs. 12.22 A & B).

The horse learns to brace with the stimulus of restriction, and the trainer has now engaged associative learning, stimulus response learning, and motor learning instincts in the horse. Once the horse has learned to hold his body in a particular way to protect himself, it is difficult to undo as all associated triggers and stimuli have to be undone and redirected toward correct, balanced, and comfortable carriage. Some people even jump in draw-reins, which causes worse *compensation syndrome* as the horse tries to make up for his inability to see or use his body properly and can fuse the seventh cervical vertebrae from the continued compression to the neck when landing.

★ **TIP:** / Never use devices that unnaturally restrict your horse from movement and ability to sense his environment. /

Just as you consider the horse as an individual when determining bit and noseband choices, before even thinking of using any specialty training devices, assess your horse's conformation, balance, muscle correctness, and mental ability. All these can influence whether your horse has the capacity to benefit at all from them.

— Using Whips and Spurs

Whips and spurs have a history of their own, which originates from humans not having the ability to communicate with their horses and get the desired response

↑ **[12.22 A & B]** It is common to see hyperflexion in the jumper, dressage, and reining arenas. While this technique is accepted by many it can limit breathing, cause unnecessary strain on neck muscles, limit the horse's ability to see, and cause mental stress.

using just their bodies and energy. A proprioceptively aware horse who is comfortable with his tack and aware of his body and his rider's, should have good ability to respond to what is asked of him, without the need for a whip or spurs. Too many riders automatically put spurs on as part of their riding attire without ever evaluating whether their horse needs them or ever learning how and when to use spurs correctly.

★ **TIP: / Spurs and whips should only be used as an aid in communication and only by educated riders who know exactly where, how, and when to "touch" their horses for added communication reinforcement. /**

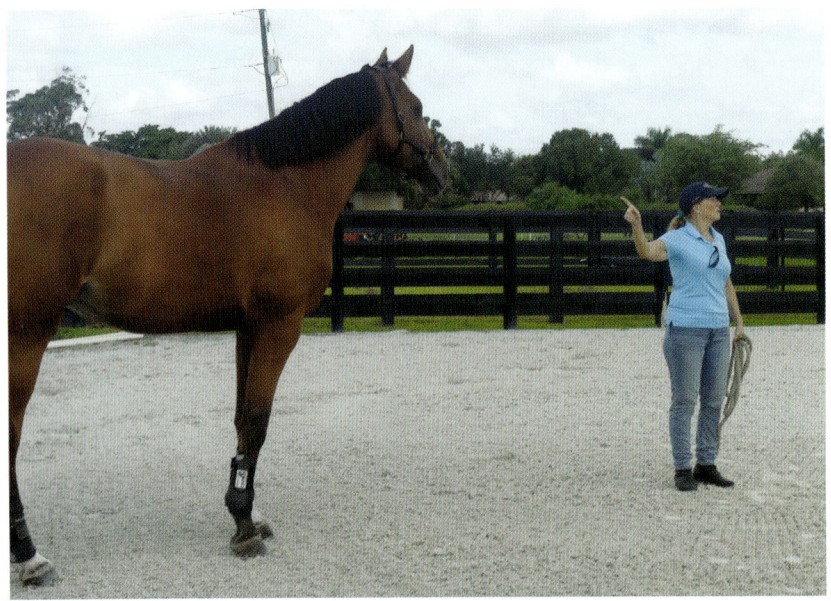

[**12.23**] Coco has learned to be aware of me, "listen" to me, respect my space, and move or stand still at liberty. Here her attention is on me and she will stay in this spot, as long as I am staying in awareness and connection with her.

Spurs and whips should only be used if they aid in communication with helping your horse become aware of his body and how to move from a refined touch. Spurs should only be used by an educated rider with a good leg who knows exactly why, when, and where they are touching their horses. Whips should only be used in a way that your horse can feel the slightest flick of a whip end. Whipping and spurring horses out of anger is never appropriate and horses who have suffered these things often will become dull to both.

Good riders teach their horses to listen to the lightest of aids. Once a horse learns it, he listens for and senses the rider's energy and body movement. Horses are much more sensitive than riders give them credit for; those labeled "dull" have just tuned out their riders, do not understand, or cannot move their bodies the way a rider is asking them to move. Remember: horses *want* to cooperate.

Limiting Stress When Teaching Specific Skills
_ Stand Still, Tie, or Cross-Tie

In nature one horse can tell another horse to stand still in a particular space. If that horse moves, he may get in trouble. In the same way, humans can teach horses how to stand still and be willingly handled with lessons in spatial awareness and respect, discussed earlier in this book (see p. 142). Once a horse understands how to stand still in the space you have asked, then he can easily learn to ground-tie (fig. 12.23). Once the horse feels safe with you, then he can learn to be restrained. No horse likes to have his ability to move freely taken away, but for safety reasons, your horse may have to be tied at some point in his life.

Never start by directly tying a horse, but instead standing with a rope you can hold and control and release as needed. Start with short timeframes of 1 to 2 minutes and then reward and release. Build up the amount of time so your horse can comfortably stand still or be tied for 15 to 20 minutes (fig. 12.24). It is helpful to make the experience positive by finding places your horse likes to be touched and scratched. They can then associate restriction with feeling good. The horse has to benefit in some way from learning new things, otherwise he will resist. Horses should never be tied or restricted for longer periods than necessary as it can interrupt their basic needs to eat, drink, move, and urinate. However, most horses will defecate while being restricted and frequent defecation may indicate stress and worry.

★ **TIP:** / Horses should feel safe and learn to stand still comfortably, before you restrict them with ropes or cross-ties. /

Many stress-related behavior problems start with restricting horses who do not need restriction, have had something fearful happen to them when they were restricted, or have been made very uncomfortable while being restricted. Everyone is familiar with the horse who "pulls back" when tied or cross-tied. Typically, the horse's poll feels the pressure of the headstall,

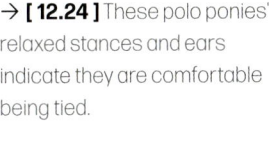

→ [12.24] These polo ponies' relaxed stances and ears indicate they are comfortable being tied.

PART TWO / CHAPTER 12 / Managing Stress in Horses

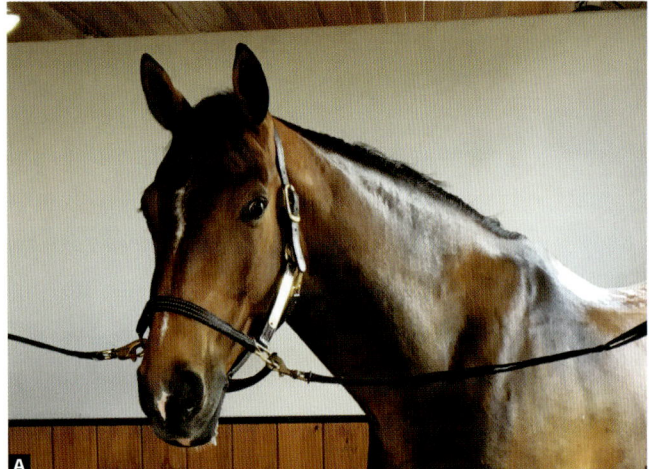

[**12.25 A & B**] In Photo **(A)**, the horse stands comfortably on cross-ties adjusted appropriately for his size. They are loose enough to allow him to comfortably move his head and neck. In Photo **(B)**, the pony strains uncomfortably in cross-ties that are too high and tight for him, compressing his spine and neck.

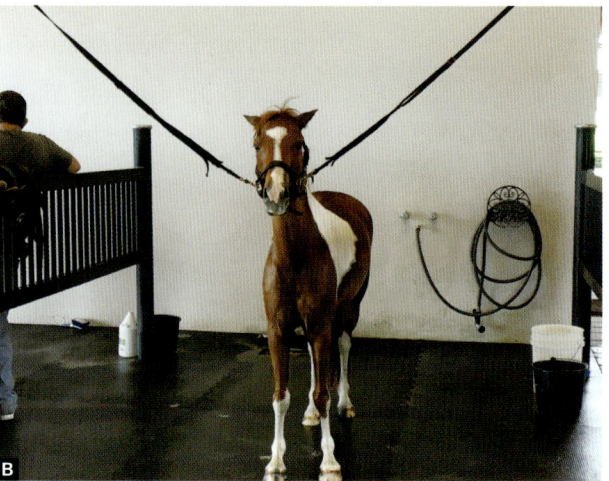

Longeing: Good or Bad?

Longeing can be beneficial when done correctly to allow a horse in training to safely find his balance or to give a horse a chance to play and buck when there is no place to turn him out; however, regularly longeing horses in place of exercise or instead of taking the time to help the horse overcome worry and stress at a show only increases the horse's risk of injury. In fact, according to several studies and veterinarians, Dr. Duncan Peters states, improper longeing "is the number one cause of death at horse shows today due to injuries."

Learn more

and that is what triggers the panic to escape. This is pretty normal horse behavior and fairly simple to fix.

Many horses actually develop tension in the poll due to various causes, such as: being asked to flex when they are worried; having a subluxated or jammed atlas, axis, and temporomandibular joint (TMJ); eating from a hay net; getting up from sleeping in a small space and throwing their heads in the same direction for balance. Gentle massage or chiropractic work can help your horse release tension in the poll. This is a common issue and will only get worse until it is fixed. Don't restrict horses who have discomfort in the head until the discomfort has been addressed. Even when restrained, make sure horses have the ability to move their heads up and down and sidewise (figs. 12.25 A & B). Keeping a horse's head tightly restrained affects his whole body and can cause extreme discomfort.

★ **TIP:** / Don't restrict horses' heads without allowing full range of motion, up and down and side to side. Head restraint affects the whole body and can cause tension, bracing, and other forms of discomfort. /

One of the most common problems I see in barns is leaving horses standing in cross-ties for too long. Often the cross-ties are too tight and force the horses to keep their heads up, preventing them from relaxing. This can cause both physical discomfort and mental stress, and sadly, most people are unaware they are making their horses so unhappy and uncomfortable. Horses in cross-ties cannot urinate easily if their heads are too high, nor can they relax their necks and backs. Then, when they are asked to go into the cross-ties again, they resist and people wonder why. Adjust the cross-ties in your barn so horses can safely and comfortably lower their heads into a relaxed position.

★ **TIP:** / Adjust cross-ties for comfort so horses can lower their heads into a relaxed posture. /

_ Ground-Driving and Longeing

Driving horses on the ground from behind (long-reining, long-lining, ground-driving) has long been an acceptable way to start young horses. This familiarizes the horse to the pressure of the bit and reins without the added stimulus of a person on his back. It allows the horse to learn the aids for turning in his mouth (the bit) and on the sides (the "reins" or "lines"), which will eventually be replaced with a rider's legs. Longeing and driving on "long lines" allow the person and the horse to get to know each other using eye contact and body language, but unfortunately, doing either correctly is becoming a lost art because people want "quick fixes" and instant results.

Managing Riding Stress

I mentioned earlier in these pages that in over 60 years of riding, one policy has never failed me—I always take time to greet my horse and help him feel safe and comfortable before I get on (fig. 12.26). That means the horse makes

> **Many stress-related behaviors start with restricting horses who do not need restriction.**

[**12.26**] Before you get on, greet your horse and talk to him in a gentle voice. Make sure he is comfortable with you on the ground before you get on his back.

eye contact, willingly nose bumps me, and wants me to touch him. If he does not want to make eye contact or acknowledge you on the ground, then you certainly are not going to have a safe ride on top. Many of the under-saddle behavioral issues we see are because the horse does not feel safe with the rider. No amount of restraining devices will make the horse feel safe. Always check for stress-related behaviors *before* getting on. Stressed and worried horses are usually unpredictable horses.

We've already talked about how stress can be reduced with a well-fitting bit and noseband—if they are even necessary for your horse. There are more mouth-and-head restrictive devices for the horse than a person can count. But I ask, how can it be that a horse learns to work cows with limited control of his mouth and face, while a jumper and dressage horse seem to need a plethora of bits, tight nosebands, and leverage devices? Is it because history and culture dictate differences rather than what is needed to communicate with the horse and rider? "Horse training" has come a long way from the abusive methods once practiced, but there is still a long way to go in implementing "horse-centric" thinking in training and management, rather than "human-centric." The irony is that there are much safer and quicker ways to get horses to do the things we want them to do when we take the time to understand the basics of how horses think, feel, and behave. I hope in these pages you gain the tools and knowledge you need to feel comfortable going outside the box of your discipline and using appropriate science to improve your relationship and performance with your horse.

As most would agree, there is nothing "natural" about the horse having a human predator on his back. But because of horses' gentle nature and willingness to work with humans, they tolerate, and in many cases seem to *enjoy*, interaction with humans. Horses want friendly people who will protect them at all times. Building a trusting relationship on the ground and then transferring this relationship to the horse's back is the best way to form a strong, trusting bond and create a general sense of safety for your horse (fig. 12.27).

★ **TIP:** / Horses need to have the freedom to communicate and move comfortably when ridden. /

Clearly, researchers need to address a number of variables when it comes to causes of stress in the ridden horse, including the effect of various riders, the horse's level of training, the horse's temperament and mood, and his

→ [**12.27**] The least stressful method for a horse is to be ridden by someone the horse trusts with his life. Kim and Dano have a deep bond, and while dressage rules require Kim to compete with a double bridle, Dano often preferred training with only a halter.

physical conformation and ability to perform the tasks asked of him. Take a good look at the tack and training methods you see in use at your barn and at showgrounds, and consider for yourself which methods are least stressful for horses. A few studies designed to measure horse learning and stress actually showed that the action of learning itself was not as stressful as the interaction with the person riding. This indicates, again, that the relationship the horse and rider have together is the key (fig. 12.28).

_ Guidelines for Minimizing Stress While Riding

• **Friendly Greeting** Before getting on, greet your horse. Horses greet each other every time they are gone or out of each other's sight for even a few minutes. Following equine protocol, always say hello to your horse to see what kind of mood he is in *before* you ride. And never get on a horse you do not know before establishing a friendly relationship with: soft eye contact while

↓ [12.28] This horse and rider are communicating and working together. Note the expression of the in the horse's eye and the relaxed ears listening to the rider, as well as his balanced posture and relaxed tail—all indicate connection and cooperation.

smiling, blinking, and talking; gentle touch of the back of your hand to the horse's nose; and a friendly touch or scratch on the withers or neck.

- **Physical Comfort** Before you get on, get your hands on your horse and do a general body check for soreness, tightness, and sensitive areas (see p. 194). Oftentimes horses will be girthy and not like to be brushed if they have ulcers or "stuck" ribs. Make sure your tack fits comfortably (figs. 12.29 A–E). Once in the saddle, continuously check for your horse's comfort in the training. If he resists doing something, it usually is because he does not understand, he physically cannot do it, or he is worried about something hurting. Horses who spook or run off suddenly with no apparent trigger may have felt a sudden pinched sciatic nerve or an ulcer pain. It is critical to not discipline your horse in these situations but rather check his eye and facial expression to see what he is feeling and thinking. The horse who is "testing" you with his behavior will have a completely different expression than the horse in pain or worry.

- **Feeling of Safety** Help your horse feel safe with you by talking to him about what he is going to experience and what you would like him to do. When we speak out loud, we visualize, and horses can then interpret our thoughts and feelings. Be just as much or more aware of your surroundings with your horse's safety in mind than your horse is. Be "horse-centric" in your thinking and feeling. Always give your horse time to:

  - Adjust to new objects and situations. Take time to listen to your horse and coach him through his concerns. Be firm, relaxed, confident, and kind.

  - Adjust to changes in light. Remember, it takes at least 15 minutes for your horse's eyes to recalibrate when going from dark to light or light to dark (see p. 49).

  - Warm up and re-establish proprioception. He needs to know where his body is in relationship to the environment, others horses, and objects. Often horses who are usually ridden alone will spook when another horse comes toward them in a ring, as the horse has no proprioceptive references for safety around others.

> **Talk to your horse. He may not understand your words, but he will understand the tone of your voice and soothing energy.**

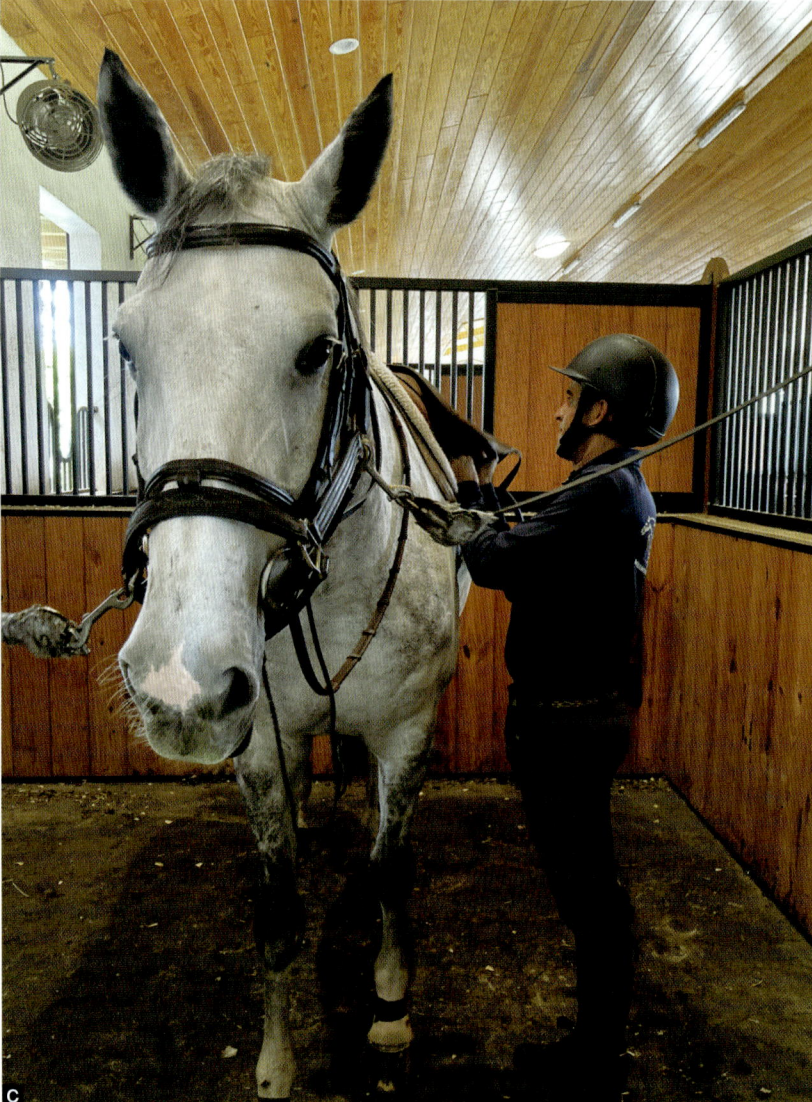

↑ [**12.29 A–E**] Gabriel checks the saddle fit **(A)** and watches his mare Cassie's eye as he tightens the girth—it shows her worry **(B)**. After a moment, she relaxes, which you can also see in her eye **(C)**. He then carefully adjusts the bridle fit **(D)**. Always ensuring his horse's physical comfort allows Gabriel to then have an amazing training and "play" session over fences **(E)**.

↑ [12.30] I put a hand on Coco's withers to acknowledge that I see what she sees and let her know there is nothing to worry about.

- **Connection and Communication** Using soft hands and seat, *feel* and connect with your horse so he knows you are there for him. Remember to be his eyes, ears, nose, and body—as long as you are on your horse's back, you are responsible for his safety. Be as light and sensitive as possible in your communication with your horse making sure to:

  - Stay focused with your horse. If you are distracted, then your horse may be, too.

  - Stay connected through your voice and regular physical touches on the withers or neck.

  - Keep an open dialogue, listening to what *your horse* needs, not just telling him what *you* want. Look for constant feedback.

_ Take the Time to Explain

It is critical to establish a positive relationship prior to riding. Good groundwork to establish trust, safe space, and leadership will benefit you and your horse by reducing stress for both of you. When your horse has learned that a hand on the withers means safety when you are with him on the ground, he will relax rather than worry when faced with a potential threat with you in the saddle (fig. 12.30).

★ **TIP:** / On the ground, train your horse to relax with a touch from you on the withers. This can have an immediate response as it does not require the horse to think once he learns that the touch means relaxation. /

Just as with children, when your horse becomes curious or worried, take time to explain what he sees or senses. Some trainers feel a horse should give his undivided attention to the rider when under saddle. For example, the horse is not allowed to smell another horse's manure when riding on the trail and isn't allowed to stop and investigate an object he does not recognize or is concerned about as a threat. Good judgment is important, but horses feel safer when they have the ability to try and understand their environment. Horses do this by processing through smell, touch, and other sensory input. Giving the horse a three-second timeout to assess and process with your support can help the horse feel safe, brave, and confident, and can minimize worry and help the horse learn.

★ **TIP:** / Take a three-second timeout to allow your horse to process, either to feel rewarded for accomplishing something or to erase and "reboot" if your horse has become worried, does not understand, or has been disobedient. /

Talk to your horse. Horses may not understand your words, but they will understand the tone of your voice and soothing energy. And staying in contact with your horse through touch is important. Using both touch and your voice to coach your horse to be his best self can help him relax and feel motivated to perform and be with you, regardless of discipline. Mental and emotional welfare in horses is often overlooked in lieu of physical performance when riding, but in fact, mental worry is often more of a stressor in horses than any physical activity. Always assess your horse for what he has

← [12.31] Horse friends in their pasture, wandering in social groups from water to feeding stations that have been spread out in different places to encourage movement.

learned or not learned under saddle, and help him feel safe, above all else. A horse who feels safe with you will want to be with you.

Habitat Management

Horses need to be out socializing and interacting with their environment daily to stay healthy. Providing a natural habitat that allows them to move and find food all day long would offset many stabled horses' stress (fig. 12.31). Ideally, all horses could live in social groups in green pastures and choose when to be ridden or interact with humans. While this model is being followed by some, it is not practical for most people. However, because of the positive impact "open design" living can have on horses, more people are developing new management concepts that also are ecologically sound and minimize labor. Systems like the Aktivstall provide solutions that can be adaptable to farms that want to incorporate feeding stations, various safe footing, and interactive gates that allow horses to stay active in their environment.

As available rural land shrinks, fewer people are able to keep horses on farms and living in natural environments and must instead house horses in stalls or paddocks. Horses living out in nature need little if any habitat enhancement, as long as there are other social creatures around. But those in confined spaces may develop stress-related disorders without enough movement, free-choice forage, and the ability to explore and make choices. If horses are generally happy, then they adapt fairly well to their housing, whatever

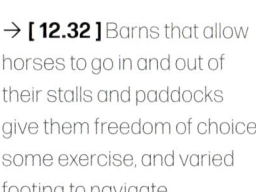

Learn more

→ [**12.32**] Barns that allow horses to go in and out of their stalls and paddocks give them freedom of choice, some exercise, and varied footing to navigate.

STORY FROM THE FIELD: //

The Same but Different

Horses are highly adaptable creatures, and as we have learned, *individually context-specific* (see p. 251). Various horses could respond differently to similar situations. For example, one horse in a barn where I worked hung his head out of his stall all day, watching everyone go by and eager for social interaction. The mare in the next stall would stand with her head hung down in the corner and did not want to interact with horses or humans. What was the difference? They both had regular food, water, and exercise. What was the underlying cause for her seeming depression? Was it simply a difference in adaptation?

To the average passerby in the barn, the mare had good muscle and a nice coat. She appeared healthy. But she showed anti-social depression-like symptoms. In taking her case history, I found she had been with several trainers, roughly treated, given a number of drugs and nutritional supplements, and had had an embryo transplant. She did not want to be touched, and when she was touched, she looked away. We treated her with vibration therapy to lift her mood and used aromatherapy in her stall. She was taken out for walks by different people who gave her new smells, food, and lots of vocal support until she was comfortable being touched, and then we offered her massage and other bodywork to help her feel physically better. Within a few days she was socially responsive.

This is a common situation for many horses who get repeatedly bought and sold, so more research needs to be done in order to establish better guidelines for ensuring the mental health of sale horses. //

it may be. But there are a number of things you can do to enhance the psychological and physical welfare of stabled horses to ensure the least amount of stress possible (fig. 12.32).

Remember, the horse was designed to be socially interactive and walk all day long, but many horses spend most of their day standing around waiting for food or waiting to be taken out of their stalls, so creating a barn life synchronized around your horse's priorities and rhythms will greatly help to reduce his stress. While most horses like routine and certainly know how to tell time when it comes to food, horses also like variety and choice. Providing stimulation through social interaction, even with other species and objects, can prevent boredom and encourage curiosity and learning. Making sure your horse has a time to eat, sleep, socialize, explore, roll, and exercise each day will help keep him mentally and physically fit (fig. 12.33).

★ **TIP:** / Get to know your horse's rhythms—when he sleeps, eats explores, socializes, naps, and so on. /

Horses who have grown up in stalls and at horse shows, may exhibit little obvious stress in confinement and may instead demonstrate stressed behaviors if turned out to pasture or in large spaces with other horses. These horses find their stalls to be their "safe space," where food and safety are located, and if turned out, they need to learn how to be around other horses. Similar to locking a child in a bedroom for years and then taking the child to the park and saying, "Go play," horses may need

[**12.33**] When one horse starts to roll, it triggers a social suggestion to other horses to roll. In nature, good rolling spots are highly desired and horses wait their turn to roll in them.

to learn how to socially interact and use their bodies in larger spaces.

The stable may be your horse's habitat, so making it as pleasant as possible is important for your horse's overall welfare. Depression-like symptoms can show up in confined horses who do not have friends and spend their days in isolation. Inactive and unresponsive horses who spend their time facing away from the stall door and not acknowledging people when they come to greet are exhibiting signs of stress.

Welfare standards for horses are highly opinionated, and thus only the bare minimum of standards used for "food animals" have been regulated. Basically, most state welfare standards simply address the minimum amount of space for a horse to stand, and indicate the horse must have adequate food, water, and shelter. And the sad fact is, a fat horse standing in a stall alone who has not been out of his stall for two months does not trigger the same awareness and sympathy as a starving skinny horse at a feed lot. But both could be emotionally stressed.

Most people who love and care about their horses know when they are stressed. The good horseman or horsewoman can tell if their horses are a "little off" or "just not right," as they know their horses inside and out. Keeping a daily journal of your horse will help you track behaviors, both physical and emotional. Whether you use a notebook in the tack room or an app on your phone, it is easy and important to write down daily notes related to your horse's life in the barn. Many people do take their horses' temperature daily, which certainly is a good indicator for health, but there are a number of other simple checks you can do. If your horse usually has 10 to 12 piles of manure per day and suddenly only has three, you will know something is wrong. Or if your horse has not spent any time lying down sleeping and he usually does, you will know to call your vet and explain the worrying behavior. It's all about relationships, and the better you know your horse, the better you can manage and prevent stressors. Go through the following eight-point daily stress check, and use it to guide your note-keeping.

Quick Daily Stress Check

- **Eating** Can your horse nibble all day on hay? How long does your horse have an empty stomach in between feedings? (Remember, the equine stomach empties about every 15 minutes.)

- **Manure** Checking this gives you a quick look at what is going in your horse's digestive system. Is it consistent in texture, color, and moisture? Is it regular?

- **Sleep** Is your horse getting enough sleep? Does your horse rest standing up or lying down or both? For how long? Horses need to lie flat out for at least 15 minutes a day to get their REM sleep.

Learn more

- **Water** How much water does your horse drink each day? How often does he urinate? What is the color and smell of the urine?

- **Movement** Does your horse lie down and roll, walk around his stall and paddock, go inside and outside? What are his movement patterns?

- **Rhythms** How many hours a day does your horse, sleep, eat, stand, move, and socialize?

- **Socialize** Does your horse have friends he can see or touch? Does he like or dislike neighbors? How much time a day does he spend socializing? This may include people and other animals, but the horse needs freedom to choose to socialize.

- **Freedom and Choice** Does your horse have any freedoms or choices? Can he go inside or out? Can he join friends or leave them? Has he changed his likes or dislikes?

Barn Design with Horses in Mind

Stall and barn design have not changed much since horses started being confined. The 12-foot by 12-foot stall is fairly standard, and while it does not allow much room for a horse to lie down or turn around, domestic horses have adapted as best they can to what humans have designed. There are a variety of stall types used in different parts of the world, from open pipe to concrete to wood panels. Keep in mind your barn and stalls are your horse's living quarters and should be designed for his comfort. Stalls that open to individual paddocks offer horses options for going inside or out and allow them to choose to interact with neighbors or not. Giving horses choice enhances their mental and emotional well-being.

★ **TIP:** / Giving horses choice to go inside or outside, to be with friends or be alone, and to eat when they want enhances their mental and emotional welfare. /

- **Air flow and windows to the outside** are key in order to keep air fresh as well as to allow your horse to see what is going on outside (fig. 12.34). Horses confined in stalls with no windows often will be spooky when they go outside as they have no idea how to associate movement, such as birds flying or wind in trees, with related sounds. Horses need exposure to nature—sunshine, dirt, air.

- **Natural light** is also important for your horse's welfare. Horses can suffer from vitamin D deficiency as well as other restrictive light disorders. Now there are light therapy systems for barns that try and regulate natural light and reduce the kinds of light waves that are harmful to horses. While natural light is good for your horse, so is natural dark—having lights on all night can cause sleep disorders in horses (fig. 12.35).

★ **TIP:** / Provide natural light during the day and darkness at night to support your horse's natural circadian rhythms. /

- **Bedding** may not seem like it would affect your horse's mental or emotional welfare, but it certainly can. If your horse has positive early foalhood memories of being on straw with his mother, then a fluffy straw bed is very inviting and secure for your horse. Straw also allows your horse to nibble on low digestible fiber, which can also be of benefit to some. If your horse was born in

Learn more

↑ [12.34] Stalls that are open on at least two sides ensure ventilation and encourage the horse to move and interact with the outdoors.

→ **[12.36 A & B]** Horses require 15 minutes of REM sleep a day, and this can only happen when they lie down flat. Make sure that your horse is comfortable with the bedding in his stall—whether straw or shavings or something else. Watch and record his rest times so you know that he is, in fact, lying down for part of the day.

↑ **[12.35]** Lights left on all night at the barn can cause horses to be sleep-deprived, as can loud music or waking them up in the middle of the night with activity. These actions are all very common at horse shows and can lead to physical and mental stress. Often, horses may act "spooky" when they are, in fact, experiencing sleep deprivation.

fluffy shavings, then shavings may be associated with comfort. Horses want to "make their bed" to lie down, and you will watch them dig in their bedding to create a pile of fluff to lie in and possibly roll (figs. 12.36 A & B). If horses do not have a comfortable bed, they may not lie down for their needed REM sleep.

- **Horses want the ability to socialize with their friends.** Horses' social nature directs them to eat with friends, sleep with friends, and play with friends. Stress builds up in horses if either they must live alone with no friends or when they are required to be separated from their friends.

★ **TIP:** / Design stalls to allow horses to see each other and at least touch noses. /

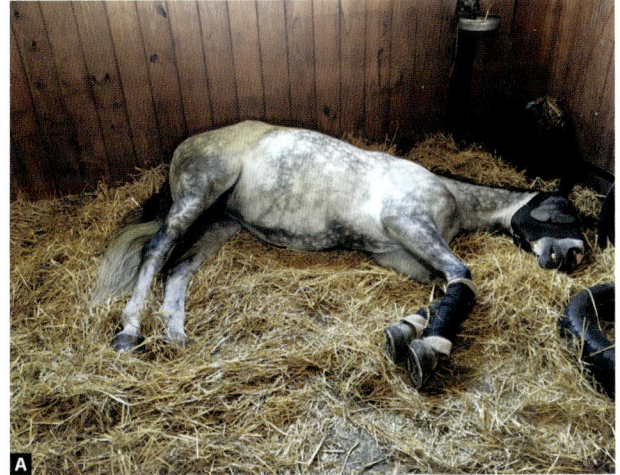
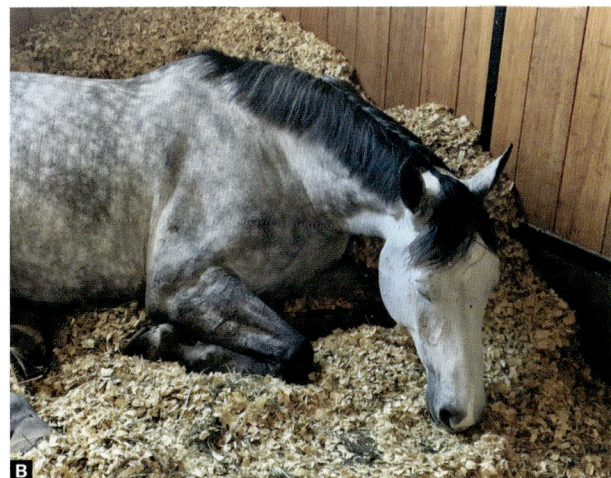

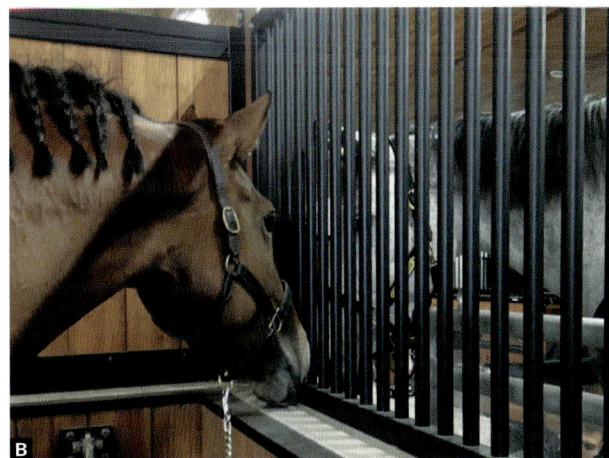

↑ **[12.37 A & B]** Horses need to be able to see other social creatures and interact with them to be happy in their home.

Horses can be happy living in stalls, assuming they have adapted and have at least one other horse or other social creatures around. Typically, the more the merrier, but horses like to pick their friends, so having an annoying gelding next to your sensitive mare could cause her more stress than being alone with her humans. In order for horses to sleep they need to feel safe and have at least another social creature watching guard for them. Many horses will stay up all night, and when people are in the barn for chores, they take their naps because they do not feel comfortable sleeping at night without someone to "keep them safe." Stalls should be set up so horses can at least can see and smell each other. Ideally, they should allow safe interaction through bars or nose holes (figs. 12.37 A & B).

STORY FROM THE FIELD: //

A Sleep-Deprived Thoroughbred

Neptune was a typical Thoroughbred. He was always up for "go." But suddenly he started being totally exhausted when I got on to ride. We would walk into the covered arena, and he would almost fall asleep. I had noticed that when I came for the late-night feeding, Neptune sometimes would be outside in his paddock, staring into the forest, while his two Mustang mare herdmates lay happily in their stalls, dozing.

Then it occurred to me that the two ex-wild mares had convinced Neptune to be "on guard" all night so they could peacefully sleep. I put him in the soft sand arena at night, away from the girls, so he could lay down and get his REM sleep. Neptune was back to my "ready-to-go" Thoroughbred after a couple of nights of sleep. After that, he always got to live in the covered arena at night so the girls would not pester him to be their night watchman. //

— Ensure Exercise and Movement

Since the perfect horse world does not exist for most horses, nor for most horse owners, mitigation and support therapies to reduce and prevent stress in horses is often needed. Think *movement, food, and fun*. Horses need to be moving all day long. As free-roaming herbivores, they stay healthiest when able to walk around, eating small bites of food all day and night, and having friends around to do it with (fig. 12.38). Social life and exploring the environment with humans can be fun, too, if the horse enjoys being with you and is comfortable.

Horses were created to walk all day on unstable and varied ground. Walking stimulates the hoof to contract, which in turn stimulates *peristalsis* and digestion. Horses with good feet usually have good digestion, and horses with poor feet often have poor digestion and are usually more prone to colic. So movement is critical to the horse's physical and mental well-being.

★ **TIP:** / Get your horse out to exercise several times a day. The more walking the better. The more movement your horse gets, particularly with friends, the healthier he will be mentally and physically. /

★ **TIP:** / Stimulate circulation in your confined horse. Magnetic therapy, massage therapy, and kinesiology tape, to name just a few options, can increase circulation similar to walking all day and help your horse's body feel like it has had exercise. /

→ [**12.38**] Horses are meant to move constantly, while searching for food and in the company of friends.

Ideally all horses should have the opportunity and choice to be either outside with friends or inside in their private stalls. Some horses will choose to hang outside, and others will choose to be inside. Some horses like to move, and others can be lazy and want to stand around and eat all day, even if they have the chance to go outside with friends. It is possible that domestication may be genetically adapting horses to be more sedentary even though physically their systems have not yet adjusted to the modern horse lifestyle.

For horses who either are confined or choose not to move around much, they may need a "gym workout" of sorts to ensure they get enough exercise. While it may be hard to motivate some horses on their own to move, having other horses walking out usually is a good motivator. Riding with other horses, going for trail rides together, being turned out in paddocks or put on walkers, treadmills, or aqua gyms can both be entertaining and good exercise for your horse.

_ Feeding: Spread it Out

Horses' digestive systems are designed to eat small bites of grasses and other plant materials for at least 15 to 20 hours a day in nature, depending upon habitat. As I've mentioned, equine stomachs are relatively small and only hold food for about 15 minutes. Hence, an empty stomach creates acid secretion, which can cause *equine gastric ulcers syndrome (EGUS)*.

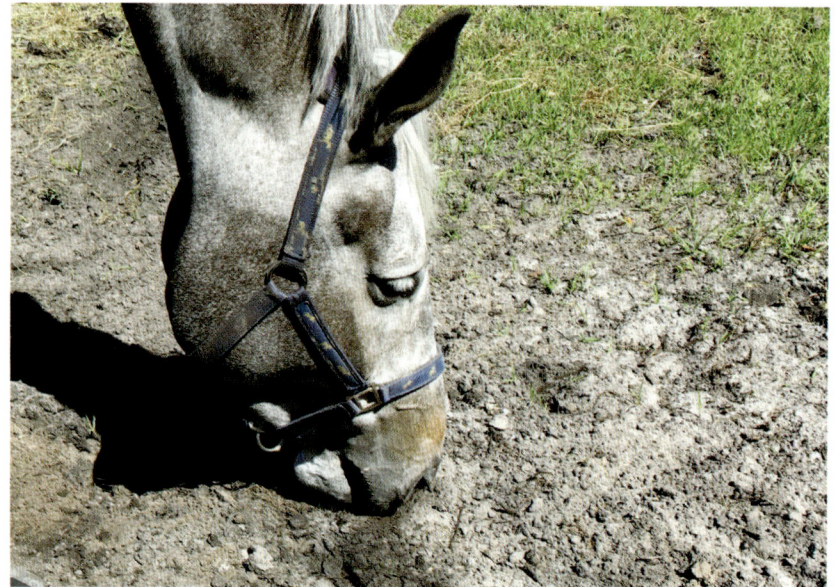

[12.39] Horses seek out what their body needs. They eat dirt for micronutrients. Here Holly searches for the right dirt in her paddock.

Unfortunately, as we've discussed, ulcers are so common that many in the veterinary community feel ulcers are "normal" for domestic horses with some studies showing at least 90 percent of performance horses having some form of them. But they do not have to be "normal" with good horse management and stress reduction.

When considering digestive issues, it is important to assess and understand your individual horse's digestive system and temperament. Sensitive worried horses will often have more ulcers than confident horses. Mares will have more ulcers than male horses. While diet and feeding schedule are key, levels of mental and emotional stress are strong contributors as well. Happy horses rarely have ulcers. Giving your horse as much choice in his diet as possible allows him to seek out what it needs.

★ **TIP:** / Test your horse and your hay to determine nutrient needs. It is money well spent to know what you are feeding your horse and where your horse may be deficient. /

Determining What Your Horse Needs

Hay grown in most countries does not require any label of the contaminants in the soil and water, so even though hay may look good, if it does not smell right to your horse, it may be missing what your horse's body needs or have toxins in it your horse does *not* need (fig. 12.39). Over-fertilization makes hay look good, but it can cause imbalances in equine digestion that can lead to disorders. There has been a tremendous amount of research on horse diets and supplements. Unfortunately, most have focused on providing concentrated feeds designed to give your horse "everything he needs" in just one scoop.

★ **TIP:** / Take your horse out for hand-grazing. Not only is it quality time for both of you, but it allows your horses to seek out plants he needs. Your horse may also want to eat dirt to gain critical microorganisms. Pay attention to what your horse consumes. /

★ **TIP:** / Allow horses to have free-choice feed all day, and particularly all night, or break up feedings into a minimum of four to five times a day. /

Feeding a low-protein grass-mix hay that allows your horse to eat at least 8 to 14 hours a day can help prevent boredom, and although it does not replace their natural grazing and browsing patterns, it can alleviate stress when horses are confined and can smell food, but cannot get to it.

★ **TIP:** / Give horses as much choice as possible. Vitamins and minerals can be set up in small self-serve containers to allow horses to choose their nutrients when they need them. /

While hay nets may slow down eating and keep stalls clean, they often cause secondary problems if hung too high by allowing dust and spores to enter the horse's nostrils. They also can cause *repetitive movement issues* due to the horse eating in an unnatural way, creating stress and tightness in his poll and neck. *Specialized slow-feeders* such as the NibbleNet® require horses take one bite at a time and can be set at a height to allow horses to eat with their heads down, but not so low they can step on it. Another option is to *spread your hay out*. Place handfuls of hay around the edges of your horse's stall so he must walk around to discover and take bites of food (figs. 12.40 A & B). "Circle feeding" hay outside encourages both exploration and movement.

Automatic feeders are gaining popularity because they save time and can be adjusted to each individual horse's needs. While these feeders are good options for many horses, those who are "always hungry" can become compulsively focused on the sound of the feeder releasing food. In such cases, it

"
Sensitive, worried horses will often have more ulcers than confident horses.

↙ **[12.40 A & B]** Hay nets can slow eating down and allow horses to nibble all day. The hay net in Photo **(A)** is set up at a correct height so Coco does not have to repeatedly twist her neck and reach up. Another way to slow feeding is to spread the hay around the sides of the horse's stall like in Photo **(B)** so the horse must move and search to eat.

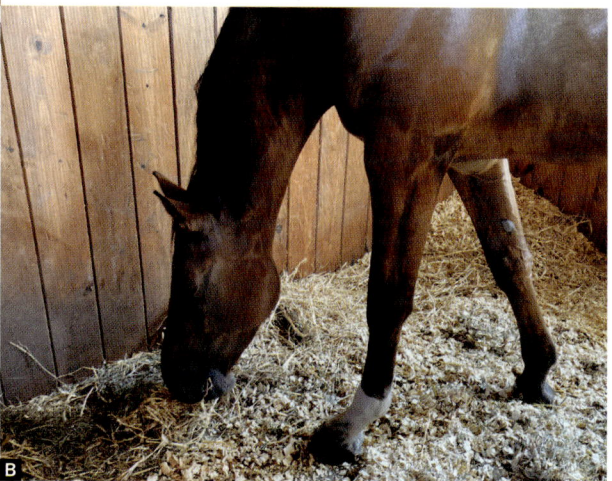

is best to mix it up—automatic feeders can just be used at night or when people are not around.

★ **TIP:** / Slow-feeding options can help horses who eat too fast have more time to chew, offsetting digestive issues as well as boredom. /

Offsetting Boredom with Habitat Enhancement ("Enrichment")

Horses can become bored, anxious, or depressed when confined in a stall all day. Making social time with friends a priority is almost always the best solution, but if a horse must be alone in a stall, then providing habitat stimulation—known as "enrichment"—is important.

Think of your horse like a preschooler on a rainy day, locked in his room. He needs stimulation for his mental and physical health. Here are few fun things to try:

- **Smell, Smell, Smell** In the wild, horses process and decipher thousands of smells a day. Place various smells—essential oils diluted in carrier oil at a 1:5 ratio, or even spices—the horse might find of interest on the wood of the stall or on another safe object (fig. 12.41). This stimulates the horse's brain and can enhance his learning ability, as well as prevent boredom. (Be careful not to put "tasty" smells on wood, or your horse may take a bite!) Mix up your smells daily. In one study that gave horses various smells to choose from, scents were rated by how much time horses spent smelling them. While in most cases horses appeared to prefer various blends of essential oils, in one case they were intrigued with the smell of fabric softener. The study showed horses have scent preferences, depending upon what they like and what their bodies might need.

★ **TIP:** / Placing various smells in your horse's stall can offset boredom and stimulate the brain. /

- **Free-Choice Food Bins** Fill several feed dishes or bins with various herbs, vitamins, minerals, and even good clean dirt. Only put a small amount—1 teaspoon will often do, at first—to see what your horse likes or needs. Put in small samples of various fruits and vegetables to see what may be of interest to your horse. (Make sure to avoid anything toxic to equines.)

STORY FROM THE FIELD: //

What Choice Can Tell Us

When working with rescue horses or wild horses, I will often lay out various essential oils for them to choose from. The exercise stimulates curiosity, allows the horses choice, and gives insight into what each horse may need. When they have ulcers, they will often choose chamomile, peppermint, or fennel. If they are worried mentally, they may choose lavender, tangerine, or clary sage, and if they have parasites, they may choose neem oil or clove. //

[12.41] The Nose-It® treat ball is a non-breakable horse toy that you can use for dispensing different scents for your horse to explore and enjoy.

- **Freedom to Explore and Socialize** If possible, allow your horse to wander out of the stall and enjoy a stroll down the barn aisle. Make sure everything is safe for both horses and people. In circumstances when the horses cannot go out to play, this can be a very social activity, but like a group of preschoolers, they need supervision.

- **Mix It Up and Change Stalls** While horses like their "safe space," and some mares may not like having someone else in their stall, one easy and good exercises is to have horses exchange stalls for an hour. Both horses usually enjoy checking out all the new smells and features before returning their own "homes."

- **Paddock Play** Every barn should have at least one paddock used for giving horses a place to roll, socializing from a safe distance, and introducing novel and interesting objects, sights, sounds, and smells. While horses enjoy interacting with other other horses as they explore, even a person standing nearby will suffice.

- **Provide Novel Objects and Textures** Using novel objects with different sizes, shapes, and textures can encourage your horse to feel and smell. Young horses, particularly, need things like this to do. Even soft stuffed animals can be pleasant friends in the stall (fig. 12.42). Horses will become bored if anything is in their stall for long, so only leave an object for a couple of days before changing it out.

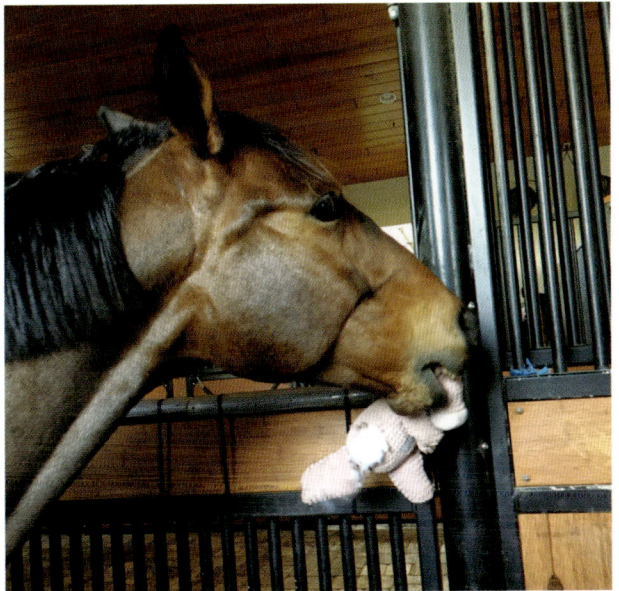

[12.42] Novel objects to chew and smell can help alleviate boredom through tactile stimulation. Radar loves biting his stuffed bear.

★ **TIP:** / Do not leave the same toy or object in a horse's stall for more than a few days as the horse will tire of it. Mix it up. /

- **Go on a Weed Hunt** Take your horse out, not just for hand-grazing but for a "weed hunt," as together you look for your horse's favorite plants. Your horse will be impressed if you know where they are hiding, and then will follow you on the hunt the next time you go.

- **Bring in New Potential Friends** Barns that have miniature goats or Miniature Horses that run around the barn all day causing havoc keep stalled horses alert and interested.

- **Give Your Horse a Massage or Stretching Session** Learning these kinds of bodywork is a great way to bond with and help your horse feel good.

- **Play Nature Sounds or Music** Music can be enjoyable stimulation to many horses, while others have little interest. As I discussed on p. 184, some horses clearly show a preference in the music they like and will navigate closer to the musical source when they enjoy it.

★ **TIP:** / Some horses enjoy listening to music for parts of the day, but played too long, loud, or often, it can transform from enjoyable stimulation to a stressor. /

Shipping, Showing, and Selling Horses

For many people, a horse is lifetime partner, and they would never contemplate selling him. But for others, horses may be a business, a lifestyle, and a sport. The reasons for having horses may also change throughout a human's life. The reality is, at some point or another, most people will ship, show, or sell a horse.

STORY FROM THE FIELD: //

Music to Improve Performance

A study conducted by Polish researchers investigated the effect of music on performance and stress in 70 three-year-old racing Arabians. The horses had music played for five hours every afternoon for three years. The music genre was "new age guitar." The 30 horses in the control group had no music or other variables. Heart rate and overall emotional state of the horses were measured six times every 30 to 35 days. Within the first 30 days, researchers noticed a noticeable difference—the horses listening to music were more relaxed and their level of relaxation and positive emotional state increased over the next several months. The researchers also monitored performance. Horses listening to music noticeably out-performed the control group. The researchers concluded that music played in barns positively affects horse welfare and improves performance. //

— Shipping

We talked a bit about the stress caused by shipping on p. 262. Horses who were shipped as youngsters with their mothers or with friends usually take later transportation in stride, but horses who have never been separated from friends and then are shipped experience a high degree of stress. The same guiding principles apply to shipping as other areas where stress is experienced: if there are other horse friends involved, then horses usually adapt well. If they have to adjust to an experience on their own, then it may take time for a horse to learn how to travel by himself.

★ **TIP:** / Ship horses with others, preferably with one being an experienced traveler, to help alleviate stress. /

Shipping is an unnatural activity for horses, so the more you make the trailer and traveling a "safe space," the less stress your horse will experience (fig. 12.43). Provide fresh water and hay in the trailer when possible to help the horse associate his home stall with a trailer stall.

★ **TIP:** / Allow your horse to explore and feel safe in an enclosed space by feeding him in the trailer and allowing him to go in and out at will. /

Before shipping your horse, practice loading him. Some horses do not like confinement and so they have to learn to adjust to the small space of a trailer. When possible, it is helpful to feed your horse in the trailer while allowing

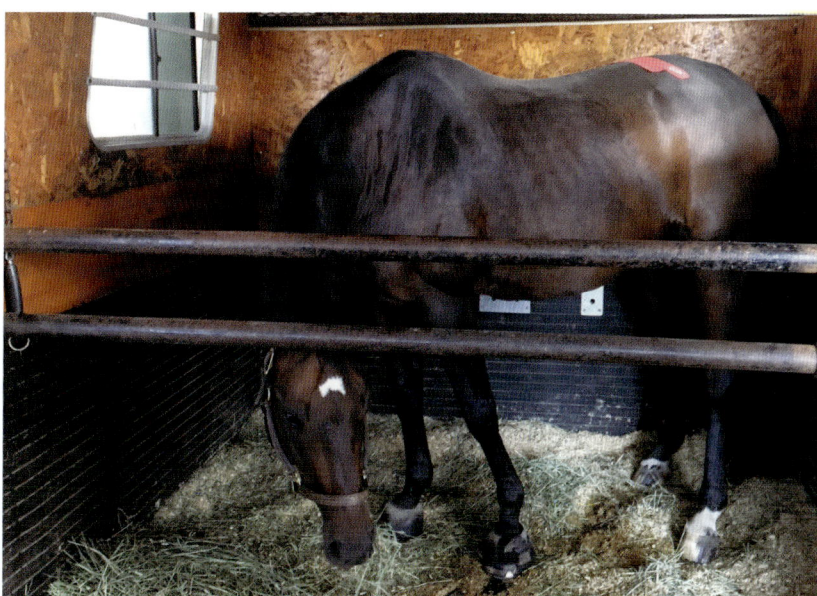

→ [**12.43**] Ascot has an entire "box stall" in the trailer with bedding and food so he can turn around and lie down if he wants.

him to enter and exit as he wishes. This allows your horse to explore and become familiar with the trailer on his own while developing a positive association with it.

★ **TIP:** / **Use familiar smells when trailering your horse. Even a brush or a towel with his horse friend's smell on it can help him relax.** /

Using familiar or relaxing smells can help horses feel safe in trailers. Many times when hauling new horses, wild horses, or horses alone, I take some manure from the horse's stall, along with a friend's manure when available, and place it in the trailer. Like following breadcrumbs, the horse would track the familiar smell of safety right up the ramp. Remember how powerful smell is for determining safety: even when it looks suspicious, if it smells safe and no other negative stimulus takes place, it is likely your horse will feel safe.

As I've recommended in other instances, using aromatherapy can also be effective. When smells are programmed into a pleasant and relaxing activity, such as massage, grooming, or eating, then when the horse smells that smell, he cannot help but have an involuntary reaction and relax.

_Managing Performance Stress at Shows, Races, and Events

Horse-human sports should be fun for horses as they enjoy socializing with new horses, running around in new habitats, and generally engaging in physical activity. Unfortunately, money and the need to "win" often drives the riders, not the horses' enjoyment or even their own. Some horses are better adapted to performance than others, but the lifestyle they must endure even under the best care is often stressful. Modern veterinary medicine tries to address the *physical* stress placed on sport horses, but little help has been available to manage the mental and emotional stress they experience.

Horses do not dream of being stars, but many people dream of having champions. There is little data that demonstrates horses would rather be doing anything other than eating grass in a pasture with their friends. However, with modern breeding and genetics, top performance horses are being created. Cloning, although relatively new and mostly being done with polo ponies, may be opening the door to producing top equine athletes that are both mentally and physically adapted to a sport. Time will tell.

A horse's temperament and physical ability can affect the type of discipline in which he can excel and the level of stress he will endure. However, the most important criteria affecting performance stress in horses is how they are managed, who is training them, and how they are trained. Horses living with friends and being trained with friends (and this includes human friends) will experience less stress than those living and training alone or with people (or other horses) who cause them to worry. Performance horses allowed free time to roll, play, and run around will have less stress than those who spend all day in a stall. Horses trained with very few devices restricting their ability to communicate, express themselves, and move their bodies will have less stress than horses trained with restriction. Horses who remain with one rider or trainer and travel with friends to familiar showgrounds or tracks will have less stress than horses who are frequently sold and moved from stable to stable.

Preparing horses for showing is similar to other types of learning. Assess your horse and know his personality and how he handles other sources of stress. Expose your horse to competitive environments prior to entering him in a show, race, or event. Make sure your horse feels safe traveling. You can simulate many of the activities your horse will experience in performance, such as loudspeakers, strange smells, horses passing from both directions, and strange people milling around to help prepare him.

Horses who have learned to feel relaxed at shows, races, or events often enjoy the competition. It is like a party for horses! Remember that growing large social networks is wired into horses, but it assumes they have social skills and can feel relaxed around social activity. Domestic horses often get so excited and overwhelmed by the stimuli of strange smells, other horses, new noises, and unfamiliar environments that they do not know how to behave. There are many potentially top performance horses who are talented but cannot handle the stress of competing because they were forced to compete when they were not ready mentally. Giving your horse frequent opportunities to experience such stimuli *without* the added stress of competing can help him adjust.

★ **TIP:** / Simulate, sights, sounds, smells, and activities from competitive environments to help prepare your horse for shows, races, and events. /

If you horse feels safe with you, then wherever you go together, "safe space" goes too. Hence the importance of your relationship really comes to the surface in new environments. Build your relationship at home and then take it on the road, always keeping in mind your horse's welfare. Since horses pick up our emotions, your horse can sense your own level of stress as well. So as we discussed on p. 285, it is key to manage your own stress.

You cannot undo a bad experience if something happens to make your horse worry and feel unsafe.

★ **TIP:** / Allow your horse to learn how to handle stress and feel relaxed at shows or events prior to competing. /

Tips for Keeping Stress Low When Competing

- **Travel and Stay with Friends** By far the best stress-reducer for horses is social distraction from friends. If your friends are okay, then you're okay. It is always best when experiencing a new situation for the first time to have your friends around, and you want to have at least one calm one.

- **Practice and Prepare Your Horse at Home** Horses who learn to be by themselves and feel safe in various environments experience less stress. Taking short trips around the neighborhood in a trailer and always returning helps your horse have a positive experience and learn he will come home. Many shows only have small tent stalls. When you can practice spending time in a tent stall at home, it will eliminate one variable of potential stress.

- **Use Smells from Home and Aromatherapy** As we have discussed for lessening stress in other scenarios, comforting scents can help your horse feel relaxed in new places. Bringing shavings or straw from home that smells like your horse or their friends can immediately relax your horse if placed in a new stall. Relaxing aromatherapy blends or other essential oils your horse may like can help override worrisome smells at shows or new places.

- **Limit Stressors Until Your Horse Can Adapt** When introduced to one stressor at a time, horses adjust well to unfamiliar noises, smells, and sights. Limiting sensory input until your horse learns to associate each stimulus with a positive experience can help him manage his stress.

- **Ensure Your Horse Can Get Enough Sleep** Control lights, noise, and distractions as much as possible. Sleep deprivation is a common problem at competitions, causing horses to become spooky and be more likely to be injured. Make sure your horse has a comfortable place to lie down and get his necessary REM sleep.

> **Horses who have learned to feel relaxed at shows, races, or events often enjoy the competition.**

- **Keep Forage and Clean Water Available** It is best if your horse can eat all day and night. Drinking and eating when they feel like it can offset worry as well as help relax horses, preventing acid build up and ulcers. Bring hay from home. Make sure your horse has time to eat before competing.

- **Use Calming Signals Around Your Horse** Reinforce feelings of safety with soft eye blinks, yawning, and relaxed posturing. Think "sleepy time" energy—the state of energy right before you fall asleep. Horses pick up on the strongest energy around, so keep yours relaxed.

- **Allow Your Horse to Roll, Stretch, and Buck** When turnout isn't available at a venue, let him play on a longe line before you get on and warm up (although do so safely and only for short periods of time—see p. 326). Being confined in small spaces like trailers and tent stalls can limit normal movement.

- **Spend "Hang Out" Time with Your Horse** Just be together and let him graze or explore. It will help your horse relax to just be with you and not being asked to do anything. Practice breathing together.

- **Stretch and Massage Your Horse** Muscles are muscles and horses store emotional stress in their bodies, too.

- **Use Relaxation Techniques** Try acupressure, myofascial release, and the Masterson Method to make your horse feel good and help him relax.

- **Keep Your Horse's Routine Similar** When on the road, traveling to competitions, try and keep your horse's routine as similar as possible to home.

— Helping Horses Adjust and Adapt When Sold

Horses do not have any choices when they are in human care. Humans can buy and sell horses like cars with little regard for the horse's friends or welfare. For a species whose core is social relationships, this often causes the most emotional stress as well as welfare issues. While many people are careful to whom they sell horses and try and track their horses throughout their

↑ [12.44] New friends share hay on the ride home. The horse on the left was a slow developer and had difficulty with the fast-paced world of show jumping. He found a home with a dressage rider willing to take time with him. When his new owners came to pick him up, they brought another horse from their farm so when the former jumper arrives at his new home, he will already have a friend.

lives, there are lots of horses who are bred for sports and are moved through numerous sales and brokers, with the original breeders and trainers never knowing where their horses end up. Because there is no tracking for and oversight of the sale of most horses, the best protection a horse has is to be able to adapt to variety of people, training, environments, and tasks.

Horses who have had a good upbringing to prepare them for changes usually adjust well. But those who did not learn basic horse or horse-human interaction skills early on often have a difficult time learning them later. They often remain worried and reactive. And because they cannot meet human needs for performance, they get sold frequently or end up in a horse rescue.

Horses do not understand being separated from family and friends when they have done nothing behaviorally, from their standpoint, to warrant leaving the herd. So it is critical if you are ever going to sell a horse, that you teach your horse how to be separated from friends and transfer friendship to humans to minimize his stress (fig. 12.44). Horses who have grown up with people and have had many different humans in their lives during their young horse education seem to acclimate to having various owners and moving around. But horses who have been bred and raised on a farm and then perhaps as four- or five-year-olds are sold and shipped to new homes with new people endure various degrees of stress.

★ **TIP:** / **Teach your horse how to feel safe being alone and in new situations in the company of people to help prepare him for someday going to a new home.** /

When I ran an equine sport horse brokerage business (Equines LTD) for many years in California, I set it up to be as horse-friendly as possible to minimize stress in horses and educate buyers and sellers how they could also assist their horses in adapting to change. My belief was and is that the better you prepare your horse for changes, the easier it will be for your horse to adapt should he need to. The following 16 considerations will help reduce your horse's stress should a sale come to pass:

1. Allow your horse to greet and establish friendships with other people besides you.

2. Practice loading in and out of different trailers with different horses and alone.

3. Go to shows and events with different horses, even if you are not competing.

4. Put your horse in a different stall or paddock than he is used to with the same food, but with different horse neighbors.

5. Give your horse smells from a new place before going, if possible, so when he gets there, he will feel like he has been there before.

6. Leave your horse alone for short periods of time to get him used to being alone. Gradually increase time alone. Feed him alone so there is something positive to do when he is without company.

7. Ride your horse in various tack to get him used to what other people may use.

8. When possible, have the person who has bought the horse come ride him at a familiar facility.

9. Make sure your horse is comfortable with veterinary procedures and with the farrier.

10. Send familiar smells or objects with your horse to the new barn or owner.

11. Write down your horse's likes and dislikes so the buyer knows your horse's personality and how he responds to various stimuli.

12. If possible, visit your horse during the first two weeks following a sale to see how he is adjusting, and help the new owner with any issues.

13. Prepare your horse for any differences in lifestyle, such as living in a large open space when he has been confined or learning how to live in a stall if he has been in a pasture. Supervise all transitions and use gradually increasing time blocks to see how your horse adapts and prevent injury until he is comfortable.

14. If a horse is going to be living in a group or herd, as might be the case at an equine retirement home, then make sure he has learned spatial awareness and spatial respect so he is less likely to get hurt by other horses.

15. Once a move or sale is imminent, send familiar hay and feed with your horse so he has a similar diet.

16. Microchip your horse and register with a microchip tracking app so you can check up on your horse's welfare throughout his life.

Buyers and sellers should be encouraged to talk with each other, as well as to the horse's veterinarian and farrier. A horse that is for sale should have a history sheet that outlines as much information known on the horse as possible, in case for any reason the buyer and seller cannot directly communicate. This should include an evaluation of temperament type and any potential behavioral issues. Full transparency and disclosure of all issues help to ensure that a buyer knows exactly what she is getting and how best to help the horse adapt to his new home.

★ **TIP:** / Give buyers a "guidebook" to the horse you are selling. Let them know what the horse likes and dislikes, as well as his history. Good buyers should want to know as much as possible about a horse they are going to purchase. /

Today, technology with online videos and advertising has linked buyers and sellers together, giving people ample opportunity to easily view videos of horses prior to seeing them in person. Some people even buy horses off video alone, never having actually met them. It is not recommended to ever buy a horse you have not met, unless you implicitly trust the person you are buying the horse from and know everything about the horse. On the flip side, horses like to meet prospective owners. The perfect horse for one person may not be the perfect horse for someone else. Horses are not cars, so meeting any horse before you buy is important for both of you.

Sadly, the horse industry is dominated by economic incentives for buying and selling horses. Whether as investments or for pleasure, horses have become a commodity that can be bought, sold, and traded like cars

> **The better you prepare your horse for changes, the easier it will be for your horse to adapt, should he need to.**

with little regard for the animal's mental and emotional welfare. Horses must tolerate separation and travel over and over again, so prepare your horse for these challenges to help reduce his stress in the long term.

Solving Common Behavior Issues Related to Stress

Having worked with thousands of horses and people over the years, it still saddens me when I arrive to solve a "behavioral issue" and find a horse who is simply communicating his worry or pain. Most behavioral issues have some form of current or past physical pain at their core. What follows are a few of the most common behavioral issues related to stress, with causes and some simple solutions.

_ Issue: Walking on Top of People (Inability to Lead Safely)

- **Primary Cause:** The horse has not learned spatial respect with people. The horse is overeager or worried about getting where he wants to go.

- **Solution:** People are quick to put chains over the nose of a horse in this scenario when simply teaching the horse spatial respect and awareness of where he needs to place his body next to the person is safer and easier. A simple method is to keep eye contact with the horse and twirl the end of the lead rope in front of the horse when he starts to walk in front of you. Circle frequently. Practice spatial respect and awareness exercises as shown on page 142.

- **Secondary Cause:** The horse has had a bad experience or is frightened walking in a particular area. The horse may not see well and seeks to get away as quickly as possible. This is often the case walking through doors or into stalls, particularly if the horse has been hit by a person or has difficulty going from light to dark, or vice versa.

- **Solution:** Have the horse's eyes checked by your veterinarian, and be patient in leading situations to give your horse confidence and build positive associations to replace negative ones.

Issue: Spooking/Bolting

- **Primary Cause:** Feeling unsafe or has discomfort.

- **Solution:** Allow the horse to understand and learn what stimuli triggered the reaction; check tack for discomfort, particularly restrictive tack and saddle fit; check for ulcers, body soreness, and pinched nerves.

- **Secondary Cause:** Testing the rider to see if she is paying attention, or the horse does not see well or is deaf. Horses who are deaf will often spook when something comes from behind them, while horses who do not see well often overreact to movement and cannot clearly see the objects that moved.

- **Solution**: Pay attention to your horse and think ahead of him. People must be as or more aware of staying safe than their horses. Test your horse's vision and hearing. You can do this by using various sources of noise around your horse, and see if your horse can stand still and track the sound with his ears (see p. 227). Allow your horse to explore novel objects when he has the ability to escape if he feels he needs to.

Issue: Running Off, Getting Quick, Bucking

- **Primary Cause:** Physical discomfort. Horses run from pain, whether it is in their mouths or their backs. Their natural reaction is to "escape."

- **Solution:** Check body for discomfort: tight muscles, soreness, subluxated vertebrae. Have a veterinarian, chiropractor, and a good bodyworker go over the horse. Check your tack for fit and comfort and appropriateness (see p. 316). Check for ulcers. Horses suffering from them can be fine some days, then suddenly they can start getting "quick" under saddle when the ulcers act up from stress or not eating regularly.

- **Secondary Cause**: The horse feels good and needs more exercise. Too much feed or the wrong feed can make horses anxious and overly energetic. Remember, horses are designed to move all day long.

- **Solution**: Give you horse turnout time to run around and the opportunity to buck and move his body in a way that unlocks discomfort and stiffness.

Issue: Pulling on the Reins, Raising Head Up or Pulling Head Down, Shaking Head

- **Cause:** Discomfort in the mouth and face. In these cases, the poll and temporomandibular joint (TMJ) is often tight and locked. The neck may feel "stuck" in the horse that shakes his head to find comfort. Poorly fitting bits, nosebands, and other tack, or saddles that pinch the withers will cause this behavior.

- **Solution:** Have a good equine chiropractor and dentist check your horse. A horse is not being "bad" when he resists in these ways, but rather trying to seek comfort. Assess and treat the tight jaw, TMJ, poll, neck, and shoulders. Adjust saddle fit to ensure the withers are clear.

- **Secondary Cause:** Some neurological disorders can cause a "nervous tic," and horses can also sometimes learn compulsive avoidance behavior such as head-shaking, grabbing the reins, and pulling the head up or down as a past association with worry from being asked to do something the horse either cannot do or experienced pain when doing it.

- **Solution**: Have your veterinarian check your horse.

★ **TIP:** / There is no such thing as a "hard-mouthed" horse. All horses have soft mouths—unless a human has caused scarring and damage. /

Issue: Grinding Teeth

- **Cause:** Worry and stress. Horses learn coping behaviors to handle their stress. Grinding teeth can be like chewing gum in humans. Horses use their mouths constantly to communicate, touch objects and other horses, and eat. When their mouths are restricted so they are not able to move their tongues, jaws, and mouths, many horses feel worry and will grind their teeth instead.

- **Solution:** Give your horse freedom in his face. Loosen or do away with the noseband. Try a bitless bridle, as many horses who grind their teeth are experiencing discomfort and worry from the bit. Help your horse learn to relax. Use aromatherapy and other stress management solutions like breathing together, calming signals, and body relaxing cues, such as using touch on the withers.

Issue: Pulling Back in the Cross-Ties or When Tied

- **Cause:** Horses naturally do not like restriction. If they have ever been scared while tied or feel discomfort when tied or in cross-ties, the instinctual reaction is to pull and get away. Tension and discomfort in the poll often trigger a horse pulling back.

- **Solution:** Make sure your horse is comfortable in the cross-ties or being tied—he should be able to hold his head in a relaxed position. Often cross-ties are set too high and tight (see p. 326). Check the poll and TMJ for tightness and discomfort. Have an equine chiropractor check for subluxations. Massage tension in muscles. Use kinesiology tape over the poll. Magnetic therapy or other modalities can also help relax the muscles behind the ears and over the poll. Allow your horse to choose to learn to override tension in the poll: Loop a rope around a secure post and hold onto it so you can control the tension. Put enough tension on the rope that it triggers a pulling-back reaction, but at the same time, offer your horse a carrot or other treat in front of his nose, but do not give to him. Release the rope immediately if the horse starts to pull back, but only give him the treat if he steps forward. You want your horse to make a choice and change the negative trigger into a positive trigger. The pressure behind the poll has triggered pulling back and worry, but now your horse can make a decision: step forward and take a carrot or pull back. Horses like choices. Once your horse makes the choice to eat the carrot in response to pressure on the poll, after a few repetitions, the pressure on the poll now means "carrot."

- **Secondary Cause:** Smart horses sometimes learn they do not have to be restricted if they don't want to be and can pull back to get away.

- **Solution:** Don't restrict and tie your horse, but rather teach him to stand comfortably while you handle him on a loose lead. Usually, if the horse is held loosely enough so he can put his head down and relax, he will not pull back. This can then eventually be translated to tying loosely.

Issue: Biting

- **Cause:** Gender and age have a lot to do with this behavior. It is common for stallions and young horses to bite as they use their mouths to play and have not learned "horse etiquette" yet. Mares use their mouths more for "talking." Identify the underlying cause: Is it playful nipping? Aggressive or protective "get out of my space" biting? Pain will cause many horses to bite as they do not want to be touched when they hurt. Horses also naturally want a bigger-than-usual "space bubble" around them when they do not feel they can "escape" quickly if necessary because they hurt.

- **Solution**: When dealing with playful gender- and age-related biting, give the "boys" oral toys to mouth—various textured objects in the stall and paddock. You can use your voice to show your disapproval when this behavior is exhibited, and that can be effective with many horses. Work with your horse on spatial respect and paying attention to you. Keep a squirt bottle with water at the ready when a horse is particularly aggressive, as some horses think it is a game to bite when a person is not looking. Particularly with stallions, you must be careful to always pay attention to mouthiness. While a chain over the nose can keep a stallion occupied, it will often create a worse problem when it comes to nipping and biting. Try using soft stretchy tape (such as kinesiology tape or electrician's tape) and placing it on the horse's nose or taping his lips gently so the whiskers pull if the horse tries to bite. With most male horses, this simple technique will keep them entertained and not biting you but rather learning that if they try and bite, it is uncomfortable. This allows the horse to make a decision, which he likes. Horses who bite from pain or anger must be treated differently. Be careful to make eye contact with your horse and talk to him. Find out why your horse is biting—pain, association, anticipation, worry? Many mares and sensitive horses will bite when getting girthed or seeing something that has caused them worry. In these cases, as long as they direct their biting to the air and not to you, allow them to express themselves. Trying to stop a horse from communicating his feelings—which may be expressed by the mouth—is like saying, "Shut up!" to a person. When a horse has to rebuild trust with humans, treat biting as communication and determine the cause, which will help with the solution.

_ Issue: Pawing and Kicking

- **Cause**: Pawing and kicking are usually only seen in stabled horses. This is a sign of impatience and lack of having the ability to move around at will and choose friends, food, and location. Mares will often kick when the horse

> **Fear and worry can easily shift to curiosity and play when you keep learning enjoyable and light-hearted.**

next to them does not respect their space. Just because a stall wall is there does not mean the space is enough for many mares. Pawing while tied often indicates discomfort, boredom, and frustration. Smart horses learn how to get people's attention by pawing. Like kids, they think that some attention, even if it is a correction for "bad" behavior, is better than no attention.

- **Solution:** Don't leave horses in stalls, cross-ties, or tied up for long periods of time. Make sure horses have activities to occupy them and are able to move around enough and get enough exercises to be comfortable. Use enrichment activities and objects to offset boredom in the barn (see p. 246).

- **Secondary Cause**: Watch your horse's eyes to see what they are expressing, as pawing and kicking can mean they do not feel good. Many horses will paw and kick if they are starting to colic or have digestive discomfort. Keep an eye on any horse who usually does not paw or kick and suddenly does. Pawing can also mean your horse wants to roll or lie down. Horses like to make a soft bed, and when they do not have the space or the right material, they will try by pawing to dig up dirt or other material to make a bed or a place to roll.

- **Solution**: Call your veterinarian if your horse is pawing and kicking and it is not normal for him, as you do not want to risk colic. Get him out of his stall and walk him around—keep him moving. Horses are designed to walk all day long, so often just taking your horse out for some exercise can help.

Issue: Overreactive to Stimuli

- **Cause**: "Vigilant" horses may be genetically predisposed to be overreactive to noise, touch, smells, movement, or novel objects by human standards. But they can learn to relax. Horses who have experienced trauma can react in a way similar to humans with PTSD. Horses have good associative memories, and it takes time, understanding, kindness, and patience to undo negative associations.

- **Solution:** Assess your horse's level of reaction to various sounds, smells, and sights. Determine your horse's "safe space" boundaries. Some horses feel they are much bigger than they are, and while one horse may be fine with an object or noise with 6 feet, another horse may require 25 feet to feel safe. Always start with your horse's safe space. Horses who have poor vision will often be more sensitive to sound. A horse who is deaf may spook or be aggressive when something comes from behind when he did not see it coming—because he did not hear it either. While some horses can be "desensitized," most sensitive horses prefer learning how to be confident and handle various stimuli. Make learning fun for your horse. Use positive rewards so your horse learns to associate loud or unusual noises, sights, and smells with positive reinforcement. Fear and worry easily can be shifted to curiosity and play when you keep learning enjoyable and light-hearted.

Our Humane Responsibility

Happy horses are usually healthy horses. So it is important to understand and manage your horse's psychological welfare. Those people who spend the time to just "be" with their horses on "horse time"—doing the things a horse likes to do—often have the most rewarding relationships (fig. 12.45). Mental and emotional health go hand in hand with physical health. Science is proving this to be true with horses and humans. Humans must also manage their own stress if they are to be interacting with horses, because as emotional social beings, horses will pick up on a person's state of mind.

Understanding your horse's natural instincts and behaviors, being able to accurately assess your horse's personality, and having the ability to clearly communicate with your horse and allow your horse to "talk back" will provide the foundation for a positive horse-human relationship. Being empathetic as well as having a good scientific understanding of stressors for your horse allows you to better manage your horse's mental and emotional stress in all phases of horse-human interaction. From the breeding shed to raising a foal, and from starting a young horse as a riding companion to competing in high performance sports, the horse is just a horse and seeks the same safety and comforts.

Horses have entrusted us with their lives and given us their cooperation. It is our responsibility to minimize their stress and oversee their welfare in every aspect of our interaction.

↑ [12.45] Spend time hanging out, walking around, and exploring the world with your horse. Talk to your horse—yes, carry on a vocal conversation about what you see and sense.

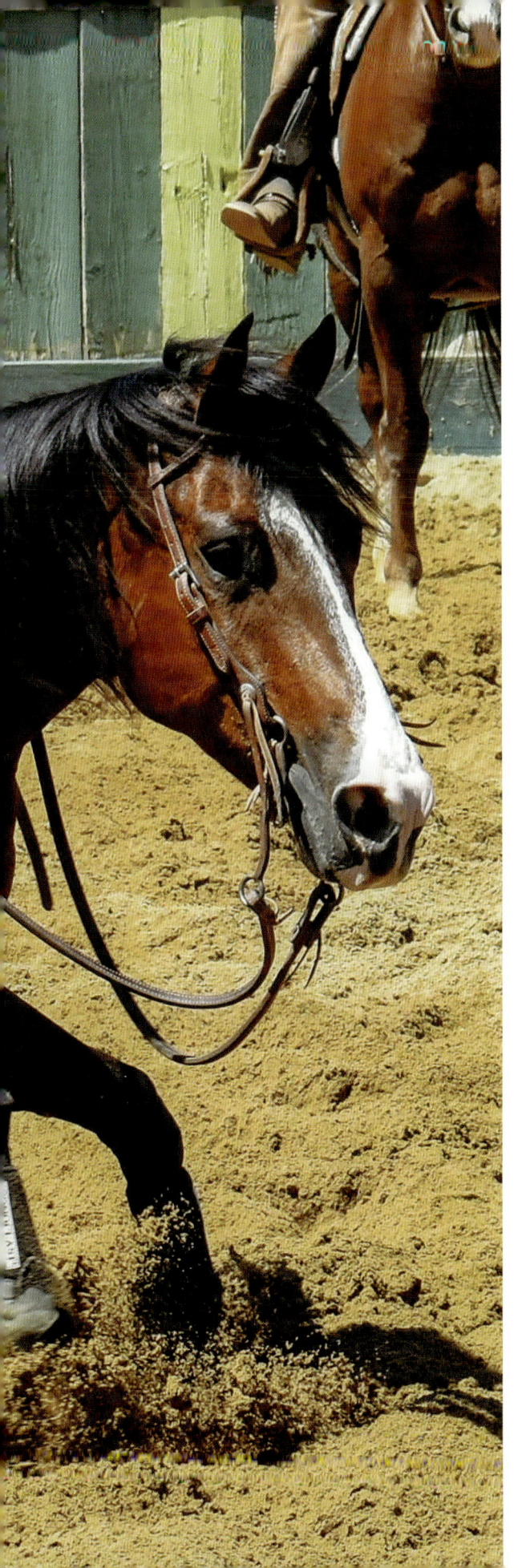

PART THREE

WHAT WE DO WITH HORSES

ETHICS, ECONOMICS & WELFARE

← [13.0] Horses can enjoy competitions and activities with us, but their welfare lies solely in our hands. It is our responsibility to pay attention to the physical and mental health of our equine companions.

INTRODUCTION
TO PART THREE
BY
DUNCAN PETERS, DVM, DACVSMR

Dr. Duncan Peters is co-owner of East-West Equine Sports Medicine, LLC, in Lexington, Kentucky, and a member of the United States Equestrian Federation (USEF) Veterinary and Drugs and Medication Committees. He is a USEF evaluation veterinarian for the eventing, dressage, and endurance discipline international teams and a multi-disciplined Fédération Équestre Internationale (FEI) veterinarian.

> **Our choices in our dealings with horses moving forward will ultimately affect how we are going to be perceived by future generations.**

I met Mary Ann Simonds almost 40 years ago. Interestingly, the circumstances then were similar to now with my review of this book—at that time, we were talking about the ethics of the use of horses in our society, and here we are again. A lot of the same questions are still facing us. At times it has felt like we are moving forward, yet at others, like we are treading water.

The way this book is laid out allows us to examine aspects of how we deal with the interaction between what horses are, what they need, and how we "use them" in what are mostly human pursuits. Mary Ann used previous chapters to explore what horses are (their foundation), how they have developed, how they interact with humans, and how we have dealt with that in a chronological sense as times have changed. We now move on to the ethics of what we are doing with horses, the cost involved, and how we attend to and maintain their welfare. This part of the book is very important, because our choices in our dealings with horses moving forward will ultimately affect how we are going to be perceived by future generations. We talk about dogs and cats as "pets," and in a lot of ways, we talk about horses as pets, too, but on the other side of it—the competition side—there are economic and ethical questions that we must confront.

Duncan Peters, DVM, DACVSMR

There is a balance, a marriage, that we must manage when it comes to caring for the sport horse. The horse just wants to be out there grazing and be peaceful and interact with his own kind, and we take him out of that comfortable environment and often put him into other environments only to fulfill our perceived needs or competitive drive. We, in many ways, "force" ourselves on the sport horse for the types of activities we are interested in pursuing, be it endurance riding, be it polo, be it show jumping, be it eventing. (I should note that I believe that we do this to sport horses in a way that is different from racehorses. I think racing, how you view or how you consider the sport's relationship with the horse, is a kind of separate commodity.)

Recreational riding has its own ethical and welfare issues where we should ask ourselves the same questions: To what ends are we utilizing horses in this manner? To what balance? What is it that we want and need and how does that match up with what the horse needs?

Mary Ann and I have very similar viewpoints on these subjects, which is how we came together to begin with. We don't decry horse sport in general or vilify specific disciplines. We believe some horses can certainly be perceived to enjoy sport horse activities, as well as the partnerships they may develop with humans. The matter to note is that while we can use horses in competition, we have to remember they are not strictly a vehicle for us; they are not a machine for us. We have the responsibility to pay attention to their welfare and their needs *as horses*. It is pretty common to talk about abuse in horse sports—but there are so many different types of abuse that go way beyond "beating on" a horse. And I think we need to be able to appreciate those more subtle abuses and stand up against them as strongly as we do the obvious ones.

Consider some of the topics Mary Ann has covered in these pages: Should competition schedules be questioned? Should transportation to a season's worth of events over many months and many miles? What about putting horses on airplanes for 24 to 40 hours? We have to look at ways to improve every aspect of horse sport and competition, be it scheduling, be it transportation, be it housing, be it turnout.

I'm not against competition at all. But I think we have to incorporate ways to allow the horse to reboot and refresh and to make the experience pleasant for him, because no horse would do it on his own. No horse would go out and play polo on his own. No horse would go on an event course on his own or perform a four-minute freestyle. I think we have to always keep that in mind. I bring it up to clients all the time. If it was up to the horse, *he wouldn't do it*. I don't think that means he can't do it or we shouldn't ask him to do it, but I do think we have to be aware that we are asking him to do some extraordinary things.

I do see some changes already taking place that can improve sport horse welfare in the ways Mary Ann and I think are important. Some of the showing "circuits" out there now allow horses to spend more time in one place while still being able to compete. It used to be that shows were week after week of trailering from one venue to another. In some ways, having horses compete on a circuit alleviates those problems. Of course, there are also drawbacks: Horses maybe don't get the turnout time they would at home or have the ability to be in that home environment that allows them to truly refresh in between competitions. But I do think that circuits minimizing traveling is a good thing for horses.

I think the stewardship put in place by the US Equestrian Federation and the FEI in terms of looking out for the welfare of the horse have been beneficial. Some circuits are now implementing light-dimming in all barns and mandatory quiet time after certain hours, which is enforced with security. How well it is working remains to be seen, as certain show activities do tend to go on at night, like braiding and other preparations. Of course, these need to be accommodated somehow, but at the same time, these changes to offer horses a more rest- and sleep-friendly environment when on the road are a good step in the right direction.

Transportation is improving, too, with new rules that require shipping companies to give horses more room in trailers and on airplanes—things that were not options in the past. Stewards are in the practice area and show ring, looking for things that might be questionable in terms of tack and equipment, and horse handling and riding practices. Rules are now in place for how much flexion a horse can have, and for how long. There are rules for how rails and jumps can be set in the schooling area before a horse goes in to compete, and for how long a horse can be warmed up. I also know, being on the veterinary committee for the United States Equestrian Federation, that there are rules that are going to go into effect to enforce good practices for longeing and the use of specific kinds of equipment. All of these are positive changes. They may be baby steps, but they are more steps in the right direction, nonetheless.

There is certainly more of a perception now that we need to maintain the welfare of the horse, that we need to pay attention to it, versus some of the things we've done and practices that have been accepted in the past. Change comes from both public pressure and from inside the sport. There is no doubt the loudest voices are outside the sport, and often they demand the most attention. But I think people *within* the sport are taking note. People like Mary Ann are saying, "Hey, look. If we don't do

Duncan Peters, DVM, DACVSMR

something, if we don't self-regulate, if we don't self-govern, we're not going to have a choice."

The problem is those baby steps we have to take within the competitive horse world to change things, while the loud people outside the sport rant and rave about how we need to take giant steps or how horses shouldn't be doing sports at all. "Horses shouldn't be in stalls!" some critics might say. Well, our answer within the sport now is, "Okay, we won't keep horses in 8 by 8 or 10 by 10 stalls anymore. All stalls will now be a minimum of 12 by 12." Certainly, if you compare a 12 by 12 area to a horse being out on the range, running across the mountains, that's nothing. But those horses in stalls don't have to worry about the mountain lion coming up behind them, either.

Every discipline has its own history and set of guidelines it has followed, and bridging all those disciplines in order to establish a set of standard horse welfare guidelines is very difficult. It is interesting because even among the different equestrian disciplines, there are comparisons where people from one sport feel it is kinder or more natural for a horse than another. People with Saddlebreds will look at a jumping horse and say, "Why should a horse ever jump? You talk about being cruel to horse? Why should they be pointed at obstacles and forced to jump them?" But others will look at a Saddlebred and ask, "Why should a horse wear those big shoes? Why can't he travel naturally?" Across the board, it is very difficult to get a consensus.

And that is where this book comes in. It speaks to the issues in generalities, rather than sport by sport, and it provides suggestions and some directional ideas of how we can all move together for the good of the horse, which I think is so important.

There are still more questions than answers when it comes to the economics and welfare aspects of competitive horse sport. That is the tough part. How are we going to deal with the environment of each particular horse in each particular sport and every particular thing they are being asked to do? How are we going to deal with the fact that professionals make money training and showing and selling horses and that impacts choices made? The answers will come over time and will develop but it is going to take a lot of work from a lot of people, and a lot of balancing the good and the bad…and compromise along the way.

[13.01] The choices we make to reduce stress in domestic horses today will ensure we will be able to enjoy riding and competing with them for years to come.

PART THREE / Introduction by Duncan Peters, DVM, DACVSMR

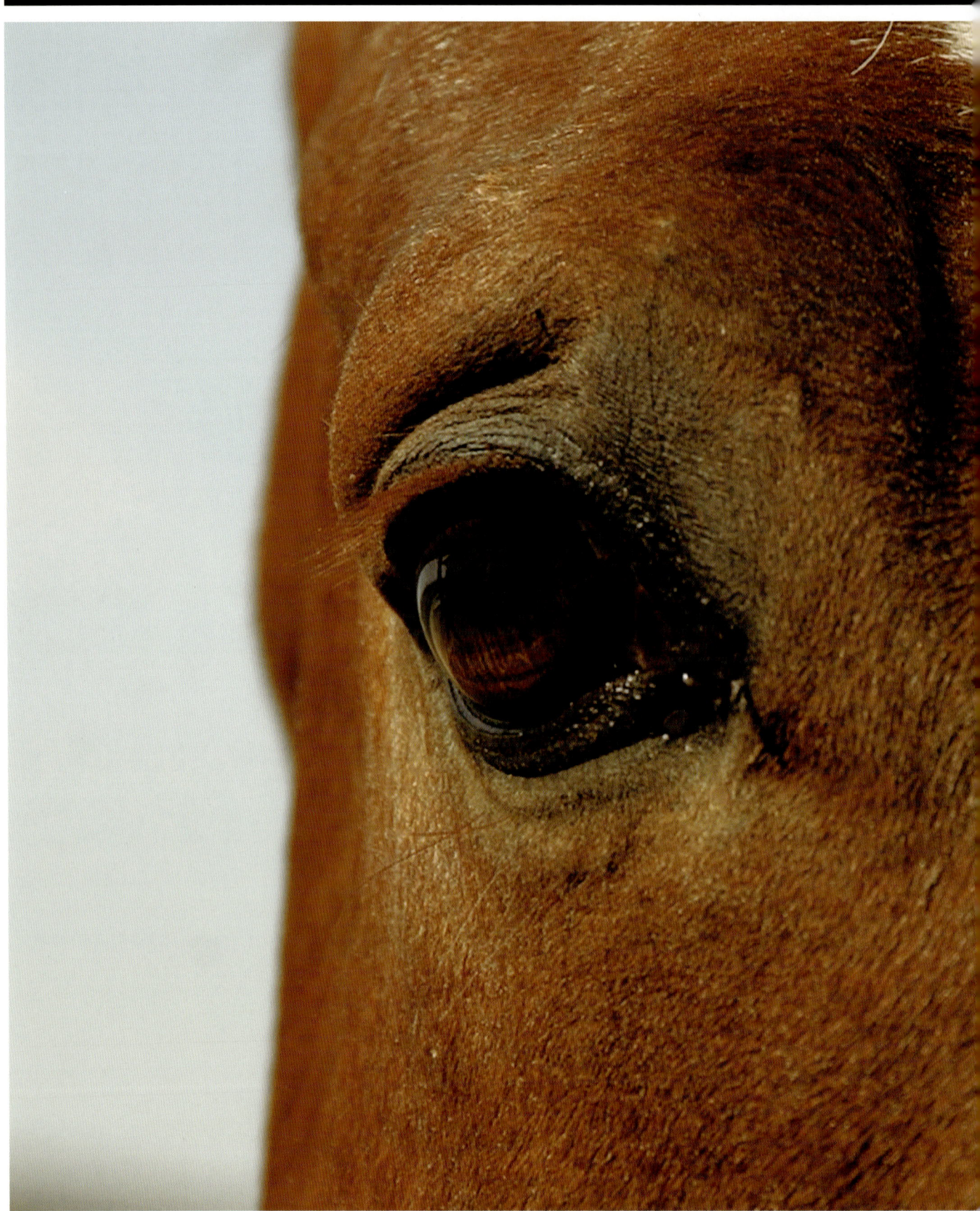

← [**13.1**] Land restrictions, increased costs related to ownership and care, limited feed, and increased prize monies in competition are all influencing the ethics of owning, riding, and competing horses. Because of horses' cooperative natures, they often do not demonstrate welfare issues until they have a physical problem.

CHAPTER

[**13**]

THE FUTURE OF HORSES

Generally speaking, the horse industry has resisted national-level horse welfare laws that cover all breeds and disciplines in lieu of a belief it can self-regulate.

[13]

Horses and humans have traveled through history together and been mutually dependent upon each other until modern times. Today, horses are more of a luxury than a necessity, and because they are such willing creatures and want to please their human companions, it has been assumed that horses enjoy what we modern-day humans do with them (primarily sport and recreation). But the tides are changing as the public becomes more knowledgeable about not just the physical, but also the mental welfare of animals. Consider that zoos must now have enrichment activities for their captive occupants, and even the pet world is growing accustomed to requirements for socialization and entertainment—"doggy daycare" and "cat TV" are now common as humans learn more about the cognitive and emotional needs of other animals.

But what about horses? Where do they fit in with these changes?

If you ask people who have horses whether they "love horses," most will say, "Of course." For the average horseperson, keeping a horse is akin to being married to someone you love and plan to care for forever. But on the other side of the horse world spectrum is an economically driven industry germinated partially by the attraction of horses and an equestrian lifestyle and partially by the attraction of money and sport. Although these two sides are not mutually exclusive, the various equestrian disciplines have primarily, because of the influence of money, often ignored what is good for the horse (fig. 13.1).

While the various equestrian disciplines have never focused predominantly on the horse's welfare, there has been an underlying assumption that horses are generally "well cared for" as companion animals rather than just treated as "livestock," and so the industry has sidestepped public scrutiny for a long time. But because the horse industry is so disjointed when it comes to opinions and accepted training methodologies for the various disciplines, it has done little to research, educate, and communicate—internally and externally—what makes a horse "happy" to partner with humans in our various sports and endeavors.

Effort has been made to grow equestrian competitions into "spectator sports," thus attracting sponsors with prize money and investors interested in the sale of high-performance equine athletes. These factors, unfortunately, tend to emphasize "winning" over what is good for horses. Regulations and what is deemed "acceptable" can vary tremendously from sport to sport and country to country. While the public in many places has pushed for better horse protection laws—and in some countries, they are developing—generally speaking, the horse industry has resisted national-level horse welfare laws that cover all breeds and disciplines in lieu of a belief it can self-regulate.

Today the horse "ping pongs" between the public that wants better welfare for horses and the public that wants to "invest" in horses. Without strides made to better welfare protection, the horse is at risk to be just a "commodity"—as horse prices and competition purses increase, some trainers and owners treat horses like cars that "lose value" when they reach a certain age. Walk into almost any high-performance barn—whether racehorses, jumpers, or reining horses—and if you ask which horses are for sale, they will tell you "everything has a price."

In addition to the rise of the horse as a commodity, the local stables where anyone could once take riding lessons are disappearing into housing subdivisions, increasing the already high cost of horsekeeping. As exclusivity increases, it encourages the development of an affluent lifestyle "brand" around horse ownership and showing, and the trainers and sales agents who can produce top performance horses that "win" became the "gatekeepers" to a life to which others aspire.

So the welfare of the horses has long been left to the whims of trainers, traders, and owners. Organizations do not differentiate between how horses win or what trainers have done to win as long as no one is caught breaking any rules or performing unethical or inhuman actions that people can observe. But thanks to a growing awareness and public question of "social license"—an activity's approval or acceptance by society—horse sports and horse welfare are now very much in the spotlight.

Education, better oversight, establishment of best management practices, collaboration among horse groups, and some form of accreditation to ensure both horses and public are treated fairly will be needed in the future if horses sports are to remain and grow with public support. Better ethics means better welfare, which impacts how people feel about being involved with horses and the horse industry. Regardless of your role or level of involvement with horses—whether you have them for pleasure or sport, whether you are an owner, investor, trainer, coach, or professional—the well-being of the horses who touch your lives should be *your responsibility*. Every one of us should be asking ourselves, "What is the ethical thing to do to ensure the welfare of the horse in this situation?"

And as I have demonstrated in these pages, it is no longer a matter of opinions. As equine and welfare science continues to conduct research on the emotional lives of horses and their relationships with humans, it is showing that horses who are mentally and emotionally comfortable—what we might call "happy"—perform better. Hopefully the future will bring productive discussions and the adoption of standards with this evidence in mind.

How Do We Best Help the Horse?
— From "Equestrian Lifestyle" to "Horse Lifestyle"

The horse industry is spread out over a vast landscape. As diverse as the habitats in which horses have flourished, are the disciplines in which humans have engaged with them. With traditions and culture woven into many of the popular disciplines, horse welfare is blurred by human-centric justification. The "equestrian lifestyle," from ranching to fox hunting, savors a memory for simpler times and days gone by when horses and humans spent most of their days together. Our histories are so entwined on many levels that there is hesitancy to give up the cultural components *regardless of how horses fare.*

For some, the development of competition was an excuse to keep training and riding horses over a lifetime. For others, traditions run deep on the human side of sporting events, such as racing at Churchill Downs in Kentucky or the steeplechase course at Ascot in England. Filled with fun, excitement, and socializing, the races are culturally significant, even marking each year with specific times for celebration. Picnics, parties, and affluence blind us to the dark side of the backside and gambling industry interests. Good trainers often don't come out the winners, caring more about a horse's welfare than money. Sadly, I have known several trainers who have lost business when a client wanted them to compete a horse the trainer did not feel was ready or felt would not hold up.

Every equestrian activity will be scrutinized in the future as it is now possible for media to cover and broadcast everything we do with horses. Public perception may, in fact, dictate the future of horses and humans if the various national and international organizations do not come together and establish acceptable welfare practices for promoting the mental and emotional health of horses in all they do with us.

Hopefully, we will evolve a culture that puts the welfare of the horse above the enjoyment and profit of humans. We can have both, but it will require a collaborative effort. Horses need to be seen as equal athletic partners

> **The future and welfare of horses with humans is not a question of need, money, or space, but rather a question of the heart.**

STORY FROM THE FIELD: //

Stop Pointing Fingers

In the 1980s, I was consulted as an equine behavioral ecologist to comment to the United States Congress on the practice of "soring"—the practice of deliberately inflicting pain to exaggerate the leg motion of gaited horses. The Tennessee Walking Horse industry requested their accepted practices be compared to practices used in other horse sports. So I conducted a year-long study on stress in horses, comparing various disciplines, using behavioral observations. At the time, polo ponies showed the least amount of mental and emotional stress when compared to Western pleasure horses, dressage horses, and hunter/jumpers. According to my study, Tennessee Walking Horses did show a high degree of mental and emotional stress compared to other disciplines. However, *data showed the strongest variable affecting equine welfare was not the discipline, but the person training the horses.* Those trainers who loved and cared well for their horses had horses who exhibited very little stress *regardless of discipline.* //

in sports, and they need to be considered companion animals rather than commodities and livestock. Our horse culture needs to be refreshed so the much desired "equestrian lifestyle" begins to focus more on the "horse lifestyle," taking the horse's welfare into consideration at every stage of his life.

_ Empowering Veterinarians to Make Welfare Decisions

In the best situations, a horse has a team of caring people that guides his lifelong care. Veterinarians, trainers, owners, farriers, and other horse professionals work together in making decisions that are best for the horse. But in many cases, particularly in show barns and racing stables, veterinarians deal directly with trainers. Trainers, often under pressure from owners, request horses be treated in certain ways to keep them competing at specific levels. Often, when this happens, there is no discussion concerning the horse's mental (or long-term physical) welfare.

In some cases, it is trainers, not horse owners, that make veterinary care decisions for horses. While owners who care for their own horses usually have a good relationships with their veterinarians and would discuss any care decisions with the horse's welfare in mind, the sport horse industry can be different. Owners often give "care, custody, and control" of their horses to their trainers, allowing the trainers to make decisions on the owner's behalf, and many times without even consulting the owners. This obviously can lead to welfare issues when trainers are either ignorant of care (for example, many trainers give routine injections to horses without knowing the risks) or would prefer to get under-performing horses out of their barns and replaced with "better" (more expensive) ones.

In the sport horse industry where money is pushing certain trainers to "win," pressure is experienced by the trainer, who then puts pressure on the veterinarian. I have had veterinarians tell me that they only deal with issues directly related to a trainer's request or question. When they start asking about training abuses or other welfare questions, trainers will not use them again and they lose business.

Clearly, the horse industry needs to sort out the ethics around horse welfare and who has the responsibility and expertise to help guide a lifetime of care for the horse. Here are just a few of the subjects and areas of concern I most commonly hear from veterinarians, farriers, trainers, owners, grooms, and others who are involved in a horse's care:

- **Pre-purchase veterinary exams:** In the case of a sale, both the owner and the trainer should be present to discuss the vet's observations regarding the suitability for intended use. Reports should be available to both buyer and seller and follow the horse's complete health history. Open discussions should take place between all parties in order to make the best decision regarding a sale on the horse's behalf. Health records should be transparent and veterinarians should be able to evaluate and comment on the mental health of the horse as part of a pre-purchase exam.

- **Competition therapy**: Injecting the horse's joints for the purpose of alleviating stiffness or pain in order to train or compete is very common. A number of vets have commented to me that at times when they would recommend rest and rehab for a horse, a trainer has requested injection therapy instead so the horse could continue competing. When a vet declines to provide the therapy requested, the trainer will find another vet to perform the treatment. As one vet told me, "If I let another vet do what I do, he could injure or kill the horse, and at least I can keep him alive." There should be agreements among veterinarians to communicate with previous vets before proceeding with the care of an individual horse.

- **Training-related welfare issues:** Some vets have communicated to me they "look the other way" as they do not have the power to comment on training-related welfare issues, such as a horse's lack of social interaction or turnout. As I've said, they are concerned about keeping their clients. While all 50 states in the United States permit or require vets to report suspected animal cruelty, and it is mandatory in 19 of those states, vets are hesitant to make welfare decisions regarding "cruelty" regarding horses because there are varying degrees of "abuse"—from confinement and social isolation, to using pain-causing devices, to failure to provide adequate nutrition. Veterinarians should be educated in identifying welfare issues regarding behavior, pain, and nutrition, and they need to be empowered to report welfare concerns related to their equine clients to a separate body responsible for the oversight of competitive horse sports.

_ Retirement and End of Life

Let us imagine one day you have a horse who is no longer capable of performing, is injured, old, or dangerous, or you no longer can support him financially. How do you ensure what is best for your horse when you can no longer care for him? These are tough but real discussions that every equestrian should consider and for which they need a process for decision-making. There is no one right answer, and many people involved in horses don't even know their options.

Ending a horse's life is never an easy decision. Many people so wish to avoid death that they would rather give their "suffering" or unwanted horses away and not know what happens than deal with the harsh reality. If we consider these issues from a horse's perspective, we need to remember they do not think about death or suffering. They just exist in the moment. Horses are present in what they feel, sense, and experience *in the now*. It is understandable why people wish to avoid keeping a horse they feel may be suffering or dying. But suffering is a subjective observation when it comes to horses, and this is part of the reason why there needs to be more ethical welfare standards of practice concerning what constitutes suffering and when end of life should be considered.

Over the years, I have been asked to "find homes" for many horses, and I have been part of many euthanasia procedures. I know that we as horse caretakers need to develop better options and guidance for people facing the reality of long-term and end-of-life care for horses. There are a number of growing areas providing new choices for owners who can no longer keep a horse, such as horse "retirement homes" and human-horse interaction programs. Both have opened opportunities for horses in their later years, and hopefully, these can help promise that in the future there will be *no* unwanted horses.

Improving Means of Reporting Welfare Issues

By becoming more knowledgeable about the horse, starting with the known sciences of basic biology, physiology, and behavior, and then staying up to date with the evolving sciences of horse-human interactions, people can have a solid foundation upon which to base welfare decisions, rather than doing so on emotion. Because we compassionately sense the feelings of horses, people *want* to help, but often do not know what to do.

Through the years I have been involved in horses and horse sport, people have called me, asking for guidance when they did not know how to best help the horse. These situations have covered a range, from unwise to inhumane. I have had grooms and working students tell me they have been asked to put electric devices on jumpers, to give magnesium drenches to sedate hunters, to tie horses in their stalls with side-reins so tight they cannot eat or drink for hours, and to withhold water before classes so horses will be calm.

While there are many people who do care about horses, the equestrian world, because it is so divided, does not yet have a way to adequately report, take action, and limit inhumane treatment to horses. And more subtle

STORY FROM THE FIELD: //

The Question of Veterinary Ethics

Dr. Barb Crabbe, a long-time equestrian and sport horse veterinarian, is focusing on bioethics—the study of the ethical issues emerging from advances in biology, medicine, and technologies—within the horse industry and veterinarian profession. Here she asks, "Who decides?" when it comes to medical decisions for the horse.

A significant problem we face in managing performance horses is determining who should be the medical decision-maker or "surrogate." Autonomy, or patients' rights to make decisions regarding their own medical care, is a guiding principle in the field of human medical ethics.

When a patient is determined to be incapable of making those decisions, a "surrogate" decision-maker is assigned. The surrogate makes decisions first based on a "substituted judgment standard" (what patients would choose for themselves), followed by a "best interest standard" (what the surrogate would choose for a patient). In equine medicine, both autonomy and the substituted judgment standard are bypassed, leaving us with a best interest standard as the only available option.

Who should play the role of surrogate? All too often, a trainer with financial interest in the horse's performance takes on this role. I would argue that if owners took on the responsibility of making fully informed decisions based on the best interest of their horses, many of the welfare issues we face in the world of the performance horse might be overcome. //

Barb Crabbe, DVM
Owner and Veterinarian
of Pacific Crest Sporthorses, Inc.

welfare issues often observed at horse shows and events, such as the use of drugs, do have scrutiny and regulation, but without direct evidence of rules violation, a large number of needles and syringes in a show barn simply identifies a trainer who uses a lot of needles and syringes. Proof of abuse can be difficult. Other issues, such as constant confinement and lack of any social stimulus or interaction, can go unnoticed until a horse begins to display a behavioral or health problem.

Further discussions need to take place within breed and discipline organizations, as well as veterinary associations and welfare organizations, to address the construction of a *collaborative approach* to determining what is best for horse welfare and how to facilitate communication and decision-making that ensures it throughout the industry, but particularly in the competitive sport horse world. The "social license to operate" (SLO) will be dictated by the public if the horse industry does not come together to justify and show that horses are "happy" competing and engaging in activities with people and that their welfare is not just a statement on paper, but has implementable actions in each sport to ensure sport horse welfare.

⌐Integrating Science and Welfare

In the area of animal emotions and welfare, science has launched ahead tremendously in the last ten years, but little has made its way into the mainstream horse industry. There are now *thousands* of studies conducted and published in a variety of academic journals on topics spanning from animal behavior to equine psychology. And with nothing more than an internet search of a particular topic, any person interested can have access to these studies and become knowledgeable.

Scientists are taking a greater interest in the welfare aspects of horsekeeping and horse sports and investigating if personal opinions and long-held beliefs about training and caring for horses hold up under scientific investigation. But the equestrian world has adopted little if any of this knowledge. Meanwhile, we see great advances in mental enrichment for other captive mammals. For example, there are few zoos today that do not have enrichment activities for their animals or closely monitor the mental and emotional health of their animals.

The International Society for Equitation Science (ISES) evolved out of a horse welfare and behavior workshop and the International Society for Applied Ethology (ISAE) has been holding international conferences since 2004. Their focus is on research and education to improve horse lives with humans. And the Fédération Équestre Internationale (FEI), which governs the Olympic horse sports globally, has assembled the Equine Ethics and Wellbeing Commission composed of industry and third-party equine behavior experts for the purpose of assessing the science and determining ways to improve sport horse welfare. While all of this is a step in the right direction, there is resistance within various countries and disciplines to regulation by scientists when many people are part of cultures that have been training horses in certain ways for generations. The cultural aspects of changing attitudes by raising awareness for the mental and emotional health of horses needs an organic paradigm shift, from the inside out.

Understanding that the world equestrian industry is diverse and tends to be skeptical when it comes to accepting science over tradition, some researchers focused on how to implement equine welfare and equitation science education within the horse industry. The study (conducted in 2018) analyzed "equestrian chat" in an online thread about equitation science. The researchers determined there were four prominent beliefs about science that prevented equitation science from being accepted:

← [13.2] A key point of this book is not to critique the activities we may enjoy pursuing with horses, but to examine the quality of our horses' relationships with us, our attention to each horse as an individual, and our ability to guide our interactions with horses in a way that benefits their natural drive to feel safe and have comfort by being with "friends." Consider this image of the same activity, two different horses. The horse on the left is an *externalizer* (see p. 222), licking his metal bar, ears back listening to the driver, eyes alert, and looking forward with relaxed facial muscles and an expression of engagement and communication. The horse on the right is the opposite—an *internalizer* with wrinkles above his eyes, a "dead" or "empty" expression, tight lips, sideways ears, and a lack of interest in what is going on around him. Because so many horses look the way the horse on the right does, I have found most people think it is normal because the horse does what he asked. This is a good example of the simple fact: some horses enjoy getting their gear on and going out with you for the day, while others dutifully obey and suffer as they have no other options.

Learn more

1. **Science discounts "feel."** People who had intuitive and emotional connections with their horses felt they knew more than science.

2. **Science is "overrated."** People felt science is often wrong when applied to the real world.

3. **Science is a gimmick.** People felt science may just manipulate ideas (data) to have it "say" what they want it to.

4. **Science is reductionist.** People felt science does not take into account the whole picture with all the variables.

Only one comment indicated that science "is useful and progressive." The researchers concluded that in order to gain acceptance of science to help ensure horse welfare, we need a sensitive, welfare-centered communication

STORY FROM THE FIELD: //

Wild vs. Domestic, Freedom vs. Control

Freedom, the value humans seem to care about the most and which is often symbolically represented by horses, lies deep inside people's belief systems. Thus, what we believe about horses and how we feel they should live and interact with humans is an emotional and cloudy subject. There are those who feel "the horse owes me because I pay the bills." And on the other end of the spectrum there are those who feel every horse, no matter how old or injured, should be cared for until the horse dies of natural causes.

In the 1970s and 1980s, I was trying to better understand why and who was killing groups of wild horses in both Nevada and Oregon. I interviewed a number of ranchers in those areas. These were deep and often profound discussions. One rancher told me, "I don't hate wild horses, but every time that stallion with his mares comes down on my land and looks at me, he reminds me of what I don't have and what he does—freedom." The bank owned this man's ranch and he had no idea what he would do, as he was a third-generation rancher and his whole life as he knew it was coming to end. He apologized for shooting the horses, and then asked me why they should survive when he and his family could not.

Another rancher's son in Eastern Oregon told me why he had shot a different group of horses. I was staying at his family's ranch, helping them develop a tourism option for their property. The son knocked on my door after dinner, and said he needed to talk, and then, in a nervous voice, he said: "Those wild horses, walking with their heads down, single file cross the Alvord Desert, their tails dragging and their manes all in knots...I just felt sorry for them. So I shot them." He was in tears.

In both cases, the rationale for killing the horses came from a deep-seated under-explored belief about life and death, freedom and control. It is no different with the decisions we make in the management and training of our horses. Horse-human stories and beliefs are complicated and reflective of how individuals view themselves and their world. And so the welfare of all horses lies in individual beliefs about horses and where they fit into our personal lives and into society. //

strategy that is tailored for nonacademic audiences who are not familiar with equitation science.

Many universities who have equine science programs now offer both *equitation science* and *equine welfare courses*. There are conferences all over the world now on these subjects, and yet little of the knowledge and applicable skill has made its way into the top ranks of most horse sports. There remains skepticism between equestrian professionals and academics.

But the paradigm of horse training, and animal welfare in general, is changing. As horses have migrated from primarily work and transportation animals to partners in pleasure and sport and now to "therapists" and "educators," their role the human world is changing. More people are discovering that horses and humans are both social mammals and that we share many emotions and behaviors. With this awareness, more people are realizing the mental and emotional needs of horses are perhaps not so different from their own.

The next generation of horse people will be key influencers for the future of horses. Once, the only way to have a career with horses was to be a horse trainer, farrier, or veterinarian, but today there are multiple careers available with horses, and many new horse enthusiasts are coming out of universities and colleges with degrees in equine science, sport science, equine behavior, ethology, training and equitation science, welfare science, and equine-assisted therapy. And although there are factors that are increasingly making it more difficult to affordably own and care for a horse, humans clearly still want horses in their lives. While few of these younger, up-and-coming horse people have yet made it into the industry at a level of influence, their continual involvement will eventually change the horse world as new careers in various scientific fields, equestrian marketing and media, horse tourism, equestrian psychology, and diverse horse-human interaction programs disciplines evolve.

As science continues to produce numerous studies focused on horse-human interactions, the key will be to develop communication channels to get the information to the right people, where it can effect lasting change for the good of the horse. Helping equestrians overcome their fear of science through better education will also improve horse welfare. Transferring the knowledge of horses as a species living naturally (see Part One of this book—p. 7) and understanding the domestication process and what it means in conjunction with the "wildness" that remains in all horses (see Part Two—p. 113) will help all equestrians better understand their horses and what can keep them healthy and stress-free.

Industry Ideas for Improving Horse Welfare in Equestrian Sports

I polled professionals in the North American horse industry in search of new ideas we can all get behind to improve the life of horses without sacrificing competitive aspirations. What follows is a summary of comments that represent the collective counsel offered by breeders, owners, trainers, judges, stewards, course designers, professional riders, jockeys, racetrack and horse show officials, farriers, veterinarians, and others involved with horses. Often, the same comment was made by different people. Because so many professionals were concerned that their comments would influence their ability to work in the horse industry, what follows is listed anonymously and in no particular order.

_ Breeding and Training for the Sport or Discipline

- *Breeders need to breed responsibly and only breed quality horses that the market will support.*

> **Resisting change is one thing; resisting knowledge when it can benefit the horse is denial. Science is producing the information and tools that should be embraced by the horse industry to help improve the life of every horse.**

- *Genetics should be used to guide the breeding of quality horses to increase performance, limit physical and mental disorders, and decrease "throw away" horses.*

- *Microchipping should be required for all registered horses, and all breeders should be encouraged to microchip horses, regardless of registry. A standardized database should be established for various microchips. Microchips should be required for interstate and international travel.*

- *The industry should limit the number of horses they allow to be registered each year to create better demand in the market and make breeders have to apply for registry. This will help control unwanted horses.*

- *A "whole life horse fund" should start with breeders' contributions when they register their horses. This fund should be contributed to by all registries and disciplines and available to programs and facilities who care for older, injured, and end-of-life horses.*

- *Horse sales should reflect standards similar to real estate. Transparency and honesty should be required. Veterinary records should follow the horse, not the owner. Sales agents should be licensed or at a minimum registered and operate under a code of ethics to protect the welfare of the horse.*

Care of Horses—Educate and Regulate

- *Require all horse professionals to pass similar knowledge tests on horses and horse welfare. The difference among countries is tremendous. Trainers and officials should have to take continuing education (CE) courses each year to stay current on horse welfare, similar to other professions.*

- *The industry needs to justify each sport with good welfare science. Partner with universities and fund the development of better guidance to protect the welfare of horses in sport. Train veterinarians and other horse professionals in welfare science, and incorporate this expertise throughout the industry.*

- *Restrict the use of abusive tack in all sports. Horses should be able to compete feeling comfortable and being able to breathe and move freely. Horses should be allowed to have full physiological function to compete. There should be standard training and tack use across disciplines.*

- *Require accountability and enforcement of ethical welfare treatment of horses. Lobby for laws to hold horse abusers accountable.*

- *Quarantine care needs to consider the mental health of horses. Depending upon the facility, horses can be highly stressed under current standards.*

Use of Drugs in Sports

- *There should be a national drug policy governing all horses in sports. Racing varies state to state, while each discipline controls their drug use, which is not fair to the horse. The horse is the same creature, regardless of the kind of sport he competes in.*

- *Educate horse professionals on nutrition and drugs. They should have to know what they are feeding and how it can impact drugs and their horses.*

- *Follow the American Endurance Ride Conference (AERC) rules for zero tolerance at top levels of competition when money is involved. It will cause trainers to train horses better and breeders to breed better athletes.*

- *Allow older competition horses to have safe levels of anti-inflammatory drugs to keep them in the sport longer. Older horses, like older people, need to feel good.*

_ Horse Shows, Racing, Events, and Competitions

- *Implement a process for defining sport horse welfare, how to measure it, and how to regulate it within the global field. Most organizations say they "protect" horses, but there is no agreement on what constitutes good welfare or how to enforce violations.*

- *Limit the number of competitions a horse can participate in per year, and require time off in between competitions, relative to discipline and intensity.*

- *Limit the number of classes a horse can compete in at an event, relative to discipline and intensity.*

- *Require competitors to have successfully completed a certain number of competitions at lower levels before allowing them to compete at higher levels. Or require them to take a horsemanship test relative to their discipline to compete at a higher level prior to showing. This will protect horses and riders as too often competitors are "buying up" above their level to ride and the horses get injured because of lack of experience and poor judgment.*

- *Don't race three-year-olds. Many horses never make it to be four-year-olds. Lower racing purses for young horses, and raise purses for older horses to encourage people to wait to race.*

- *Horse shows and competitions need to implement horse welfare standards and have them enforced in order to have approval. Facilities housing horses for more than two days need to provide turnout paddocks and grazing areas. Stall sizes need to be 12 feet by 12 feet for horses.*

- *Expand opportunities for young horses (three to seven years old) and old horses (16 years and older) that focus on age-appropriate training, and reward people for having happy and healthy horses. Offer classes that allow horses to demonstrate their positive relationships with humans that reinforce good "age-appropriate" training.*

> **Not recognizing horses have feelings is *ignorance*.**
>
> **Saying only humans can express emotions is *arrogance*.**
>
> **Not addressing the emotional welfare of horses is *negligence*.**

- *Tack should be comfortable for the horse, and minimum use of tack required for communication should be rewarded. For example, allow bitless bridles or no nosebands for dressage and other competition sports.*

- *Take the money away and look for support via sponsors and people who love horses and horse sport. Prize money has made equestrian competition about money and not about horses. Put the money into better event grounds and experiences for both horses and people.*

- *Limit conflict of interest in sports. Officials should not be able to train and sell horses. Judging should be a neutral profession.*

- *Require equine welfare education and testing for all horse organization members in order to participate in equestrian sports.*

Everyone Is Responsible

While the main focus for horse welfare improvement may target those involved in the business of horses—from breeder to trainer to investor to corporate sponsor to feed manufacturer to pharmaceutical company—anyone involved with horses, whether for business or pleasure (backyard horse owner, tack manufacturer, veterinarian, farrier, body worker, horse show organizers, breed and discipline associations, welfare organizations, event spectators) has the ability to impact the lives of horses. Question what you see if you feel a horse is worried, in pain, or does not understand. Do not assume that because it is "culturally accepted" in a particular discipline that it is good for a horse.

Toy manufacturers cannot make toys that are not safe for children, so why can tack manufacturers make tack that is abusive to horses? Would sponsors be so willing to sponsor a sport horse competition with high-dollar purses if they knew that many of the top trainers competing may be using abusive methods to win? As purses go up, regardless of the sport, horse welfare goes down.

The sport horse industry, while it is making efforts to improve, is not set up organizationally to monitor horse welfare, and "reports" on the subject often have no consequences. Because each discipline has its own regulations, training methods may not be monitored, allowable drug use varies, education is not required for trainers, rules related to "professional conduct" do not exist, and while associations may, in fact, encourage reporting cases of abuse, often nothing is done, and the horse is the victim.

Everyone can be instrumental in change, but first they have to be aware of the issues. I hope this book both raises awareness and provides a foundation for education and further exploration. What follows identifies positive action you can take to help evolve the horse industry into an ethical, horse-centric, and welfare-friendly future.

Buyers, Sellers, Agents Horse sales are where the real money is and are a target for welfare issues. Since, in the United States, horse sales are not regulated like real estate, nor is there oversight or accountability, most sales are handled privately through direct money transfers. There have long been such bad associations with the term "horse dealer" that, when I bought and sold sport horses, I called my agency a "horse brokering firm" so as not to be misclassified. Since there were no professional appraiser organizations in the early 1980s, I reached out to the International Society of Livestock Appraisers and developed guidelines for conducting equine appraisals. Although there are certainly ethical horse sales agents and trainers, and today we do have the American Society of Equine Appraisers (ASEA) in the United States, there is no requirement to use appraisers when conducting horse sales. There has always been a hesitancy in the horse industry to disclose information concerning horses if it might limit their salability, and this lack of transparency often causes horses to be at risk.

Whether buyer, seller, or agent, be honest and open. Matching horses and people is more of an art than a science, and you cannot make relationships happen. Always try a horse before buying. Research and get references on anyone you are either selling to or buying a horse from. Always have a legal agreement disclosing important information about the horse and the money involved. Customize agreements and bills of sale to include information you feel is important about the horse or the conditions of purchase. Seek an equine attorney if you need help.

Verify papers, microchips, show records, and any other information represented about a horse you are purchasing. Know the laws in your state or country for horse sales, and do not feel intimidated when asking questions or requiring full disclosure of related commissions and money transfers. As a seller, you should want your horse to succeed in life, so inquire about management practices, and even include how the horse will be kept, trained, or shown in the sales agreement. I always required buyers to sign an agreement stating they had "honestly represented themselves" to prevent problems in the future.

As we have discussed in this book, prepare any horse for the possibility of a new home in the future. Give him the skills he needs to make friends and adapt to new environments. Include a diary of information about the horse in any sale, and stay connected with the buyer to assist with any issues.

Breeders Microchip all horses, regardless of whether they are registered or not. Along with required papers and passports, provide additional information you feel would be helpful to a new owner about a particular horse. Note in writing if you wish to have the option to buy back the horse in the future. Only breed good-quality horses for the market.

Trainers Assess a horse both physically and mentally. Learn about his history and needs before beginning training. Report abuse when you are sent a horse you know has been in an "at-risk situation." By allowing "bad" trainers to continue to solicit clients, or by just being silent, you are supporting further welfare issues. Hold other trainers accountable when you witness a potential welfare issue, whether mental or physical. Lobby your discipline

> **Everyone can be instrumental in change, but first they have to be aware of the issues.**

organizations to develop professional "codes of conduct" for horse welfare and business practices. Teach your clients not just how to ride, but how to develop positive relationships with their horses.

Veterinarians and Health Professionals Recommend microchipping horses. Make mental health assessments part of pre-purchase exams, and learn to spot welfare issues related to training or care. Discuss issues you see with your clients—many may not know they are putting their horses at risk. Lobby regional governments and associations to expand the role and the responsibility of veterinarians to report welfare issues. Develop partnerships with horse organizations and governments to implement a welfare board for horse sports. You are the primary voice for horse welfare. Speak up. Good trainers and owners will appreciate it.

Manufacturers of Tack and Horse Supplies Be responsible and develop products that are horse-friendly and limit abusive use. Learn the science behind everything you make and the potential effect it can have on horses. Label how to use products and put warnings on products that could be used inappropriately. Support horse welfare education.

Sponsors Do your due diligence before the public does it for you—do not support organizations, sporting events, or professionals that do not provide a guarantee of welfare standards for horses. Do not assume that if the event is "approved" by an association that it is "welfare friendly" or if a person is a top trainer that the horses in the person's care are treated well. Since money is driving many of the welfare issues we are confronting today, sponsors have the most power to make change by guiding how they allow their money to be used.

Horse Show and Event Organizers Provide facilities that are horse-friendly—comfortable stabling and large stalls for big horses; turnout paddocks and grassy areas for hand-grazing. Implement light and noise restrictions in barns for six hours every night to allow horses adequate sleep. Design classes and schedules so horses are not overused. Give incentives to reward good training and offer classes that are age-appropriate. Lobby for better welfare standards within your sport.

Shippers, Farriers, and Other Equine Professionals If you see a welfare situation, speak up. Ask why a horse acts or looks a certain way if you suspect an abusive situation. Often just saying something can cause people to change their behavior. Reports can be made to local animal control for serious situations, as well as relevant breed, discipline, and veterinary organizations. "Red-flagging" at-risk situations for horses can alert the right people to a potential problem.

Organizations and Associations Breed, discipline, sport, veterinary, academic, and horse welfare organizations and associations need to work together to establish welfare standards for horses, and professional ethics for members, using current science as a guide. Educate your members and the public on horse welfare. Establish professional codes of conduct for your members and set up a way to report welfare issues that can have accountability, both to your members and to the public.

Public Take a video of any situation you suspect is a welfare issue and submit it to breed, discipline, and welfare organizations. Ask questions and learn about different disciplines. Sometimes a situation is not what it seems, and simply asking questions can educate you, as well as hold those engaged directly with horses more accountable.

Horses are not objects to be simply bought and sold, but rather sentient, feeling creatures who have evolved complex emotional intelligence toward social cooperation. Thus, **we need to adopt a "whole life cycle" approach to horse care, understanding where the horse has come from and where he may be heading.** Regardless of our role or level of involvement, we can positively influence both horses' lives and the horse industry in general by seeking to improve and protect equine mental and physical welfare. We can give the horse essential horse skills to adapt in the modern human world, while at the same time, we can work for better oversight and transparency to follow and protect our horses throughout their lives.

Keys to Remember

Regardless of the riding discipline you prefer or the breed of horse you favor, the more you understand a horse's nature and the differences and similarities he has with humans, the better you will relate not only to your own horse, but to all horses. While many books tell you "what to do" with a horse and outline various training systems, I hope that in these pages you gained the awareness that the most positive relationships do not start with systems or "doing" things, they start with just "being"—being able to slow down your thoughts and energy, open up your heart, and engage your horse's curiosity to get to know you.

While humans may continue to debate "What is good horse welfare?" while trying to come to some consensus, the horse remains in limbo, caught between worlds and cultures. We must remember that the racehorse sold at yearling sale is no different than the yearling out on the range or the yearling in your barn—horses are still horses, regardless of how well they have adapted to domestication. As science weaves its way into public acceptance and people discover more about the rich emotional lives of horses, hopefully we will grow more aware of how to minimize the stress they experience in a domesticated life and enhance their mental health, while continuing to evolve with them as partners. The more we collectively learn about horses, the more we embrace science, the more willing we are to change our own thinking as well as our culture, the more likely horses are to have a positive future with us.

On the following pages I've listed important keys to remember in order to reinforce positive horse-human relationships and limit emotional and mental stress in your equine partner.

Keys to Remember

1. **Be present in the moment. Be "horse-centric."** Horses are "in the moment" and cannot process too many stimuli at once. So slow down and *be present*. Spend some "horse time" with your horse. Whether eating together, resting together, or "playing" together, find your horse's rhythm and spend time doing what horse friends do together. This allows your horse to get to know you.

2. **Listen to your horse and communicate clearly.** Be a good observer and a good communicator. Understand the neurophysiology of how horses process sensory input and communicate. Learn to look "with" not "at" your horse and understand all the ways your horse communicates.

3. **Know your horse and yourself.** Understand horses as a species and how humans and horses are similar and different. Be able to assess your horse and develop a long-term care and education program to ensure his welfare and his mental health. Teach your horse the things a horse needs to learn as well as how to be with humans.

4. **Create safety and comfort.** Be consistent and recognize the importance of spatial awareness for both horse and human safety. Create "safe space" wherever you go with your horse. Reinforce awareness and respect.

5. **Limit restriction.** Restriction—whether confined through cross-ties, ropes, draw-reins, bits, or other devices—is unnatural for a horse and limits thinking and freedom. While restriction may be necessary in certain cases, the restriction should always allow the horse to move comfortably and communicate safely.

6. **Give horses freedom to choose.** Horses are used to choosing their food, friends, where to go, and what to do. Provide enrichment by giving horses choices in small things, such as forage, vitamins and minerals, and time to wander where they want to investigate and explore (even with you on their backs). This can improve their mental and emotional welfare.

↑ [**13.3 A & B**] Horses are social creatures. The friendship they share with each other in the wild can be ours to share with them, as well, if we learn to recognize and understand their nature.

❝

It is not the strongest of the species that survives, nor the most intelligent, but rather the one most responsive to change.

Charles Darwin

7. **Talk to your horse.** Instead of being on your cell phone, tune in to your horse, and ask him what he is thinking and feeling. Horses may not understand words but they understand your intentions.

8. **Be hands-on with your horse.** Allow him to tell you where he wants to be groomed and scratched. Give your horse freedom to "talk back" and show you where he hurts or where it would feel good to have a rub. Horses spend a lot of time each day, grooming with their friends, reinforcing social bonds as well as keeping their bodies feeling good. We can play a role in this instinctual behavior.

9. **Be a horse advocate.** Don't be afraid to speak out for a horse if you see a condition that needs to change. Get involved in humane or other organizations that are working to affect positive change.

Regardless of why you have an affinity for horses—whether it's about a sport, or just enjoying being with them, or even just "horse watching"—become horse-centric in your interaction. Feel and sense what the horse is experiencing. The horse's wild nature, which still lies deep in every domestic horse, has made him highly sensitive to safety, but it has also driven him to make social bonds with strangers—whether new horses or humans (figs. 13.3 A & B). The horse's innate ability to be curious about others and to explore and learn who we are, perhaps as much as we have learned about them, is what continues to bring us together.

PART THREE / **CHAPTER 13** / The Future of Horses

> **The love for all living creatures is the most notable attribute of man.**
>
> Charles Darwin

ACKNOWLEDGMENTS

Thank you to all the people and horses who contributed to this book, and to the photographers and friends who have allowed me to use their photographs.

A very special thank you to Barbara and Marty Wheeler who I met many years ago because their amazing wild horse photographs so aptly showed what I had observed in my field research. This book would not have been possible without their beautiful images and the years they spent in the field, following the natural history of various herds. Their photos captured many behaviors and moments of communication and cognition so often missed by others. Sadly, many of the horses pictured in this book are no longer with us, but their stories will live on because of Barbara and Marty.

Thank you to all the people and horses who contributed to this book, and to the photographers and friends who have allowed me to use their photographs. Thank you, Kim Suddaby, Diane DeLano, Gabriel de Matos Machado, Ava DeCaster, Raquel Mazur, Lisa and Kevin Sink, Nicole Harrington, Emma Whillans, Lori Stewart, Wendell Stockdale, and others I may not have named but who were just as important to the end result. This book is truly a cooperative effort of support to help educate and improve the welfare of horses across all disciplines.

A special thank you to my dear friend Dr. Dorothe Meyer, who questioned the craziness of chasing horses in a round pen and "demanded" I share my

Acknowledgments

spatial awareness research with the German equestrian world. Dr. Meyer worked hard translating my first two books into German and introduces Part Two of this book—Horses with Humans—with her own wise observations.

A grateful "thank you" to Julie Goodnight who kindly wrote the foreword. As a gifted horsewoman and educator, Julie has traveled North America, inspiring people to think more like horses, and she has helped thousands better understand themselves and their horses.

Thank you to Dr. Dan Rubenstein, who through the years has taken the time to share his research on wild equids and brilliantly wrote the Introduction for Part One—Horses without Humans. Thank you for your endless energy and mentorship, guiding generations to better understand and conserve our wild equids and their habitats.

Thank you to Dr. Duncan Peters, a longtime friend and colleague who has worked within the sport horse industry as both a veterinarian and equestrian to improve equine welfare and who wrote the Introduction for Part Three on the ethics and economics of what we do with horses.

An acknowledgment of Fred Wyatt (deceased), Director of the BLM Palomino Valley Wild Horse Facility, who asked me to help him learn how to talk to horses and became a "born-again horseman," as he stated after learning a more "horse-centric" way to work with wild horses. Fred bettered the lives of thousands of Mustangs with his gentle approach, demonstrating how to rope a wild horse by allowing the horse to become curious and stick his head through the rope by choice. May he rest in peace.

To my friend Robert Vavra, who shared all his field knowledge, photos, and film with me from his lifetime studying and photographing horses around the world—thank you.

I am so grateful to all the horses and people over the last 50-plus years who have contributed in some way to this book by allowing me to share their stories and develop methods and models to help others better understand, communicate, and care for horses.

And a special thank you to my publisher Trafalgar Square Books, managing director Martha Cook, and editor Rebecca Didier for believing in the importance of this project and allowing this book to evolve over the three years it has taken to be a much bigger and robust book than we first imagined. Thank you for all your hard work on this project.

Appreciation to my husband Bill who has supported my horse passion for over 50 years and only once complained that I smelled like horse manure.

Thank you, Bill, for never questioning my adventures off into the field to study wild horses or being gone for weeks at a time, giving clinics while you held down the farm. And thank you to my son Chase who spent his childhood going from barn to racetrack to BLM facilities—your gentle nature was always calming to horses as they could easily read your authentic kindness.

Further Reading

In no particular order, here are a few other books I recommend (you can find more on my website maryannsimonds.com):

- *Stress bei Pferden: erkennen und behandeln* by Mary Ann Simonds with Dr. med. Vet. Dorothe Meyer
- *Mental Health and Well-being in Animals* (Second Edition) edited by Franklin D. McMillan
- *Equine Welfare* by Marthe Kiley-Worthington

Was Pferde wirklich brauchen by Mary Ann Simonds

- *Horse Wisdom* (The Henry Blake Omnibus) by Henry Blake
- *Fraser's The Behaviour and Welfare of the Horse* edited by Christopher B. Riley, Sharon Cregier, and Andrew Fraser
- *Horse Behavior* (Second Edition) by George H. Waring
- *The Horse's Mind* by Lucy Rees
- *Wild Horses of the Great Basin* by Joel Berger
- *Horse Brain, Human Brain* by Janet Jones, PhD
- *Minding Animals: Awareness, Emotions, and Heart* by Marc Bekoff
- *Equitation Science* by Paul McGreevy and Andrew McLean
- *Equine Behavior: Guide for Veterinarians and Equine Scientists* by Paul McGreevy
- *Language Signs and Calming Signals of Horses* by Rachaël Drassisma
- *Connecting with Horses* by Margrit Coates
- Anything by Linda Tellington-Jones!
- Anything by Jim Masterson!
- *Beyond Words: What Animals Think and Feel* by Carl Safina
- *Kinship with all Life* by J. Allen Boone
- *Tug of War: Classical Versus "Modern" Dressage* by Gerd Heuschmann
- *The Welfare of Horses* edited by N. Warren
- *The Compassionate Equestrian* by Allen Schoen, DVM, and Susan Gordon
- *What Horse Really Want* by Lynn Acton
- *Animals in Translation* by Temple Grandin
- *Where Does My Horse Hurt?* by Renee Tucker, DVM
- *Such Is the Real Nature of Horses* by Robert Vavra

INDEX

**Page numbers in *italics*
indicate illustrations.**

A

Abuse, 249, 254–55, *254–55*, 368, 380–82, 386
Adaptability
    of diet, 53–54
    in horse-human relationships, 5, 116
    in horses, generally, 11, 27–29, 42–43, 87–88, 107, *108*
    of social networks, 33
    stress and, 2–3, 263
    temperament and, 133, 135, 336
    of training approaches, 290
Advocating, for horses, 389, 391
Aggression, 37–42, 203, 231, 253
Aids, responsiveness to, 324
Air flow, in barns, 338, *339*
Alarm signals, 67–68
Alertness, 43, 72, *72*. *See also* Vigilance
Alfalfa, 54
Alpha roles, 40, 75, 90
Altruism, 36–37, *36–37*, 124. *See also* Nurturing
Animal cruelty. *See* Abuse
Anthropomorphism, 152, 155
Aromatherapy
    in communication, 186–87, *188*, 189
    in habitat enhancement, 347
    for stress management, 336, 351, 353
    in training, 232, 290
Associative learning
    overview, 99, 290, 294
    case histories, 234–35
    motivation in, 310–11
    negative experiences in, 181, 188, 221, 263, 292, 360
    pitfalls of, 322
    in testing intelligence, 236, *237*
Assumptions, 210, 328–29, 384. *See also* Bias
Athlete personality type, 157
Athletes, horses as, 133
Attention. *See* Awareness
Autistic-like behaviors, 233, 235
Automatic feeders, 234–35, 345–46
Autonomy. *See* Choice, freedom of
Awareness
    attention as, 313
    in equine culture, 109, 228
    in horse-human relationships, 146, 163, 206–7, 286, *324*
    in OFFER Technique, 204–8
    in riding, 210, 212, 215, 333–34
    in SAICC Evaluation, 228–30, *229*

B

Bachelor bands
    characteristics of, 79, 88, 93, 108
    disruption in, 35–36
    domestic equivalents, 302
    friendships in, 38, 125, 146
    play in, 80, 83
Back
    in body check, 196
    tension in, 192
Balance
    in movement, 60, 97
    in riding, 191, 194, 210, 212, 312–13
    stress and, 192–93
Barn design, 339–41, *339–41*
"Be still" signals
    among wild horses, 101
    in training, 142–43, *142–43*, 146, 180, 229, *242*, *324*
Bedding, for stalls, 339–40
Behavior
    assessment of, 337
    in domestic vs. wild horses, 10, 11–15, 21–22

INDEX

functional baselines of, 107–11
innate vs. learned, 21–22, 28–31
prosocial, 66
in response to observation, 22–23
stress and, 117–19, 264
tack and, 317–21
trauma and, 247–55
unwanted, 305–6, 325–26, 328
Behavioral despair, 263
"Being with" horses. *See* "Horse time"; Looking "with"
Beliefs. *See also* Bias
about control, 231, 316–17, 322
about horses, 126–27, 212, 384
Beyond Horse Massage (Masterson), 192, 196
Bias
about dominance, 81
about horse-human relationships, 125–27
about wild horses, 18–19, 20
in intellectual testing, 236
minimizing, 5, 259
Bighorn sheep, 101
Biting, 203, 304, 360–61
Bitless bridles, 213, *214*
Bits and bitting
communication role, 210, *212–13*, 213–14
stress and, 261, *321*, 360
training considerations, 309, *309*, 316–21, *319–20*
Bladder meridian, 196
Blake, Henry, 57–58
Blind spots, 49
Blindness, 51
Blockages, 191–92
Blowing, as communication, 68, 71
Body awareness, 308–9. *See also* Proprioception; Spatial awareness
Body checks, 193–97, *193*, 331
Body language
in communication, 69–70, *69*, 178, 180–83
context in, *182–83*, 183
interpretation of, 39, 63, 281
Bolting, 233, 243, 250, 252, 318
"Bombproof" horses, 239, 240
Bonding, 56–57, *56–57*, 124, 298. *See also* Friendship, among horses; Horse-human relationships
Boredom, 223, 336, 345, 346–49
Brain function
in horses, 60, 202, 203–4, 288–90, 321
in humans, 23, 152, 179, 202, 203–4
Brambell, Roger, 266
Breathing
disruption of, 251, 272, 323
by human/handler, 201–2, 212, 286
synchronization of, 59, 202, 202
Breeding
among wild horses, 93–99, *94–95*

ethical considerations, 386
human preference and, 131–32, 231–33, 352
personality types and, 158
as social activity, 94, 296
stress management for, 294–96
Bridleless riding, 213, 214, *319*
Bridles, 213–14, 279. *See also* Bits and bitting
Bucking, 243, 359
Buddy-scratching
among horses, 34, *35*, 78, *78*, 140
in horse-human relationships, 190, *190*
Burrows, Ann, 275
BusyBody personality, 223
Buying horses, 257–58. *See also* Selling horses

— C

Calm horses, 51, 238–39, 309
Calming signals, 180, *181*, 354
Cambridge Declaration on Consciousness, 169
Caregiver personality type, 156–57
"Carrot focus" exercise, 230
Carrot stretches, 196, *197*
Cattle, 19, 31
Cautious temperament, 221
Choice, freedom of
denial of, 254
in feeding, 345, 347, 349
habitat management for, 338, 343, 389
in self-expression, 135
training considerations, 360
Circle feeding, 345
Circles, 314, 326
Circulation, stimulation of, 342
Clicker training, 185, 290
Cloning, 294, 352
Colic, 125, 264, 342, 362
Color, perception of, 49
"Combo" personality type, 224
Comfort
horse's desire for, 260, 267
human responsibility for, 147–48, 316–17, *317*, 331, *332*, 389
touch in, *57*
Commodities, horses as, 3, 5, 125–27, 257–58, 376–77
Communication, among horses. *See also* Communication, horse-human
body language, 39, 69–70, *69, 71*
defined, 65, 67
emotional dimensions of, 37, 39–40, 42
energy perception in, 76–77
in equine culture, 65–67, 125
motivations for, 171
play as, 80–83
in prosocial behavior, 65, *66*
smell in, 71–75
spatial awareness as, 77–80
touch in, 75–76

vocalizations, 67–69
Communication, horse-human. *See also* Communication, among horses
channels of, 177–79, 208–15, *208–9, 211*
energy in, 197–208
horse-centric, 3–4, 66, 146
as interspecies communication, 163–65, 169
OFFER Techniques for, 204–8
practical considerations, 169–71, 174–75
in riding, *161*, 210–15, *211–15*, 316–17, 324, 333
safety considerations, 166–67
smell in, 186–89
sound in, 184–86
touch in, 189–97
two-way conversations, 175–77, 207, 212, 389
visual, 180–83
welfare considerations, 388–89
Companion animals, 349, 376
Compensation syndrome, 322
Competition/events
ethical considerations, 116–17, 367–70, 380, 387–88
relaxation techniques for, *181*
rules and standards for, 283, 369–70, 383
socioeconomic factors, 154–55, 376
stress and, 256, 263, 271, 351–54
training considerations, 300–301
Competitiveness, among horses, 102
Competitor personality type, 157
Compulsive behaviors, 233. *See also* "Vice" behaviors
Concentrated feeds, 345
Concentration, in riding communication, 212
Conceptual learning, 291
Confidence, 222, 237–41, *238*, 243
Confinement, 116–17, 151, 301–2, 310, 319, 343
Conflict, among horses, *16*, 40, 90, 93
Conflicts of interest, 388
Connection
development of, 56–57, *56–57*, 229, *330*, 333
heart rate variability measure, 199–200
in riding, 212
Consistency, 137, 148, 292, 294
Contact, 319, *319*, 320
Control, expectations of, 213, 316–17, 322
Controller personality type, 156
Cooperative horses, 222, 241–43, *242*
Cooperative sensing, 55
Coping behaviors, 251
Courtship, 70, 94. *See also* Reproduction
Crabbe, Barb, 381
Cross-ties, 324–27, *326*, 360
Culture, equine
context in, 101
evolution of, 43, 87–89
food/water and, 96–97
in fostering intelligence, 231–33
"functional" horses in, 235, 244–47, 378–79

human influence on, 93
interpretation of, 22, 147
key characteristics, 78, 88, 100–101, 103
learning/education in, 99–103, *102*, 124, 157, 244–45, 304, 310
rolling, 97, *98*
sleeping, 98–99, *99*
social structures in, 11, 27–29, 31–35, *85*, 89–90, 93–99, 304
space games in, 91–92, *91*
variability of, 29–30, 88–89, 90, 123
Curiosity
lack of, 249
in learning/training, 99, *136*, 221
as motivator, 41, 42, 147
selective breeding and, 131–32
shared by horses and humans, 151, *164*
vs. vigilance, 42–43

_ D
Dalla Costa, Emanuela, 274
Dangerous horses, 143, 144–45, 254
Darkness
sleep and, 339, 342
vision and, 49–50
Darwin, Charles, quoted, 14, 35, 151, 391
Deafness, 359, 363
Decision-making, 75, 92, 111, 147
Deductive reasoning, 235, 236
Dehydration, 264
DeLano, Diane, *113*, 253
Demonstrative temperament, 222
Depression/depression-like symptoms, 263, *268*, 336–37
Depth perception, 48
Desensitization, 239, 240, 291–92, 363
Developmental slowness, 221
Diet. *See* Feeding
Difficult horses, 231, 251
Digestive system, 152, 343–44
Dirt, eating of, 21, 30, *344*, 345
Disassociation, 240. *See also* "Shut down" horses
Discrimination learning, 291
Distractions, 212, 221
Dogs, 115, 119
Domestication
behavioral adaptations to, 3, 10, 15–16, 109, 231–33
benefits, 28, 131–40
history of, 115
horse-human relationships in, 165
as mutualism, 123
tradeoffs of, 2, 163, 210, 236–37, 343
Dominance
dogs and, 119
expressions of, 73–74, 75, 80, 89–90
human models of, 81, 136–37
vs. leadership, 145

Doorways, 358
Draw reins, 321–23
Dressage, training and judging of, *255*, *283*, 388
Drugs, 382, 386–87
"Dull" horses, 166, 225, 250, 324

_ E
Eagerness to please, 125–26, 131–32, 221, 223, 225, *261*, 263
Ears, 22, 40, 55, 70, 175, 270. *See also* Hearing
Education. *See* Learning; Training
Elephants, energy perception in, 76
Elk, 19, 31
Emotional intelligence, 37, *85*, 87–89, 200, 233, 235, 236
Emotions
in behavior, 108–9
in communication, 67–68, 76–77
energy dynamics in, 32, 59
in horse-human relationships, 149–50
horse's sensitivity to, 200
in learning, 100
limited scientific study of, 167, 169
neurology of, 289
stress and, 59, 125, 193–94
Empathy, 57–58, 119, 207
End-of-life considerations, 381–82, 386
Energy
calming signals, 180, *181*, 354
as communication channel, 179, 197–208
horse's perception of, 47, *48*, 57–59, 91
human/handler presentation of, 41, 203–8, *204–5*
music and, 185
Enrichment practices, 346–49, 375, 383
Environment, as cause of stress, 266. *See also* Habitat management
Equestrian industries. *See also* Competition/events; Socioeconomic factors
case histories from, 118–19, *254*
early training for, 300–301
ethical considerations, *283*, 383, 385–89
pitfalls of, 125–26
Equid Research and Conservation, 16
Equine, as term, 16
Equine Comfort Assessment Scale, 275
Equine etiquette, Golden Rules in, 100–101, 103. *See also* Culture, equine; Spatial awareness
Equine Facial Action Coding System (EquiFACS), 16–17, 150, 275
Equine gastric ulcers syndrome (EGUS), 343
Equine Mental Health Assessment, 275–80
Equines LTD, 155, 356
Equipment, introduction of, 306, *306*. *See also* Tack; Training devices
Equitation science, 384
Equus caballus, 27–29
Essential oils. *See* Aromatherapy

Ethics
breeding, 386
competition, 116–17, 367–70, 380, 387–88
equestrian industry, 383, 385–89
habitat management, 369, 386, 387–88
of horse-human relationships, 368, 375–82, 384–89
in racing, 368, 386–88
Euthanasia, 382
Exercise
breeding considerations, 296
at competitions/events, 354
habitat management for, 342–43
locations for, 313–15
longeing as, 326
need for, 359, 361–62
for young horses, 305
Exploration, by horses, 102, 111, 292, *293*, 348
Externalizer types, 222, 239
Eye contact
among horses, 66, 70, 78, *78*
in horse-human relationships, 144–45, *170*, 182–83, 234–35, 240, 250
Eyes
in body language, 70, 182–83, 250
deadness/dullness in, 166, *250*, 254, *255*
signs of stress in, 268–69, *268*, 272
"soft," 70, *141*, 183, 234–35

_ F
Facial Action Coding System (FACS), 16–17, 150, 275
Facial expressions
in horses, 70, *172–74*, 269–70
in humans, *170–71*, 185
Facilities. *See* Habitat management
Fascia, 193–95, *195–96*
Fear
expression of, 152, 272
learning and, 100, 294
suppression of, 41, 42
in vigilant horses, 221
Fédération Equestre Internationale (FEI), 369, 383
Feeding. *See also* Food
guidelines for, 343–46
"horse time" during, 174, *174*
stress checks during, 338
during travel, 257, 350–51, 354, 357
Feeling. *See also* Emotions; Energy; Looking "with"
in assessing individuality, 221
author's use of, 118
communication and, 197–98
lagging scientific recognition of, 2, 22–23, 150–51, 166, 220, 383
Feet, 308–9, 342
Females, subtle behaviors in, 18. *See also* Gender differences; Mares
Feral horses, 27. *See also* Wild horses

INDEX

Fighting, *16,* 40, 269
Fight-or-flight response, 32, 58–59, 264–65, 285
Financial considerations. *See* Socioeconomic factors
Five Freedoms, 266
Flehmen response, *52, 73, 186,* 187
Flexibility, tests for, 195–96, *197*
Foals
   care/nurturing of, *94,* 95–96, *96, 99,* 101, *124,* 125
   communication and, 68, 70
   curiosity in, 41
   friendships, 34
   nervous system development, 59–60, *61*
   play in, 80, 83
   protection of, 36, *39,* 40, *40*
   spatial awareness in, 79, 143, 304–5
   stress management for, 297–302, *297, 299,* 305–7
   teaching of, 53, 99–103, 124, 304
Focus, of horse, 207, 230
Follow reflex, 110, 146, 309
Food. *See also* Feeding
   contaminants in, 31, 53, 296
   in evaluating intelligence, 235–37, *237*
   foraging for, 47–48, 51, 96–97, 111
   free-choice, 345, 347, 349
   as motivator, 139, 145, 147, 234–35
   wild horse dietary preferences, 19, 21, 30–31
Footing, varied, *335,* 342
Foraging, 47–48, 51, 96–97, 111, 349
Forward movement, 318
Free time, 133. *See also* Choice, freedom of; "Horse time"
Freedom, as human value, 384
Friendly attitude, in OFFER Techniques, 205, 207
Friendship, among horses. *See also* Bachelor bands; Horse-human relationships; Social networks/interactions
   communication in, 52
   empathetic pairing in, 57–58
   horse's desire for, 260
   importance of, 33–35, *34–35, 105,* 107, 109, 125
   models of, 136–37
   training considerations, 352–53
   welfare considerations, *255, 334*
Functional adaptive behaviors, 108–9, 304. *See also* Culture, equine

—G
Geldings, 146, 191, 223, 303
Gender differences, 146, 190–91, 223
Genetics
   ethical considerations, 386
   learning and, 99
   temperament and, 3, 42–43, 131–32
   in wild vs. domestic horses, 9, 115
Gestures, in body language, 70

Girthy horses, 192, 331
"Golden Rules" of equine etiquette, 100–101, 103
Grandin, Temple, 178
Grazing, in-hand, 147–48, 345, 349. *See also* Foraging
Greetings and greeting rituals
   communication in, 67–68, 78, *78,* 189–90
   examples of, *50, 138, 182, 208, 209, 211, 328*
   as functional skill, 245, *246*
   in horse-human relationships, 139, 209–10, 327, 330–31
   universality of, 100
Grooming, *76,* 111, *278,* 308–9
Ground-driving, 327
Ground-tying, 324, *324*
Groundwork, 147–48, 313, 326, 327, 333
Grunting, in communication, 68

—H
Habitat, of wild horses, 30, *30*
Habitat management
   overview, 335–38
   barn design, 339–41
   at competitions/events, 351–54
   early training and, 301–2, *303*
   enrichment, 346–49
   ethical considerations, 386, 387–88
   feeding, 343–46
   land availability, 5, 16, 335
   stress management for, 16, 266
   transportation, 349–51
   when selling horses, 354–58
Habituation, 225, 239, 291–92
Hackamores, 213, *214*
Half-halts, 214–15, 290
Halters and haltering, 213, *214,* 277, 298–300, 309
Hand signals, 180, *180*
Hand-walking, 147–48, 345, 349
Hanging out. *See* "Horse time"
"Happy-Go-Lucky" Surfer Dude personality, 224
Hay, 54, 344–45, 350
Hay nets, 345, *346*
Head, in body language, 70, 91, 318, 359
Headache anecdote, 176
Healer personality type, 224
Healing energy, 198
Health
   human responsibility for, 381
   indicators of, 73–74, 83
Hearing, 54–55, *54,* 152
Heart rate
   coherence, 59, 198, 199, 215
   music and, 350
   potential to increase, 152
   synchronization of, 59
   variability, 59, 199–201, 238, 260–61
HeartMath Institute, 200–201
HeartMind Speak visualization, 201–3

Herbivores, horses as, 19–20, 29–30, 88, 342
Herding behavior. *See also* Social networks/interactions
   body language for, *71*
"Move" signals, *101,* 142–43, *242*
Hierarchies, social, 89, 90
Hjortsjo, Carl-Herman, 275
Hormones, 52, 244, 264
Horse Brain, Human Brain (Jones), 288–89, 318–19
Horse Grimace Scale (HGS), 16–17, 274
"Horse time"
   benefits of, 138, 247, 287, 363, 388
   at competitions/events, 354
   practicing, 140, 147–48
Horse-centric understanding, 3–5, 19, 22–23, 138–40, 267, 331
Horse-human relationships
   based in equine culture, *113,* 116, 149–50, 154–55, 288, 307–10
   bonding in, 137, 165–69, *165,* 256, 389
   communication in, 169–215
   equestrian psychology, 148–50, *149*
   ethics of, 375–82, 384–89
   examples of, *113, 121, 126, 129,* 166
   greetings in, 209–10, 327, 330–31
   importance of, 119, 287, 388–89, 391
   as mutualism, 123
   observational learning in, 298
   personality type and, 155–59
   under saddle, 327–35
   trust in, 166
   working toward, 119, 137–45, 159, 168, 185, 285–87
Horsekeeping. *See* Habitat management
Horsemanship. *See* Horse-human relationships
Horses
   behavioral attributes, 107–9
   compared to humans, 150–54
   domestic vs. wild, 115–19
   evolutionary history, 27–29, 87–89
   roles and classifications for, 125–27
   skills assessment for, 244–47
Human-animal interaction field, 149
Humans
   compared to horses, 150–54
   horse's willingness to approach, 277
   personality types, 155–59
   responsibilities of, 363, 375–82, 384–89
   self-assessments for, 154–55, 259–60, 388–89
Hyperflexion, *255, 261,* 323

—I
Imbalance, physical, 191–92
Immune system, 262, 264
Imprint training, 298
Individuality, in horses. *See also* Personality
   overview, 217–21
   assessment of, *216,* 221–22, 246–47, 280–81, 336

in equine culture, 11, 107–8
learning and, 313
personality types, 222–24
in SAICC Evaluation, 224–43
training considerations, 290, 319–20, 323
trauma and, 251
Inherent traits, 222
Injuries, 264, 326, 386
Insecurity, signs of, 190–91
Instructors, 157–58
Intelligence
emotional, 37, 85, 87–89, 200, 233, 236
evaluating, 224, 228, 230–39, *230*, 237, 243
Intentions, communicating, 140, *140*, 184–85, 200–201, 207
Internalizer types, 222, 239
International Society for Applied Ethology (ISAE), 383
International Society for Equitation Science (ISES), 383
Interspecies relationships, 2, 133, 166, 199–200, 349, 376. *See also* Horse-human relationships
Intuition, in OFFER Techniques, 205
Isolation, 296, 340–41. *See also* Separation/separation anxiety
"I-Want-to-Please" personality, 222, 223. *See also* Eagerness to please

__J__
Jimenez, Carlos, *193, 209*
Joint injections, 380
Jones, Janet, 288–89, 318–19
Juvenile behavior, in domestic horses, 231–33

__K__
Kemmer, Heike, 118, *319*
Kicking, 110, 144–45, 236–37, 361–62
Knowledge
vs. feeling, 2
horse-centric, 3–5, 22–23, 138–40, 267, 331

__L__
Land, availability of, 5, 16, 335
Larkspur, 53
Leadership
as collaboration, 146
in equine culture, 9, 28, 75, 89–90, 92, 108
in horse-human relationships, 145, 146–48, 155–59
models of, 136–37
personality types in, 224
by social facilitators, 37, 39–40, 42
in training systems, 146–47
Leading, 277–78, 298–300, 307–9, *308*, 358
Learned helplessness, 253–55, 263
Learned traits, temperament and, 222
Learning, by horses. *See also* Training
disabilities in, *230*, 231–33, 235

in foals, 53, 99–103, 124, 304
group vs. individual, 221, 232
pace of, 313
under saddle, 330
styles of, 290–94
in wild herds, 101–3, *101–3*
Learning, in humans, 5
"Licking and chewing," 183
Lifestyle, equestrian, 377–78
Lifestyle Enthusiast personality type, 158
Light
in barns, 339
changes in, 49–50, 331, 358
Lips, of horse, 57, 70
Listening, by horse, 171
Listening, by humans, 66, 141–45, *145*, 171, 176–77, 388–89
Livestock, 127, 376
Logger anecdote, 206–7
Longeing, 326, 327
Looking "with," 19, 41, 118, 207, 259, 388–89
Love
Darwin on, 36, 389
for horses, 117, 157, 198, 375
horse's response to, 259
physiology of, 200, 202
vs. respect, 208
Lung function, 152

__M__
Male-dominated models, 18. *See also* Geldings; Gender differences; Stallions
Manners, 305–6. *See also* Culture, equine
Manure
in communication, 52, 73–75, *74*, 111, 187
eating of, 60
in monitoring stress, 337–38, 351
Mares
altruism in, *37*
breeding considerations, 294–96
characteristic behaviors, *34*, 137, 190, 223, 244, 281
communication by, 70, 74, 170
roles of, 88, *89*
as social facilitators, 125, 146, 228
spatial awareness/respect in, 77, 79, 91–93, 141, 143, 203
stress and, 302–5
tack and, 213
as teachers, *21*, 298, 300–301, *301, 303*, 304–5
in wild herds, 10, 20, 21–22, 90, 97
working with, 117–19, *139*
Massage, 349, 354
Masterson, Jim, 192, 196
McComb, Karen, 275
Media coverage, of welfare issues, 377–78
Memory, in horses, 222, 236, 307
Mental stress

case histories, 280
emotional stress as, 59, 125, 193–94
lack of adaptations to, 32
in people, 204
vs. physical, 3, 32, 37, 125, 193–94
Meyer, Dorothe, 114–19
Microchipping, 386
Mimickry/mirroring, 102, 170, 180–81, 200–201, 305, 311
Mindfulness, 167, 169, 179, 201–2, 286
Morality, Darwin on, 36. *See also* Ethics
Motivation
case histories, 134
to communicate, 177, *177*
curiosity as, 147
determining, 221
food as, 139, 145, 147
friendship as, 33–35
of human/handler, 118
for learning/training, 310–12
Motor learning, 291, 322
Mounting (for riding), 277, 279
Mouth
in communication, 70, *176*, 190–91, 213
sensitivity of, 57, 318, 360
signs of stress in, 271, 272
Mouthy horses, 190–91, 232, 360–61
"Move" signals, *101*, 142–43, 242
Movement. *See also* Proprioception
in equine culture, *32–33*, 111, 308–9, 334
habitat management for, 342–43, *343*
in horse-human relationships, 215, 245
monitoring, 338
neurology of, 289, 321
perception of, 48–49, *48*
Muscles, 194–95, 264, *272*
Music, 178, 185, 349, 350
Mutual grooming, 76, 91–92
Mutualism, 123
"My space, your space"
training/reinforcing, 142–43, *142–43*, 236–37, 315
for young horses, 298–99, 307–8
Myelin, 59–60

__N__
Neck
hyperflexion of, *255, 261,* 323
"snaking" of, 18, 70, 91, *100*
stretches for, *197*
tension in, 192, *323*, 345
Negative reinforcement, 291
Negligence, 387
Neolocality, 89
Nervous system, 59–60, *61*, 264–65, 318–19
Neuroendocrine system, 265
Neurological disorders, 359
Nickering, 42, 68–69

INDEX

Night vision, 49
Nose. *See also* Olfactory system; Smell, sense of
    communication role, 75
    length of, in mammals, 51
    sensitivity of, 57, 110, 153, 213, 318
Nose bumps, 175, 190–91, 209
Nose chains, 299–300, *308*, 358, 361
Nosebands, 261, 317–21, *317, 319–20*
Nostrils, 51–52
Novel objects
    as enrichment, 348, *349*
    exploration of, 309, 312, 331, 359
    introduction of, *242*, 306
Nurturing
    by humans, 156–57
    personality type, 224
    in social networks, 124
    touch in, 56–57, *56–57*
    in wild herds, 36–37, 39–40, 42, 66, *66*
Nutrition, 243, 266–67, 296. *See also* Feeding

O

Obedience, expectation of, 210, 283, 316–17
Observational learning. *See also* Looking "with"
    overview, 291
    examples of, *42*, 302, *312*
    training considerations, 294, 296, 297–98, *310*, 311
O'Connell-Rodwell, Caitlin, 76
OFFER Techniques, 204–8
Older horses, 386
Olfactory system, 51, 152, 178, 186–89, 289
Operant learning, 291
Opportunistic foragers, 31, 53
Opportunistic thinking, 235
Overreactive horses, 362–63
Ownership, responsibilities of, 376–82, 386–87
Oxygenation potential, 152

P

Paddock Play, 300, 348
Pain
    behavior issues and, 241, 269–70, 358–60
    as cause of stress, 265
    learning effects, 294
    locating, 190, 191–93, *191*
    in "shut down" horses, *254*
    somatic, 260–61
    from tack, 318
Passive temperament, 222
Patroling behavior, 101–2, *102*
Paul (cowboy), 166
Pawing, 68, 361–62
Peace-keeping behavior, 90, 100
Perceptual learning, 291
Performance (athletic), 3, 264, 334–35, 350–54. *See also* Competition/events
Personality. *See also* Temperament
    assessment of, 3, 219–21
    learning and, 99
    types of, 107–8, 222–24
    working with, 117–18, 244
Peters, Duncan, 276, 326, 367–70
Physical stress
    evolved tolerance for, 31–32, 37, 259–60
    vs. mental, 3, 32, 37, 125, 193–94
    signs of, 191–97
Plants, toxic, 31, 53
Play
    vs. acting out, 269–70
    as communication, 80–83, *82–83*
    creating opportunities for, 300, 348
    in equine culture, 102, *103*, 111
    examples of, 40, 361
    gender differences, 303
    importance of, 83, 354
    learning and, 297–98
    as self-expression, *135*
    training considerations, 292, *293*, 305
Poll, tension in, 192, 272, 326, 345, 359, 360
Positive reinforcement, 147, 291, 292, 311
Posturing behavior, *16*, 69–70, *69*, 74, 78, 110
Praise, 185, 294. *See also* Rewards
Preconceptions, about horses, 210, 328–29, 384. *See also* Bias
Preferences, in horses
    changeability of, 221
    examples of, 53, 185, 256, 347
    expression of, 118–19
Pre-purchase exams, 276, 380
Prey-predator relationships, 23, 32, 150
Prisoner analogy, 116–17
Problem-solving, 11
Projection, of human feelings onto horses, 152, 155
Pronghorn antelope, 31
Proprioception
    energy sensing and, 33, 197
    in spatial awareness, 92, 245
    synchronization of, *32–33*
    training/riding considerations, 210, 215, *293*, 331
Prosocial behavior, 66
Psychology, human, 2, 148–50, 248–49
Public perception
    of scientific research, 383–85
    of welfare issues, 261, 376–78, 382
Pulling back, while tied, 325–26, 360

R

Racehorses, 227, 302, 311, 350, 368, 386–88
Range of motion, 195–96, *197*
Reaction
    vs. response, 152, 259, 289–90
    as trait, 43
Rearing, *16*, 273, 304
Record-keeping, 257
Recreational riding, 116, 368, 375
Red blood cells, production of, 152
Rehabilitation, 156
Reins, use of, 210, 359
Relaxation
    in horses, 187, *188*, 195, *195*, 350, 354
    in humans, 286
Repetitive movements, tension from, 345. *See also* "Vice" behaviors
Reproduction
    in domestic horses, 267–69
    in equine culture, 70–71, 74, 93–96, 111, 124
Rescue horses, 156, *164*, 248, 263
Resistance, 222, 236, 241, 325
Resources, natural
    land availability, 5, 16, 335
    wild horses and, 28, 89, 124
Respect, 148, 207–8, 244, 245, 304. *See also* "My space, your space"
Response, vs. reaction, 152, 259, 289–90
Responsibilities, of horse ownership, 376–82. *See also* Ethics
Restriction
    acceptance of, 309–10, 325
    behavior issues around, 360
    minimizing, 294, 296, 322, 389
Retirement, of horses, 381–82
Rewards
    in motivation, 310–11
    positive reinforcement, 147, 291, 292, 311
    "timeout" as, 294, 313, 316, 334
    uses of, 147, 185, 296, 363
Rhythm keepers, 90, 97, 125
Ribs, tension in, 192, 331
Riding
    behavior and, 279–80
    communication ABCs, 210–15, *215*
    ethical considerations, 368
    first rides, 135, 168
    horse-human relationships in, 166–67
    safety considerations, 277
    stress management for, 327–35
Rolling, 97, *98*, 111, 354, 362
Round pens, 313–14, *314*
Routines, 73, 336, 338, 354
Rubenstein, Dan, 8–11, 18, 58
Running away, 243, 318, 331, 359

S

Sacroiliac (SI) joint, 193
Saddles
    fit of, 191–93, 265, 313, *332*, 359
    horse's response to, 278, 306
Safety, horse's sense of
    barn design for, 336–37
    in equine culture, 32, 59, 100, 109
    friendships in, 133, 135, 152

in horse-human relationships, 117–18, 147–48, 331–34, 352, 389
during sleep, 98–99
stress and, 260, 263, 267
trauma and, 248–51
Safety considerations, for people, 137, 148, 277
SAICC Evaluation
overview, 224–25
awareness in, 228–30
confidence in, 237–41
cooperation in, 241–43, 242
intelligence in, 230–37
sensitivity in, 225–28, 226
Scent. See Smell, sense of
Schiller, Warwick, 319
Sciatic nerve pain, 243, 331
Science/scientific findings
awareness of, 261, 382–85
models in, 81, 167, 169, 197–99
narrow focus of, 18, 22–23, 47, 150, 166, 220
Scratching
in locating pain, 192
of sweet spots, 194–95, 195
Self/self-expression, in horses, 117, 135, 241, 270
Selling horses, 117, 336, 349, 354–58
Senses and sensation. See also Smell, sense of; Touch, horse's sense of; Vision
overview, 45, 47
hearing, 54–55, 54, 152
reliance on other horses, 55
taste, 53–54, 295
Sensitive horses
case histories, 134
confidence and, 239
evaluating, 225–28, 226
genetics and, 43
as personality type, 224
stress in, 262
Sensitization, learning and, 292. See also Desensitization
Separation/separation anxiety, 249, 255–56, 260, 262, 296, 303–4
"Sharing awareness," 163, 199–200, 201, 206–7
Shipping. See Transportation
Shows. See Competition/events
"Shut down" horses, 166–67, 222, 248, 273
Side-reins, 321–23, 322
Sight. See Vision
Signature smell, 187
Simonds, Mary Ann
Equines LTD, 155, 356
Meyer on, 118
Peters on, 367
research projects, 15, 17–21, 23, 88–90, 134–35, 201
Simplicity, in training, 292, 294
Skin, 55–56, 57, 195
Skin roll technique, 195, 196

Sleeping
barn design for, 339, 353, 369
deep/REM, 98–99, 100, 153, 153
in equine culture, 110
nocturnal/diurnal cycles in, 49, 98
relaxed energy in, 41, 203
in stress checks, 338
vigilance considerations, 342
Slow-feed containers, 345
Smell, sense of
overview, 50–53, 50, 52
in blind horses, 51
in breeding, 295
in communication, 71–75, 73–74, 179, 186–89
in equine culture, 78, 110, 124, 140, 152
in habitat enhancement, 347
in horse's sense of safety, 132
neurology of, 51, 71, 152, 178, 186–89, 289
sensitivity evaluation, 226, 227
in stress management, 351, 353
in training, 300
vs. vision, 188
Smiling, importance of, 170, 170, 210, 286
Snorting, as communication, 68
Social facilitators
conflict management by, 90
mares as, 125, 146, 228
nurturing by, 39–40
role of, 37, 97, 108, 141, 146, 221
"Social license to operate," 382
Social networks/interactions
altruism in, 36–37
among wild horses, 9, 13, 29–31
benefits of, 32, 59–60
communication and, 77, 177
Darwin on, 35
disruption in, 35–36, 254
in equine culture, 11, 27–29, 31–35, 85, 88–90, 93–99, 304
evaluating, 32, 224, 338
habitat management for, 334–42, 341, 346, 349, 350
human need for, 2
importance of, 18, 107, 116, 276–77
nurture vs. aggression in, 37–42
training considerations, 311
Socioeconomic factors
in competition/sport, 352, 388
in horse-human relationships, 136, 154–55, 349–50, 357–58
welfare considerations, 127, 375–82
"Soft" eye, 70, 141, 183, 234–35
Soreness, evaluating, 191–92
Sound. See also Hearing
in associative learning, 55, 185, 290
in communication, 178, 184–86
in horse-human relationships, 133, 135, 141, 349

sensitivity evaluation, 227
Space bubbles, 203
Space-taking, 77, 80, 91–92, 297, 315
Spatial awareness
case histories, 232
communication and, 70, 77–80, 77
developmental aspects, 233
in equine culture, 81, 91–93, 100–101, 110 11, 304
in horse-human relationships, 137, 141, 143, 148, 203, 315
in intelligence testing, 236
as learned skill, 33, 244–45, 246, 297–98, 302, 304–5
Spatial respect, 245, 252, 358. See also "My space, your space"
Species, horses as, 27–29
Spooking
causes of, 146, 236, 322, 331, 353, 359
evaluating, 238
as sign of stress, 273
vigilance and, 221
Sport. See Competition/events; Racehorses
Spurs, 213, 323–24
Square pens, 314–15, 315
Squealing, 68
Stallions. See also Bachelor bands
breeding considerations, 294
communication by, 69, 69, 70–71, 73
domestic, 146, 244
friendships, 34, 38, 38, 40, 42
leadership behavior, 9, 20, 90
in logger anecdote, 206–7
nurturing by, 36, 40, 96, 125
posturing by, 16, 74, 74, 78, 94
space games, 91–93
stress and, 302–5
teaching by, 99
training considerations, 223
Stalls. See also Habitat management
changing, as enrichment activity, 348
size of, 257, 338, 370, 387
Stance, significance of, 238, 239
Standing
adaptation for, 153
habitat management considerations, 336–41
Standing still. See also "Be still" signals
as functional skill, 245, 307–8, 314–15, 324–27
as test of confidence, 240–41
Status, 77, 80, 244
Stereotypic behaviors, 253. See also "Vice" behaviors
Stockholm Syndrome, 253–55, 254
Straw bedding, 339–40
Stress. See also Stress management
assessment of, 191–97, 259–60, 273–81
behavior and, 117–19, 358–63
causes of, 3, 151, 257, 260–61, 265–68, 338, 378

INDEX
[407]

INDEX

in domestication, 2
effects of, 264–65
good vs. bad, 263
in humans, 194, 204, 285
internalization of, 231, 238–40, 244, 260, 292, 307
mental, 32, 37, 125, 193–94, 280
physical, 31–32, 37, 191–97, 259–60
signs of, 183, 262, 268–73, *268, 270–73, 321,* 325, 337
types of, 261–63
in wild horses, 131–32
Stress management. *See also* Stress
in breeding, 294–96
equine learning styles and, 290–94
equine neurophysiology for, 287–90
for foals, 297–302, *297, 299,* 305–7, *306*
gender and, 302–5, *303*
habitat considerations, 335–58
for humans, 204, 285–87
music for, 350
Stretching exercises, 349, 354
Striking, 304
Such is the Real Nature of Horses (Vavra), 15
Suddaby, Wes, *299*
"Sweet spots," 194–95, *195*
Sympathetic nervous system, 264–65

_T
Tack
as cause of stress, 261, *272*
fit of, 192, 213, 316–17, *317, 332,* 388
in horse-human communication, 210
introduction of, 306
rules and standards for, 369, 386
training considerations, 294
Tactile communication channel, 179, 189–97, *190–91, 193*. *See also* Touch, horse's sense of
Tail
in body language, 70
signs of stress in, *272*
Take-Charge Leader personality, 224
Talking, to horses, 186, 331, 334, 389
Talking with Horses (Blake), 57–58
Taste, sense of, 53–54, 295
Teacher personality type, 157–58
Teeth
clacking of, by foals, 68
grinding of, 271, 360
Temperament
adaptability and, 2, 133, 135
assessment of, 219, 298
case histories, 134–35
differences in, 107–8
performance and, 352
vs. personality, 220
research regarding, 278
training considerations, 244, 313

types of, 221–22
Temporomandibular joint, tension in, 326, 359, 360
Tension, release of, 192–97. *See also* Relaxation
"Testing" behavior, 148
Therapists, horses as, 3, 127, 201, 224, 285
Thinking in Pictures (Grandin), 178
"Thinking like a horse," 168, 185
"Timeout" breaks, 294, 313, 316, 334
Timid horses, 222
Topline, in body check, 196
Touch, by people
horse's acceptance of, 277, 308–9
uses of, *151,* 192, 209–10, 214–15, 334
Touch, horse's sense of
overview, 55–57, *56*
in communication, 75–76, 179, 189–97
in equine culture, 35, *57,* 78, 110, 124, *132*
in horse-human relationships, 133, 135, *151*
sensitivity evaluation, 227–28
Toxins, in feed, 31, 53, 344–45
Toys, *136,* 349, 361
Trailers/trailering, 51, 256, *306,* 309–10
Training
assessment in, 219–21, 244–47
communication in, 169–79
discipline-specific, 300–301
ethical considerations, 379–81, 384
horse-centric, 3–4, 141–48, 154–55, 292, 294, 328–29
imprinting, 298
individuality of horse in, 146–47, 221–24, 231–33, 292–94
over-response to, 250
patience in, 294
play in, 292, *293,* 305
preconceptions regarding, 328–29
SAICC Evaluation in, 224–43
session duration, 294, 301, 312
space/location for, 313–15
stress and, 261, *261,* 352, 378
"systems" of, 136–37, 144–45, 290
Training devices, *272,* 294, 316–17, 321–23, 369, 386
Transportation
for breeding, 295–96
case histories, 51
ethical considerations, 369
stress of, 261–62, 350–51
training considerations, *306,* 309–10
Trauma, 247–55
Trust
greetings in, 139–40, *139–40,* 168
in horse-human relationships, 137, 146, 329, *329*
sense of safety and, 125, 159, 166, 313
Tucker, Renee, 192
Tying, 309–10, 324–27, *325,* 360

_U
Ulcers
aromatherapy for, 347
behavior and, 243, 331, 359
in foals, 260
gender differences, 304
risk factors, 223, 343–44
Unwanted horses, 378–79, 381–82, 386
Urine, in scent-based communication, 53, 73–75, 111

_V
Vavra, Robert, 15
Ventilation, in barns, 339, *339*
Veterinary care, ethics of, 379–81
"Vice" behaviors, 223, 233, 250, 275, 307
Viewer/observer models, 22–23. *See also* Looking "with"
Vigilance
vs. curiosity, 42–43
in domestic horses, 131–32, 133, 141, 342, 362–63
learning and, 99, 221
in wild horses, 18, *28,* 109
Vision
overview, 47–50, *48–49*
behavior and, 358–59, 363
biomechanics of, 47–48, 152
in communication, 178, 180–83, *180–83*
in horse-human relationships, 133, 135, 141
sensitivity evaluation, 226–27
vs. smell, 188
Visualizations, 201–3, *202,* 313, 331
Vocalizations
in horses, 67–69
in humans, 184–86, 331, 334, 389
Von Bredow-Werndl, Jessica, 118–19

_W
Walking, importance of, 342, *343*
Waller, Bridget, 275
Water
consumption of, 338, 350
wild horses and, 51, 81, 81, 96–97, *97*
Wathan, Jen, 275
Weaning, 117, 260, 297–98, *297,* 302
Welfare, of horse
across life cycle, 378–79
best practices, 3, 176, 382, 384
challenges of, *373,* 378
checks for, 337–38
human responsibility for, 176, 363
public opinion in, 261, 276–78, 382
reporting considerations, 381–82
scientific research role, 382–84
socioeconomic factors in, 126–27, 375–82
standards for, *283,* 337
Wheeler, Barbara and Andy, 15

Where Does My Horse Hurt? (Tucker), 192
Whinnying, 68
Whips, 213, 323–24
Whiskers, facial, 57, 75
"Whole life horse fund," 386
Wild horses
    adoption of, 134–35, 137, 168
    attitudes toward, 18–19, 20, 384
    author's work with, 15, 17–21, 23, 88–90, *167*
    equine culture in, 25, 88–90, *91*, 123–25
    human interactions, *22, 29, 113*, 227–28, *246, 253*, 276–77
    individuality of, 89–90
    intelligence of, 230–31
    population dynamics, 88, 93
    temperament and, 131–32
Wildlife species, 19, 22–23, 28, 41
Willing to please. *See* Eagerness to please
Withers
    tension in, 192
    touching of, by handler, 209–10, *211*, 333
Worrier personality, 224
Wyatt, Fred, 22, 168

Y

Young horses. *See also* Foals
    habitat enhancements for, 348
    learning capacity of, 292
    play in, 80, 83
    spatial awareness in, 297–98
    stress management for, 305–7
    training considerations, 221, 239, 298–300, 304, 307–10, 312, 327